Determinantal Ideals of Square Linear Matrices

os

ersity of Sergipe
ão, Sergipe, Brazil

Aron Simis 🆔
Mathematics
Federal University of Pernambuco
Recife, Pernambuco, Brazil

N 978-3-031-55946-4   ISBN 978-3-031-55284-7   (eBook)
s://doi.org/10.1007/978-3-031-55284-7

hematics Subject Classification: 13C40, 13D02, 13H10, 14M05

s work was supported by Conselho Nacional de Desenvolvimento Científico e Tecnológico

Zaqueu Ramos • Aron Simis

Determinantal Idea
Square Linear Mat

Springer

This book is dedicated to all collaborators and students of the senior author, past and present.

Preface

Toda caminhada começa no primeiro passo (Acioly Netto, pop composer) [A poet version of the inductive method]

If I knew where it was, I wouldn't have lost it (Street talk) [A layman notion of tautology]

The theory of determinants is a fascinating classical subject which stands as a strong facet of both algebraic and geometric theories. Much has been written on the subject, with many old and new books versing on the overall aspects of the theory. The latter has spread out to many modern parts of algebra, geometry, and singularities, perhaps a great deal through the so-called determinantal ideals (or varieties).

This book is not strictly a book on determinants. Rather, it is about facets of determinantal ideals of matrices, from the point of view of commutative algebra, with a particular emphasis on linear such matrices. Yet, non-square matrices are dealt with quite a bit throughout, starting with the first chapter. As the subject of the book evolves, they are considered as well – there is even a full chapter on $m \times (m-1)$ linear matrices. Often, non-square matrices play a strategic theoretical role. The main reason to tackle square matrices is to look at a determinantal hypersurface $V(f)$, its singular locus and related invariants, particularly at those properties not shared by arbitrary projective hypersurfaces. Geometrically, the singular locus of $V(f)$ in the generic case is given by the determinantal variety defined by the submaximal minors. However, for non-generic matrices, the next best is the variety defined by the Jacobian J_f ideal of f. Thus, understanding the nature of this ideal in its relation to the ideal of submaximal minors comes out as a challenging matter. In the book, we explore this nature, and that of related ideals, from the algebraic stand, with the understanding that one is considering square matrices whose entries belong to a (commutative) Noetherian standard graded ring over a field. The present results depend on the typical numerical invariants of commutative theory, with a strong nuance toward those available in finitely generated algebras over a field. Though inevitably interspersed with basic tools of algebraic geometry, it is mainly aimed at a commutative algebra audience with a basic training in the field. The book shares a facet of classical endeavor in that it underlines the nature and structure of the matrices at stake. Yet, it additionally aims at the ideal theoretic invariants related to the corresponding determinants, driving with some depth into the details of the related rational maps, polar maps, and dual varieties via the Plücker embedding of the Gauss map. However, it is not a book on any of these geometric issues either, so the reader should not aim at finding full theories on them. In fact, it is often the case that the book delves longer on more basic facts of special matrices and their minors.

Having an encompassing theory of the determinantal ideals of linear square matrices over a field k looks beyond current knowledge. The overall objective of the book is an effort to deal with this theory in a certain number of such classical matrices, usually known as linear sections of the generic matrix. A typical preliminary step is to understand the role of the codimension of the k-vector space spanned by the entries of the matrix. This role, additionally supplemented with subtler information on the conjugation action, has been considered and often leads to important classification results. Relevant classes of such linear sections are the so-called coordinate sections, meaning that the entries are either the original indeterminates of the generic matrix or else zero. The number of sufficiently many such zeros makes a case for a sparse coordinate matrix. Various models of such linear sections will be given a thorough treatment, perhaps justifying assigning new names apparently not used before.

Since the book evolves around such sections, along with the properties of their determinants, the nature of their ideals of submaximal minors also comes up as a major goal. Altogether, for such a linear section L, where R stands for the polynomial ring over k generated by its entries, the related algebraic structures will be primeval: the determinant $f \in R$ of L, its Jacobian (gradient) ideal $J \subset R$, the Hessian matrix $H(f)$ of f, the polar map of f defined by its partial derivatives, and the ideal $I \subset R$ of the submaximal minors of L. For all this, ideal theory and homological aspects play a dominant role.

As to the gradient ideal J, the classical geometric way is to look at it as defining the base scheme of the polar map often without additional thorough insight into its scheme nature, much less its ideal theoretic side. Here, a major point is to first understand the ideal theoretic features of J such as its height (codimension) and some of its associated prime ideals, as well as the impact of its linear syzygies. In a second step, one focuses on the intertwining between J and the ideal $I \subset R$ of the submaximal minors of the matrix. As per default, we will have $J \subset I$ and typically (but not always) of the same codimension. Of great interest is to find out when I is a prime ideal and to study its potential nature as the radical of J or of its unmixed part. The question as to when J is actually a reduction of I becomes of neat interest. For the question of primality, besides the usual method of verifying beforehand the normality of R/I, there is the approach via t-generic matrices, following Eisenbud and Harris. We draw on both, depending on the available players. A beautiful question, to our knowledge still open in general, is to determine when the singular locus of the determinantal variety defined by I is set-theoretic defined by the immediately lower minors, just as happens in the generic case.

Another major theme is the structure of the polar map and its image (freely called polar image) in terms of its birational potentiality. Its homogeneous coordinate ring is known as the special fiber (or fiber cone) of the ideal J and plays a central role in the theory of reductions of ideals. Its Krull dimension is also known as the analytic spread of the ideal J. In this regard, we face two basic questions: first, to compute the analytic spread of J (equivalently, in characteristic zero, the rank of the Hessian matrix $H(f)$ of f); second, to decide when J is a minimal reduction of the ideal I of submaximal minors.

Complete answers to the above questions are largely dependent upon the sort of linear sections of the generic square matrix one is looking at. The prevailing tone of this study is to understand the effect of the shape of these sections on the properties of the underlying ideal theoretic structures. There will be noted differences regarding the behavior of a few properties and numerical invariants, such as codimension, primality, Gorensteiness, and Cohen–Macaulayness. As known from the earlier work of Boocher, Eisenbud, Glassbrenner–Smith, Giusti–Merle, and Herzog–Trung, the precise behavior of the numerical invariants attached to sparse linear sections is rather delicate and may in fact depend on the slots of the null entries. Parts of the book will focus diligently upon various such situations. For example, one deals with classes of matrices L which are linear sections of other matrices M which are themselves linear sections of the generic matrix. This is usually undertaken in such a way as to have L preserve the matrix structure of M – e.g., symmetry, Hankel-like, Hilbert–Burch-like, and so forth.

The more precise description of a chapter is given in the abstract that precedes it, to which we refer the reader.

Historical Ramblings

Though this is not a book on determinants, we give the beginner a brief idea of the early development.

As has become common dicdatics, the beginnings of matrices and determinants go back to the second century BC. Alas, one had to wait pretty much till near the end of the seventeenth century for the ideas to crystallize into a pervasive mathematical subject.

A rough method was practiced that eventually became known as Gaussian elimination, but as the name suggests, mathematical stability was only acquired toward the early nineteenth century. Even Cramer, with its famous rule for solving a 2×2 system, does not reach the full-fledged idea of a determinant. Many standard results of elementary matrix theory first appeared long before matrices were the object of mathematical investigation.

There seems to be agreement about the fact that a notion of determinant first appeared in Japan (Seki, 1683) before it did so in Europe. The latter seemingly took place 10 years later in a letter by Leibniz to de l'Hospital concerning the solution of a certain linear system, followed by a vivid exchange around the discussion of notation. By and large the notation used by Leibniz was possibly unfriendly to most mathematicians of the period, causing his findings to be partly neglected by the subsequent practitioners of determinant theory.

In his work, Leibniz used the word "resultant" for certain combinatorial sums of terms of a determinant. He proved various results on such resultants, including what looks to be essentially Cramer's rule. He also knew that a determinant could be expanded using any column – what is now called the Laplace expansion.

In the 1730s, Maclaurin wrote *Treatise of Algebra* although it was not published until 1748, two years after his death. It contains the first published results on determinants, proving Cramer's rule for 2×2 and 3×3 systems and indicating how the 4×4 case would work. Cramer gave the general rule for $n \times n$ systems in the work *Introduction à l'Analyse de lignes Courbes algébriques* (1750). It arose out of a desire to find the equation of a plane curve passing through a number of given points, since geometry of curves was the main target of this work. The rule appears in the Appendix no. 1 to the paper but no proof is given: "One finds the value of each unknown by forming n fractions of which the common denominator has as many terms as there are permutations of n things." Cramer does go on to explain precisely how one calculates these terms as products of certain coefficients in the equations and how one determines the sign – he even introduces the terminology "dérangements" (in the sense of transpositions inducing the sign of a term). He also says how the n numerators of the fractions can be found by replacing certain coefficients in this calculation by constant terms of the system. Thus, he essentially gives the notion of a determinant as a signed summation of suitable products, which is the basic definition inspired by the work of Leibniz.

Work on determinants now began to appear regularly: (Bézout, 1764), (Vandermonde, 1771), (Laplace, 1772), (Lagrange, 1773), (Gauss, 1801). The term "determinant" was first introduced by Gauss in *Disquisitiones arithmeticae* while discussing quadratic forms. However, it was Cauchy (1812) who used "determinant" in its modern sense. Cauchy's work is the most complete among the early works on determinants. He reproved the earlier results and gave new results of his own on minors and the adjugate matrix. With Cauchy, the full interplay of matrix theory and determinants became a deep theory, soon to be extended and improved by several other mathematicians, such as Sturm, Jacobi, Kronecker, Weierstrass, Cayley, and Sylvester. Cayley was first to use the current notation of two vertical lines on either side of the array to denote the determinant.

The first to use the term "matrix" was Sylvester, in 1850. Cayley quickly saw the significance of the matrix concept and published *Memoir on the theory of matrices* (1858) containing the first abstract definition of a matrix. He gives the first proof of the known Cayley–Hamilton theorem (only proved in its total generality later by Frobenius).

In 1870, the Jordan canonical form appeared in the *Treatise on substitutions and algebraic equations*, by Jordan. It appears in the context of a canonical form for linear substitutions over a finite field. In 1878 Frobenius wrote an important work on matrices: *On linear substitutions and bilinear forms*. This paper also contains the definition of the rank of a matrix. It should be said that, around this time or slightly later, many books on determinants would appear, such as R. F. Scott, *The Theory of Determinants and Their Applications*, 1880 (First Ed.) (Cambridge, Second Ed. revised, 1904), and T. Muir, *The Theory of Determinants in the Historical Order of*

Development, 1905 (Dover, New York, 1960). The latter contains a fairly complete detailed proceedings of all previous results by the main mathematicians of the period up to 1840 circa.

An axiomatic definition of a determinant was used by Weierstrass in his lectures and, after his death, it was published in 1903 in the note *On determinant theory*. In the same year, Kronecker's lectures on determinants were also published, again after his death. With these two publications, the modern theory of determinants was in place but matrix theory took slightly longer to become a fully accepted theory. An important early text which brought matrices into their proper place within mathematics was *Introduction to Higher Algebra* by Bôcher in 1907. Turnbull and Aitken wrote influential texts in the 1930s, and Mirsky's *An Introduction to Linear Algebra* in 1955 saw matrix theory reach its present major role as one of the most important undergraduate mathematics topics.

The reader is further referred to the excellent exposé in *MacTutor History of Mathematics*, by J J O'Connor and E F Robertson, which has helped us in parts of the above short account. For a detailed bibliography of Pfaffians and their determinants, one is referred to the beautiful account in D. Knuth, Overlapping Pfaffians, J. Electr. Comb., Volume 3, Issue 2 (1996) (The Foata Festschrift volume).

<table>
<tr><td>Aracaju, SE, Brazil</td><td style="text-align:right">Zaqueu Ramos</td></tr>
<tr><td>Recife, PE, Brazil</td><td style="text-align:right">Aron Simis</td></tr>
<tr><td>Winter 2023–2024</td><td></td></tr>
</table>

Acknowledgments

The senior author heartily thanks the ICMC-USP (São Paulo, Brazil), specially Daniel Levcovitz, and the Istituto Politecnico di Torino (Turin, Italy) in the name of Letterio Gatto, for providing the support to a series of lectures by him on the preliminary chapters of the book. Thanks are also due to the Springer personnel for their kindness and resilience during the preparation of the manuscript. Last, but not least, we are indebted to the authors of Macaulay 2, and its previous version Macaulay, for our extensive use of these programs.

Contents

Part I General Oversight

1 Background Steps in Determinantal Rings 3
1.1 Some Matrix Notions 3
1.2 Generic Determinantal Rings 4
1.3 Classical Rational Determinantal Varieties 7
    1.3.1 Veronese Varieties 8
    1.3.2 Rational Normal Scrolls 11
    1.3.3 Segre Varieties 13
1.4 Theorems on Generic Maximal Minors 14
1.5 Linear Sections 18
    1.5.1 General Linear Forms 19
    1.5.2 1-Generic Matrices 21
    1.5.3 Applications 23
Exercises 25
References 26

2 Algebraic Preliminaries 29
2.1 Minors and Cofactors 29
    2.1.1 The Adjugate of a Square Matrix 29
    2.1.2 Matrices Over Polynomial Rings 32
    2.1.3 Linear Sections Endowed with Decomposition 36
    2.1.4 A Role of the 2×2 Minors 37
2.2 Ideal Theory Basics 40
    2.2.1 Special Fiber, Analytic Spread, and Reduction Number.. 41
    2.2.2 Linear Rank 43
    2.2.3 Ideals of Linear Type 44
    2.2.4 Perfectness in Height 2 45
    2.2.5 On the Symbolic Rees Algebra 47
2.3 Algebraic Aspects of Rational Maps 49
Exercises 50
References 52

3   Geometric Oversight 53
   3.1   The Polar Map 53
   3.2   The Hessian of a Projective Hypersurface 54
         3.2.1   Parabolism and the Dual Variety 55
         3.2.2   The Rank of the Jacobian Matrix 55
         3.2.3   The Grassmann Algebra 58
   3.3   Propadeutics: Generic Determinantal Hypersurfaces 63
         3.3.1   The Generic Square Matrix 63
         3.3.2   The Generic Symmetric Matrix 64
         3.3.3   The Circulant 66
         3.3.4   A Glimpse of the Inverse Problem: When Is a Homogeneous Polynomial the Determinant of a Linear Matrix? 69
   Exercises 71
   References 73

Part II   Linear Sections of Notable Structured Square Matrices

4   Linear Sections of the Generic Square Matrix 77
   4.1   Ta(i)l(e)s of Ladder Ideals 77
         4.1.1   Gorenstein Plethora 78
         4.1.2   Codimension Four Tight Gorenstein Subideals 83
   4.2   Hollow Type Linear Sections 86
         4.2.1   Notation 87
         4.2.2   Generic Hollow Sections 88
   4.3   Codimension One Linear Section 93
         4.3.1   The Submaximal Minors 95
         4.3.2   The Gradient Ideal 97
   4.4   Case Study: One Term Linear Form 101
         4.4.1   The Submaximal Minors and the Gradient Ideal 102
         4.4.2   The Polar Map 105
         4.4.3   Diagonal One Term Iteration 111
   4.5   Semi-Hollow Linear Sections 112
         4.5.1   Preliminaries 113
         4.5.2   The Structure of the Polar Image 115
         4.5.3   Narrowing the Set of Associated Primes 120
   Exercises 127
   References 128

5   Symmetry Preserving Linear Sections of the Generic Symmetric Matrix 129
   5.1   Preliminaries 129
   5.2   Codimension One Diagonal Section 130
         5.2.1   The Submaximal Minors 131
         5.2.2   The Gradient Ideal Versus the Submaximal Minors 135
         5.2.3   Complements on the Gradient Ideal 138

		5.3	Symmetric Sections of Hollow Type	141
		5.3.1	Symmetric Hollow Sections	142
		5.3.2	Symmetric Semi-hollow Linear Sections	148
	Exercises			158
	References			159

6 Linear Sections of the Generic Square Hankel Matrix ... **161**
6.1 Nuts and Bolts ... 161
6.2 Special Results for the Square Generic Hankel Matrix ... 162
    6.2.1 Related Combinatorics ... 162
    6.2.2 The Generic Polar Map ... 166
6.3 Linear Sections Preserving the Hankel Structure ... 168
    6.3.1 Hollow Like Version ... 168
    6.3.2 Semi-Hollow Like Version ... 170
    6.3.3 The Lower Minors of $\mathcal{H}_m[r]$ ... 171
    6.3.4 The Gradient Ideal of $\mathcal{H}_m[r]$ ... 175
6.4 Special Results for the Sub-Hankel Matrix ... 185
    6.4.1 Properties and Syzygies of the Partial Derivatives ... 185
    6.4.2 Homology of the Gradient Ideal ... 189
Exercises ... 197
References ... 198

7 Hankel Like Catalecticants ... **201**
7.1 Terminology ... 201
7.2 r-Leap Catalecticants Are 1-Generic ... 202
7.3 The Gruson–Peskine Paragon ... 203
7.4 Properties at Large ... 208
    7.4.1 Parabolism and the Dual Variety ... 208
    7.4.2 The Homaloidal Problem ... 212
Exercises ... 212
References ... 213

8 The Dual Variety of a Linear Determinantal Hypersurface ... **215**
8.1 Hollow-Filled, Hollow, and Semi-Hollow Linear Sections ... 215
    8.1.1 The Generic Case ... 215
    8.1.2 The Generic Symmetric Case ... 218
    8.1.3 The Generic Hankel Matrix ... 222
8.2 One Term Linear Section of the Generic Matrix ... 227
8.3 One Term Linear Section of the Generic Symmetric Matrix ... 230
Exercises ... 232
References ... 233

Part III Other Classes of Linear Sections

9 Hilbert–Burch Linear Sections ... **237**
9.1 General Linear Forms ... 237
    9.1.1 The Overall Picture ... 237

9.1.2   The Symbolic Rees Algebra 241
9.1.3   The Implicitization Case $(m = d + 1)$ 243
9.1.4   A Notable Subsumed Cremona Map: The
         Inversion Factors 248
9.2   Hilbert–Burch Matrices of Catalecticant and Hankel Nature 260
9.2.1   Generic Hilbert–Burch Catalecticant Matrices 260
9.2.2   Deformations of Sparse Hilbert–Burch Hankel
         Matrices 262
9.3   Hilbert–Burch Linear Sections in Dimension 3 264
9.3.1   The Chaos Invariant 264
9.3.2   Reduction Number and the Rees Algebra 269
9.3.3   A Notable 1-Generic Matrix 270
9.3.4   Chaos Invariant One 272
9.3.5   Paradigm Classes 277
Exercises 290
References 292

10   Apocryphal Classes 295
10.1   Mixed Catalecticants 295
10.1.1   Basic Results 296
10.1.2   Basic Conjectures 299
10.1.3   The Dual Variety 300
10.2 3×3 Matrices: Unmapped Cases 300
10.3   Stepwise Hankelization 304
10.3.1   The Especial Case $r = m - 2$ 305
10.4   Hankel Matrices Bordered by the Circulant 306
Exercises 307
References 308

A   Complement 309
A.1   Non-vanishing Hessian 309

Index 315

Acronyms

R	A commutative ring
k	A field
$\mathbf{x}$	A set of variables over a ring
$\mathbf{y}$	A set of variables over a ring
I	An ideal of a ring
$I^{(r)}$	The rth symbolic power of an ideal I
$\det \mathcal{M}$	The determinant of a square matrix $\mathcal{M}$
$\mathrm{GL}(m, R)$	The group of $m \times m$ invertible matrices over the ring R
$I_t(\mathcal{M})$	The ideal generated by the t-minors of a matrix $\mathcal{M}$
$\mathrm{adj}(\mathcal{M})$	The adjugate of an $m \times m$ matrix $\mathcal{M}$
$\mathbb{I}_m$	The $m \times m$ identity matrix
$\mathrm{Ass}(M)$	The set of associated primes of an R-module M
$H(f)$	The Hessian matrix of a polynomial f
$h(f)$	The determinant of the Hessian matrix of f
$\mathcal{R}_R(I)$	The Rees algebra of an ideal I over a ring R
$\mathcal{F}_R(I)$	The special fiber of an ideal I over a ring R
$\dim R$	The Krull dimension of a ring R.
$\mathrm{grade}\, I$	The grade of an ideal I
$\ell(I)$	The analytic spread of an ideal I
$\mathrm{ht}(I)$	The height of an ideal I
$\mu(I)$	The minimal number of generators of an ideal I
$\mathrm{r}_J(I)$	The reduction number of I with respect to a reduction J
$\mathrm{r}(I)$	The reduction number of I
$\mathrm{reg}(M)$	The Castelnuovo–Mumford regularity of a module M
$\mathrm{pd}_R(M)$	The projective dimension of an R-module M
$\mathrm{depth}(M)$	The depth of a module M over a local ring R
$\mathrm{rank}\, \varphi$	The rank of a matrix φ
$\mathcal{S}_R(I)$	The symmetric algebra of an ideal I over a ring
$\mathcal{R}_R^{(I)}$	The symbolic Rees algebra of an ideal I over a ring R
v_d	The Veronese map of order d

$\mathcal{H}_{d,e}$	The generic $e \times (d-e)$ Hankel matrix
$S(a_0, \ldots, a_r)$	The rational normal scroll of orders $a_0, \ldots, a_r$
$\mathbf{L}^{\perp}$	The orthogonal matrix to an $m \times n$ linear matrix $\mathbf{L}$
$V_p^s(f)$	The sth polar of $V(f)$ with respect to p
C_n	The generic circulant matrix on n variables
$\Omega_{A/k}$	The module of Kähler differentials of a k-algebra
$\mathfrak{P}(f)$	The polar image of f (or of the hypersurface $V(f)$)
$G(n, m)$	The Grassmann algebra of size (n, m)
$\mathrm{trdeg}_k K$	The transcendence degree of a field extension $k \subset K$
$\mathrm{Pf}_{2j}(\mathcal{M})$	The ideal generated by the $2j$-pfaffians of an alternating matrix $\mathcal{M}$
$a(C)$	The a-invariant of a standard graded k-algebra C
$e(R)$	The multiplicity of a graded ring R
$C(m, n, r)$	The r-leap $m \times n$ generic catalecticant matrix
$GP(C(m, r))$	The GP-associated matrix to $C(m, r)$

Part I
General Oversight

Chapter 1
Background Steps in Determinantal Rings

Abstract This chapter takes a general overview of determinantal rings. Our general reference for the material discussed is Bruns and Vetter (1988, Determinantal rings, LNM No. 1327. Springer, Berlin) and Simis (2023, Commutative algebra, 2nd edn. De Gruyter Graduate, Berlin), where full proofs can be found. Of these, we kept a few explicit proofs that seemed to us relevant to the general style of the book, e.g., some of the early theorems of Eagon and Northcott. In addition, we look at some important rational determinantal varieties, such as the Veronese variety, the Segre variety, and the rational normal scrolls. In these, both the algebraic and the geometric sides are emphasized. The second part of the chapter is devoted to an account of linear sections of a generic matrix in the case where these sections are sufficiently general. Thus, first one focuses on the case where the sections are general linear forms and then surveys a collection of results assuming that the section is t-generic in the sense of Eisenbud and Harris. Both approaches are useful and will be called upon in the later chapters. The chapter assumes familiarity with first principles of commutative algebra, such as contained in Simis (2023, Commutative algebra, 2nd edn. De Gruyter Graduate, Berlin), and the basic notation of algebraic geometry.

1.1 Some Matrix Notions

An $m \times n$ matrix is a rectangular array of m rows and n columns of elements (called *entries*) in a commutative ring R, delimited left and right by either ( ) or [ ]. The notation used here is $(a_{i,j})_{1 \leq i \leq m; 1 \leq j \leq n}$, or simply $(a_{i,j})$ when the ranges are self-understood, where the emphasis is on the entries of the matrix. We will not fix any preferred letter to denote a matrix as it will vary according to the convenience and emphasis of every appearance.

Picking two sets of $t \leq \min\{m, n\}$ indices each, $\{i_1, \ldots, i_t\}$ and $\{j_1, \ldots, j_t\}$, we consider the determinant

Z. Ramos, A. Simis, *Determinantal Ideals of Square Linear Matrices*,
https://doi.org/10.1007/978-3-031-55284-7_1

$$\det \begin{pmatrix} a_{i_1,j_1} & \cdots & a_{i_1,j_t} \\ \vdots & \cdots & \vdots \\ a_{i_t,j_1} & \cdots & a_{i_t,j_t} \end{pmatrix}.$$

We make the usual abuse of calling the above determinant – instead of the matrix itself – a *minor* of the matrix $(a_{i,j})$, more completely, a t-minor or a $t \times t$-minor.

If the matrix is denoted, say, $\mathcal{M}$, then $I_t(\mathcal{M})$ stands for the ideal generated by the t-minors in the ground ring R of the entries.

Say, $m \leq n$. The $(m - 1)$-minors, often called submaximal minors, play a significant role in the theory. If $m = n$, they give rise to the notion of the *cofactor* of an entry $a_{i,j}$, which is by definition the $(m - 1)$-minor omitting the ith row and the jth column, affected with the sign $(-1)^{i+j}$.

Clearly, the full nature of this ideal would not come up when taking the ground ring to be a field (except to declare whether it vanishes or not). Yet, regardless of the nature of the ground ring, one of the beautiful properties of the ideal of minors is its invariance under conjugation, namely,

$$I_t(\mathcal{M}) = I_t(\mathcal{U}\mathcal{M}\mathcal{V}),$$

where $\mathcal{U}$ and $\mathcal{V}$ are invertible matrices of order $m \times m$ and $n \times n$, respectively.

In other words, the ideal is invariant under the action of $\mathrm{GL}(m, R) \times \mathrm{GL}(n, R)$ on the set of $m \times n$ matrices with entries in the ring R. This implies that the R-linear map $\psi : R^n \to R^m$ of free R-modules induced by $\mathcal{M}$ is well-defined independently of the chosen bases. On the other hand, ψ defines an R-module E as its cokernel. Hans Fitting then proved a second spectacular result to the effect that $I_t(\mathcal{M})$, for any t, depends only on E. However, to make it work one has to replace t by $e - t$, whenever the matrix comes up by choosing a set of e generators of E. As such, the ideals of minors are often called *Fitting ideals* or *Fitting invariants* of E.

For an expanded account on Fitting ideals and their relation to the basic module theory, we refer to [30, Chapter 3, Section 3.3]. An excellent exposé about the basics of matrices over arbitrary commutative rings is [20], where the (McCoy) rank of a matrix and other invariants are defined (the notion of rank is also given in [30, Definition 3.3.5]).

1.2 Generic Determinantal Rings

Having established the notion of ideals of minors in the previous section, one is ready for the basic notion of this section.

Definition 1.1 Let R be any commutative ring, let $\mathbf{X} = (x_{i,j})$ denote an $m \times n$ matrix of indeterminates over R, and let $R[\mathbf{X}]$ by abuse denote the polynomial ring in the entries of $\mathbf{X}$. Given an integer $1 \leq t \leq \min\{m, n\}$, the R-algebra $R[\mathbf{X}]/I_t(\mathbf{X})$ is called a (generic) *determinantal ring* over R of order t.

We will soon talk about certain numerical invariants of this ring. For the moment being, note that often one needs to consider non-generic versions of it. One approach to such a consideration is typically encoded in a ring homomorphism $R[\mathbf{X}] \twoheadrightarrow S$ and consists in taking the $m \times n$ matrix M whose entries are the images of the indeterminates $x_{i,j}$. Then $S/I_t(M)$ is often called a (special) determinantal ring, coming equipped with the specialization map $R[\mathbf{X}]/I_t(\mathbf{X}) \twoheadrightarrow S/I_t(M)$.

It may of course happen that, for a given $t \geq 1$, $I_t(M)$ vanishes. For example, if S is a field, then such a nonzero homomorphism $R[\mathbf{X}]/I_t(\mathbf{X}) \to S$ implies that M has rank at most $t - 1$; hence the S-linear map $S^n \to S^m$ induced by M factors through S^{t-1}. This gives that $M = M_1 M_2$, where M_1, M_2 are matrices over S of order $m \times (t - 1)$ and $(t - 1) \times n$, respectively.

This is suggestive of a pure ring theoretic map that factors likewise, to wit, one lets $\mathbf{Y}$ and $\mathbf{Z}$ stand for new matrices of mutually independent indeterminates over R, of respective orders $m \times (t - 1)$ and $(t - 1) \times n$, and defines an R-algebra homomorphism (the same notational abuse)

$$\rho_t : R[\mathbf{X}] \to R[\mathbf{Y}, \mathbf{Z}], \ x_{i,j} \mapsto w_{i,j}, \tag{1.1}$$

where $w_{i,j}$ denotes the entry of order (i, j) of the product matrix $\mathbf{YZ}$. Similarly, the matrix $\mathbf{YZ}$ represents a homomorphism of $R[\mathbf{Y}, \mathbf{Z}]$-modules

$$R[\mathbf{Y}, \mathbf{Z}]^n \to R[\mathbf{Y}, \mathbf{Z}]^m$$

satisfying $I_t(\mathbf{YZ}) = 0$. In particular, $I_t(\mathbf{X}) \subset \ker \rho_t$. A remarkable issue is that this inclusion is an equality if R is an integral domain; hence, in this case, $I_t(\mathbf{X})$ is a prime ideal – this and other matters will be considered in Section 1.4.

Thus, in particular, if $R = k$ is a field, it implies in addition that a generic determinantal variety is *unirational* in the sense that its coordinate ring embeds into a polynomial ring of the coefficients.

Say, k is moreover algebraically closed. The *determinantal variety* of order $t - 1$, denoted here $\mathfrak{D}_{t-1}$, is the zero locus of the determinantal ideal $I_t(\mathbf{X})$ in the affine space $\mathbb{A}_k^{mn}$ endowed with its affine coordinate ring $k[\mathbf{X}]$. Thus, the affine defining ideal of $\mathfrak{D}_{t-1}$ in the embedding $\mathfrak{D}_{t-1} \subset \mathbb{A}_k^{mn}$ is the radical of $I_t(\mathbf{X})$. In this environment one can show rather easily that this radical is a prime ideal or, equivalently, that $\mathfrak{D}_{t-1}$ is irreducible in the Zariski topology of $\mathbb{A}_k^{mn}$.

For this it suffices to show that the determinantal variety is the image of a map between affine varieties. Let us identify $\mathbb{A}_k^{mn}$ with the k-vector space of all $m \times n$ matrices with entries in k. Then $\mathfrak{D}_{t-1}$ gets identified with the subset $\mathbb{V}_{t-1}$ of $m \times n$ matrices of rank $\leq t - 1$. As remarked above, any matrix in $\mathbb{V}_{t-1}$ is a product of two matrices of order $m \times (t - 1)$ and $(t - 1) \times n$, respectively. This shows that the product map of matrices

$$\mathbb{A}_k^{m(t-1)} \times \mathbb{A}_k^{(t-1)n} \to \mathbb{A}_k^{mn}$$

is onto $\mathbb{V}_{t-1}$. Since this map is algebraic and the source, being an affine space, is irreducible, one is done with the claim.

Surprisingly enough, there is no similarly trivial way of getting the dimension of $\mathfrak{D}_{t-1}$. It is possible to argue by describing a birational map of $\mathfrak{D}_{t-1}$ to a product of two affine spaces obtained by suitably chosen subspaces and their vector complements. Unfortunately, to our view, the method is a bit opaque from the algebraic stand.

Another proof, emphasizing even more the algebraic geometric content, is based on introducing a suitable incidence variety involving the Grassmann variety of subspaces – somewhat surprising the Grassmann variety and its dimension are more easily grasped than the determinantal varieties.

Notably, a lower bound for $\dim \mathfrak{D}_{t-1}$ is more easily available. In order to both state it and give a proof, we resort to a more algebraic version.

Proposition 1.2 ([9, 10]) *Let $M = (a_{i,j})$ denote an $m \times n$ matrix with entries in a Noetherian ring R, and let $1 \leq t \leq \min\{m, n\}$. If $I_t(M) \neq R$, then it has height at most $(m - t + 1)(n - t + 1)$.*

Proof The following proof is due to J. Eagon and D. Northcott and is part of the PhD thesis (1962) of the former – the method itself has been dubbed the "Eagon–Northcott indeterminate trick."

Set $I := I_t(M)$. It suffices to show that every minimal prime ideal of R/I has height at most $(m - t + 1)(n - t + 1)$. Let $P \supset I$ be such a prime ideal. One inducts on m. For $m = 1$, then necessarily $t = 1$ and I is generated by n elements; hence the result follows from Krull's prime ideal theorem.

Assuming $m > 1$ and localizing at P, it is easy to see that one may suppose that R is a Noetherian local ring with maximal ideal P and that I is a P-primary ideal. We may further assume that $I_1(M) \subset P$, otherwise some entry is a unit, and hence I will be generated by the minors of order $t - 1$ of an obvious $(m - 1) \times (n - 1)$ matrix. In this case, the result follows from the inductive hypothesis on m. One may in addition suppose that $t > 1$ as the height of a minimal prime of R/I is at most the number of a set of generators (again by Krull's prime ideal theorem).

Now one is ready for the main step (trick). Let x denote an indeterminate over R, and consider the "deformed" matrix

$$M(x) := \begin{pmatrix} a_{1,1}+x & a_{1,2} & \cdots & a_{1,n} \\ a_{2,1} & a_{2,2} & \cdots & a_{2,n} \\ \vdots & \vdots & \ddots & \vdots \\ a_{m,1} & a_{m,2} & \cdots & a_{m,n} \end{pmatrix}.$$

We consider the free extension $R \subset R[x]$. Set $\widetilde{I} := I_t(M(x)) \subset R[x]$. Since $t > 1$ and $a_{i,j} \in P$ for all i, j, it follows that $\widetilde{I}$ is contained in the extended ideal $PR[x]$, which is prime in $R[x]$ of the same height as P. It is clear by a direct inspection that

$(\widetilde{I}, x) = (IR[x], x)$, and it is known that $PR[x]$ is a minimal prime of $R[x]/IR[x]$. But, more holds:

CLAIM $PR[x]$ is a minimal prime of $R[x]/\widetilde{I}$.

To see this, let $Q \subset R[x]$ be a prime ideal such that $\widetilde{I} \subset Q \subset PR[x]$. Introducing x, we see that $(PR[x], x)$ is a prime ideal containing the ideal (Q, x); hence there exists a minimal prime Q' of (Q, x) contained in $(PR[x], x)$. One has

$$IR[x] \subset (IR[x], x) = (\widetilde{I}, x) \subset (Q, x) \subset Q' \subset (PR[x], x).$$

By contraction back to R, one gets $I \subset Q' \cap R$, and since P is the only prime containing I (because I is P-primary), one must have $Q' \cap R = P$. It follows that $PR[x] \subset Q'$, and since $x \in Q'$, one gets an equality $Q' = (PR[x], x)$, i.e., $(PR[x], x)$ is a minimal prime of (Q, x).

To conclude, we pass to the residue class ring $R[x]/Q$ and observe that its ideal $(Q, x)/Q$ is principal with minimal prime $(PR[x], x)/Q$. By Krull's principal ideal theorem, the latter has height at most 1. Therefore the chain

$$\{0\} = Q/Q \subset PR[x]/Q \subset (PR[x], x)/Q$$

must collapse somewhere, and it can only happen in the leftmost inclusion since $x \notin PR[x]$ as $1 \notin P$. This shows that $PR[x] = Q$ is a minimal prime of $R[x]/\widetilde{I}$ as was to be shown.

Now localize $R[x]$ at $P' := PR[x]$. Since $a_{11} + x \notin P'$ (again because $1 \notin P$), it becomes invertible in this localization; hence by the same argument as in the beginning of the proof, the prime ideal $P'_{P'}$ has by the inductive hypothesis height at most $(m - t + 1)(n - t + 1)$. But since P' is prime, P' and $P'_{P'}$ have the same height, and so do $P' = PR[x]$ and P. Thus, the proof is complete. $\square$

This is a marvelous piece of core commutative algebra as it uses exactly first principles.

Remark 1.3 We will see in Section 1.4 that if M is generic, then the above number is actually the height of the ideal $I_t(M)$.

1.3   Classical Rational Determinantal Varieties

Let k denote an algebraically closed field (of characteristic zero whenever required in the context).

Some typical projective varieties are obtainable as images of rational maps with source a subvariety X of the projective n-space $\mathbb{P}^n = \mathbb{P}^n_k$. For the notion of rational maps in this context, we refer to [27, Section 3.1].

1.3.1 Veronese Varieties

Note that we are for the sake of the present examples, taking the subvariety $X \subset \mathbb{P}^n$ to be $\mathbb{P}^n$ itself.

Fixing the space $\mathbb{P}^n$, given an integer $d \geq 1$, the celebrated *Veronese map* of order d is the rational map $v_d : \mathbb{P}^n \dashrightarrow \mathbb{P}^N$ defined by

$$\left(x_0^d : x_0^{d-1} x_1 : \cdots : x_{n-1} x_n^{d-1} : x_n^d \right),$$

where the coordinate forms are the monomials of degree d in the ground variables of the homogeneous coordinate ring of $\mathbb{P}^n$. Here the order of the monomials is irrelevant in that the resulting images will be projectively equivalent. The same remark holds if one replaces the monomial basis of the k-vector space $k[\mathbf{X}]_d$ spanned by the forms of degree d by any other basis. For this reason one tends to think of the Veronese map as the rational map defined by the *linear system $k[\mathbf{X}]_d$.*

The image of v_d is called the *Veronese variety* (or simply, the *Veronesean*) of order d of $\mathbb{P}^n$. The designation "order" seems more convenient to avoid confusion with the degree of the variety in this embedding.

Setting $\mathbf{X} = \{x_0, \ldots, x_n\}$, the monomials generate an $(\mathbf{X})$-primary (i.e., irrelevant) ideal; hence the Veronese map is defined everywhere.

Showing that the map is birational onto the image is within reach. Namely, one has that $N = \binom{n+d}{d} - 1$, so let $\mathbf{T} = \{T_\alpha \,|\, |\alpha| = d\}$ denote a corresponding set of $\binom{n+d}{d}$ new variables over k that will stand for the coordinate functions on $\mathbb{P}^N$. Now one considers the first $n + 1$ monomials in the above ordering, $x_0^d, x_0^{d-1} x_1, \ldots, x_0^{d-1} x_n$, and let $T_0, T_1, \ldots, T_n$ denote the coordinates on the target $\mathbb{P}^N$ defining the corresponding hyperplanes, where for reading facility we wrote $T_0 = T_{(d,0,\ldots,0)}, T_1 = T_{(d-1,1,0,\ldots,0)}$, etc. The inverse map is given by the representative $(\overline{T_0} : \cdots : \overline{T_n})$, where the overline means taking residue modulo the homogeneous defining ideal of $v_d(\mathbb{P}^n) \subset \mathbb{P}^N$. In fact, adhering to one of the previous checking devices, one gets

$$(\overline{T_0}(\mathbf{X}_d) : \cdots : \overline{T_n}(\mathbf{X}_d)) = \left(x_0^d : \cdots : x_0^{d-1} x_n \right) = (x_0 : \cdots : x_n).$$

In particular, one sees that the inverse map is given by a *linear projection* – see [29, Chapter 1, Section 1.3] for a modern account on these.

As a corollary, the Veronesean has dimension n.

The following question does not seem to have been satisfactorily addressed so far.

QUESTION For arbitrary positive integers n and d, what is the structure of the homogeneous defining ideal of the Veronesean of order d of $\mathbb{P}^n$?

What can be seen at the outset is that the variety is the zero locus of a bunch of 2-forms, in the following way: For each quadruple of multi-exponents α, β, γ, and δ such that $\mathbf{X}^\alpha \mathbf{X}^\beta = \mathbf{X}^\gamma \mathbf{X}^\delta$, one derives a quadratic defining equation $T_\alpha T_\beta - T_\gamma T_\delta$

for the Veronesean. That these binomials of degree 2 actually generate the defining ideal of the Veronese embedding is also true and has an interesting background.

HISTORICAL INTERMEZZO As it appears, the first proof is due to L. Goddard in 1943 [15, Theorem I]. More general results, obtained with elements in a regular sequence as a replacement for independent variables over a field, appeared 20 years circa later in [2, 8]. The latter results actually dealt with the problem of explaining the Rees algebra of the corresponding "monomial" ideals – these were the heroic times of the Rees and symmetric algebras in the hands of P. Samuel, A. Micali, and the English School of Rees. From a different perspective, it has been shown that the homogeneous defining ideal of the Veronese embedding is still generated by the 2-minors of certain matrices – so-called symmetric hypermatrices as in [3, Section 3] or else a certain version of catalecticant matrices as in [26, Corollary 2.4], the latter having been thoroughly studied by Iarrobino, Geramita, and others. See also [23] for yet another approach. An additional source is the PhD thesis of Parolin [25]. Other references touching the subject but mainly dealing with the associated secant varieties and the built-in tensor theory are, e.g., [4, 21].

Remark 1.4 The fact that relations are binomials is not a surprise since the parameters are monomials; the fact that binomials of degree 2 suffice brings up slight perplexity.

There is one case where the exact determinantal structure of the defining ideal is known: $d = 2$.

Here the idea is to note that the number $\binom{n+2}{2}$ of monomials of degree 2 is also the number of distinct entries in an $(n+1) \times (n+1)$ symmetric matrix with indeterminate entries. Thus, denoting the entries as

$$T_{i,j}, 0 \leq i \leq j \leq n,$$

let $\mathcal{T}_{n+1}$ denote such a matrix. For instance, with $n = 2$ we would take the symmetric matrix

$$\begin{pmatrix} T_{0,0} & T_{0,1} & T_{0,2} \\ T_{0,1} & T_{1,1} & T_{1,2} \\ T_{0,2} & T_{1,2} & T_{2,2} \end{pmatrix}.$$

Here $T_{0,0} \mapsto x_0^2$, $T_{0,1} \mapsto x_0 x_1$, $T_{0,2} \mapsto x_0 x_2$, $T_{1,1} \mapsto x_1^2$, $T_{1,2} \mapsto x_1 x_2$, $T_{2,2} \mapsto x_2^2$.

Proposition 1.5 ([28]) *The homogeneous defining ideal of the Veronese embedding $v_1(\mathbb{P}^n) \subset \mathbb{P}^N$ is the ideal generated by the 2×2 minors of $\mathcal{T}_{n+1}$.*

Proof (Sketch) First one checks that the minors really vanish on the monomials of degree 2, i.e., they are contained in the homogeneous defining ideal of the Veronesean. Next one shows that the ideal of these minors is a prime ideal of height

$\binom{n+1}{2}$. Finally, one observes (easily) that the height of the Veronese embedding in $\mathbb{P}^N$ is $\binom{n+2}{2} - 1 - n = \binom{n+1}{2}$. This does it. $\square$

(The only reason we explained this proof is that it will be a model for other upcoming situations).

In general, the homogeneous coordinate ring of the dth Veronese embedding of $\mathbb{P}^n$ can be viewed – as will be the case of several of the unirationally defined varieties here – as the special fiber of the ideal $(\mathbf{X})^d \subset k[\mathbf{X}]$. This will provide a nice interplay between the geometry and the underlying algebra.

We conclude these general preliminaries on the Veronese embedding with the following remarkable result.

Proposition 1.6 *The Veronese embedding $v_d(\mathbb{P}^n) \subset \mathbb{P}^N$ is a projectively normal and arithmetically Cohen–Macaulay variety.*

Proof Both properties are essentially algebraic, meaning that the k-subalgebra $k[\mathbf{X}_d]$ is integrally closed and Cohen–Macaulay, respectively. By Hochster [19], it suffices to prove the first of these assertions. This assertion follows by [31, Theorem 7.1] provided one shows that the Rees algebra of the ideal $I \subset k[\mathbf{X}]$ generated by the linear system $\mathbf{X}_d$ is normal, i.e., that any power of I is normal (*complete* in the older language of Zariski). For that, the following cute argument was given by Villarreal: Pick a monomial f in the integral closure of the power I^n, that is, $f^k \in I^{kn}$, for some $k \geq 1$. Taking degrees we get $k \deg(f) \geq dkn$, and consequently, $\deg(f) \geq dn$. Thus, we can find monomials $g_1, \ldots, g_n$ of degree d such that the product $g_1 \cdots g_n$ divides f; hence, $f \in I^n$. $\square$

Example 1.7 The Veronesean embedding of order d of $\mathbb{P}^1$ is called a *rational normal curve*. The corresponding Veronese map is defined by the $d + 1$ monomials $x_0^d, x_0^{d-1}x_1, \ldots, x_1^d$; hence the Veronese embeds in $\mathbb{P}^d$. Since the arbitrary Veronese embedding of order d of $\mathbb{P}^n$ has degree d^n (use Bézout theorem and the bijection between hypersurfaces of degree d in the source and hyperplanes sections in the target), then the Veronese curve has degree d.

The homogeneous defining ideal of the rational normal curve can be expressed as the ideal of maximal minors of the so-called *Hankel* or *persymmetric* or *orthosymmetric* matrix of order $2 \times d$

$$\begin{pmatrix} T_0 \; T_1 \; T_2 \; \ldots \; T_{d-1} \\ T_1 \; T_2 \; T_3 \; \ldots \; T_d \end{pmatrix}. \tag{1.2}$$

Following a common pattern in this book, one shows that this ideal is prime, is contained in the homogeneous defining ideal of the Veronese curve, and has the right height. Arithmetic normality (and Cohen–Macaulayness) is given by Proposition 1.6. Since the ideal has the "expected" height, an alternative way to show these properties is via yet another theorem of Eagon–Northcott which tells that the ideal is Cohen–Macaulay. Therefore it satisfies property (S_2) of Serre's, and since the corresponding variety is non-singular, then Serre's criterion applies, so we conclude normality as well (note this reverts the argument in the proof of Proposition 1.6).

For a different approach in greater generality, we refer to [33].

More generally, the homogeneous defining ideal of the rational normal curve can be expressed as the ideal of 2×2 minors of the generic $e \times (d - e)$ Hankel matrix

$$
\mathcal{H}_{d,e} = \begin{pmatrix}
T_0 & T_1 & \cdots & T_{d-e} \\
T_1 & T_2 & \cdots & T_{d-e+1} \\
\vdots & \vdots & \cdots & \vdots \\
T_{e-1} & T_e & \cdots & T_{d-1} \\
T_e & T_{e+1} & \cdots & T_d
\end{pmatrix}.
$$

The other minors have also an important geometric role. Namely, for any $t \le e$, the ideal $I_{t-1}(\mathcal{H}_{d,e})$ is the defining ideal of the t-secant variety of the rational normal curve [18, Proposition 9.7].

Generic Hankel matrices, with a focus on the square ones and their linear sections, will be studied in Chapter 6.

We leave as an amusing exercise to look at a few other initial cases of the Veronese embedding. More challenging problems will be given in the Exercises.

1.3.2 Rational Normal Scrolls

There are (at least) three different ways of introducing rational normal scrolls.

The first, perhaps more familiar to differential geometers, is to go ab initio with the notion of a scroll as a sort of crude ruled variety and then descend by imposing restrictions making it a rational and normal variety. Easy start, yet pay a price in the way down. As far as the underlying algebra is concerned, this is the most opaque of the three (see [7, Section 1.2]).

The second notion is by the way of projective bundles. This is very efficient but pushes the underlying algebra to the very end as much as possible (see [11]).

Finally, the third is a geometric pearl, working in terms of a construction based on earlier, easier objects. In addition it has the advantage (to the readers of this book) in that the underlying ideals and rings soon show up in their clear role.

We will not adopt the first option as the theory of general scrolls would take us far out the intended goal of the book. Let us give a tint of the second option or at least its highbrow formalism and then explain in the way of [11] how one can land on the third option and the resulting algebra.

Throughout, k is an algebraically closed field. One starts with a vector bundle (locally free sheaf) $\mathcal{E}$ over $\mathbb{P}^1$ of rank $r+1$. For a more algebraic version, think about a finitely generated graded $k[s, t]$-module locally free of rank $r+1$ on the punctured spectrum $\mathrm{Spec}(k[s, t]) \setminus \{(t, s)\}$. The Birkhoff–Grothendieck theorem says that $\mathcal{E} \simeq O(a_0) \oplus \cdots \oplus O(a_r)$ for certain integers $a_0 \le \cdots \le a_r$ that depend solely on the isomorphism class of the bundle (see [17] for an elementary proof based on simple linear algebra over the Laurent ring $k[s, s^{-1}]$, an idea going back to George Birkhoff).

We assume moreover that $\mathcal{E}$ is positive, i.e., $a_0 \geq 0$ – in fact, for simplicity, we assume that $a_0 > 0$. In this case $\mathcal{E}$ is generated by $\sum_{i=0}^{r}(a_i + 1) = \sum_{i=0}^{r} a_i + r + 1$ global sections (think about the number of monomials of degree a_i in two variables s and t over k). Next take the projective version $\mathbb{P}(\mathcal{E}) = \mathrm{Proj}(\mathrm{Sym}_R(\mathcal{E}))$ of the vector bundle ("total space"), where $R = k[s, t]$. Then, besides the structural map $\mathbb{P}(\mathcal{E}) \twoheadrightarrow \mathbb{P}^1 = \mathrm{Proj}(R)$ induced by the inclusion $R \hookrightarrow \mathrm{Sym}_R(\mathcal{E})$ in degree 0, one gets a map

$$\mathfrak{s} : \mathbb{P}(\mathcal{E}) \longrightarrow \mathbb{P}^{\sum_{i=0}^{r} a_i + r}.$$

Since $a_r > 0$, it can be shown that this map is "birational" onto its image – a matter of some clarification as it requires the idea of a birational map for arbitrary non-embedded varieties. The latter has therefore dimension $r + 1$ and will be denoted $S(a_0, \ldots, a_r)$.

Definition 1.8 $S(a_0, \ldots, a_r)$ is the *rational normal scroll* of orders $a_0, \ldots, a_r$.

(The terminology "orders" is not very convenient, but actually it turns out that the degree of the scroll in this embedding is exactly $\sum_{i=0}^{r} a_i$).

These preliminaries take us to the third alternative in the following way. The canonical projection $\pi_i : \mathcal{E} \twoheadrightarrow O(a_i)$ induces a section

$$\mathbb{P}^1 \hookrightarrow \mathbb{P}(\mathcal{E}) \tag{1.3}$$

by identifying $\mathbb{P}^1$ with $\mathbb{P}(O(a_i))$. With this identification, the restriction of the tautological line bundle $O_{\mathbb{P}(\mathcal{E})}(1)$ to $\mathbb{P}(O(a_i)) = \mathbb{P}^1$ is $O_{\mathbb{P}(O(a_i))}(1) = O_{\mathbb{P}^1}(a_i)$. Now the map $\mathfrak{s}$ is really induced by the linear system associated to $O_{\mathbb{P}(\mathcal{E})}(1)$ – perhaps the most mechanical way to see this is to note that

$$\mathbb{P}(\mathcal{E}) = \mathrm{Proj}(\mathrm{Sym}(O(a_0) \oplus \cdots \oplus O(a_r))) = \mathrm{Proj}(\mathrm{Sym}(O(a_0)) \otimes \cdots \otimes \mathrm{Sym}(O(a_r)))$$

$$= \mathbb{P}(O(a_0)) \times_k \cdots \times_k \mathbb{P}(O(a_r)),$$

so that on each factor of order i above $\mathfrak{s}$ is defined by the monomials $s^{a_i}, s^{a_i-1}t, \ldots, t^{a_i}$, and hence is induced on that factor by the linear system associated to $O_{\mathbb{P}(O(a_i))}(1) = O_{\mathbb{P}^1}(a_i)$.

Therefore, the image of the section (1.3) is the Veronese C_{a_i} of order a_i of $\mathbb{P}^1$ inside a well-defined linear subspace $\mathbb{P}^{a_i} \subset \mathbb{P}^{\sum_{i=0}^{r} a_i + r}$ (this $\mathbb{P}^{a_i}$ is quite canonical by the earlier mechanism). Moreover, these subspaces for varying $i = 0, \ldots, r$ are mutually complementary (i.e., at the level of the vector space $k^{\sum_{i=0}^{r} a_i + r + 1}$, the respective vector spaces form a direct sum and cover the whole ambient).

Finally, letting $\varphi_i : \mathbb{P}^1 \to C_{a_i}$ denote the above map (really projectively equivalent to the Veronese map), the rational normal scroll $S(a_0, \ldots, a_r)$ is the subvariety swept out by (i.e., the set union of) the r-planes (i.e., linear subspaces of dimension r) spanned by the points $\varphi_0(p), \ldots, \varphi_r(p)$, for varying $p \in \mathbb{P}^1$.

These r-planes can also be obtained as images by $\mathfrak{s}$ of the fibers over closed points of the structural map $\mathbb{P}(\mathcal{E}) \twoheadrightarrow \mathbb{P}^1$ – note that every such fiber is identified with a linear k-space of dimension r.

Wrapping up the algebraic version yields the following:

Proposition 1.9 *The homogeneous defining ideal of the rational normal scroll in the above embedding is the ideal of 2×2 minors of the following $2 \times ((\sum_{i=0}^r a_i) + r + 1)$ matrix:*

$$\begin{pmatrix} x_0 \ x_1 \ \ldots \ x_{a_0-1} & y_0 \ y_1 \ \ldots \ y_{a_1-1} & \ldots & z_0 \ z_1 \ \ldots \ z_{a_r-1} \\ x_1 \ x_2 \ \ldots \ \ x_{a_0} & y_1 \ y_2 \ \ldots \ \ \ y_{a_1} & \ldots & z_1 \ z_2 \ \ldots \ \ z_{a_r} \end{pmatrix}.$$

Proof (Sketch) Following the scheme of proof of Proposition 1.5, one first checks that the 2-minors vanish on $S(a_0, \ldots, a_r)$. This is the easy part after we have identified the explicit parameters of the map. Namely, a typical 2-minor is of the form, say, $x_l y_{k+1} - x_{l+1} y_k$, and the association between variables and parameters reads as follows:

$$x_l \mapsto s^{a_0-l} t^l,\ x_{l+1} \mapsto s^{a_0-(l+1)} t^{l+1},\ y_k \mapsto s^{a_1-k} t^k,\ y_{k+1} \mapsto s^{a_1-(k+1)} t^{k+1}.$$

The required vanishing is then immediately seen.

The second task is to check that these minors generate a prime ideal of height $\sum_{i=0}^r a_i - 1$ (note that this exceeds by r the sum of the heights of the individual Hankel pieces). The height is not so scary as soon as we show that the matrix is a nice specialization of the analogous generic matrix; primeness is the difficult issue, but we will leave it at that.

Finally, the height of the defining ideal of the embedding is

$$\sum_{i=0}^r a_i + r - \dim S(a_0, \ldots, a_r) = \sum_{i=0}^r a_i + r - (r+1) = \sum_{i=0}^r a_i - 1.$$

Thus, the proof is complete. $\qquad\square$

1.3.3 Segre Varieties

In a sense this is the easiest of the examples so far although it carries a bit of perplexity since the departing source is not a projective variety. Thus, rigorously, its discussion would require the full notion of rational maps of varieties in the abstract.

Actually we have already encountered much earlier the required gimmick of this case in Section 1.2, only here we have a special case, namely, when $R = k$ is a field and $t = 2$; hence $\mathbf{Y}$ and $\mathbf{Z}$ are $m \times 1$ and $1 \times n$ matrices, respectively. Explicitly, take

$$\mathbf{Y} = (y_0 \cdots y_{m-1})^t \text{ and } \mathbf{Z} = (z_0 \cdots z_{n-1}),$$

and consider the subring $k[\mathbf{YZ}] \subset k[\mathbf{Y}, \mathbf{Z}]$.

So, what is the Segre map? It arises by considering the fiber product $\mathbb{P}^{m-1} \times_k \mathbb{P}^{n-1}$ (called a *biprojective space*) and introducing the map

$$\mathbb{P}^{m-1} \times_k \mathbb{P}^{n-1} \longrightarrow \mathbb{P}^{mn-1} \tag{1.4}$$

$$((a_0 : \cdots : a_{m-1}); (b_0 : \cdots : b_{n-1})) \mapsto (a_0 b_0 : \cdots : a_{m-1} b_{n-1}).$$

Note that the bihomogeneous coordinate ring of the biprojective space is the tensor product over k of the respective homogeneous coordinate rings of the two factors, namely,

$$k[\mathbf{Y}] \otimes_k k[\mathbf{Z}] \simeq k[\mathbf{Y}, \mathbf{Z}].$$

Decomposing the map (1.4) in its induced map onto the image and the inclusion of the latter in the projective ambient, we find the corresponding algebraic transcription as the inclusion of k-algebras $k[\mathbf{YZ}] \subset k[\mathbf{Y}, \mathbf{Z}]$ given before and the surjection $k[\mathbf{X}] \twoheadrightarrow k[\mathbf{YZ}]$ with kernel $I_1(\mathbf{X})$.

In addition, one easily sees that (1.4) is defined everywhere. Had we introduced the definition of rational maps to encompass more general varieties, this map could be checked to be birational onto the image and, in fact, an isomorphism (biregular) onto the image in the category of all varieties. This shows that any problem concerning biprojective varieties can be translated into one about projective varieties.

As in the case of the Veronese map, we leave as an amusing exercise to look at a few initial cases of the Segre embedding. For example, the Segre embedding of $\mathbb{P}^1 \times \mathbb{P}^1$ in $\mathbb{P}^3$ is the simplest scroll surface $S(1, 1)$.

1.4 Theorems on Generic Maximal Minors

There are three fundamental results concerning the determinantal ring $R[\mathbf{X}]/I_m(\mathbf{X})$ of maximal minors of the generic $m \times n$ matrix $(\mathbf{X})$ ($m \leq n$) over the Noetherian ring R:

- If R is Cohen–Macaulay, then $R[\mathbf{X}]/I_m(\mathbf{X})$ is a Cohen–Macaulay ring of codimension $n - m + 1$.
- If R is an integral domain, then $I_m(\mathbf{X})$ is a prime ideal.
- If R is (locally) regular, then the set-theoretic singular locus of $R[\mathbf{X}]/I_m(\mathbf{X})$ is $I_{m-1}(\mathbf{X})/I_m(\mathbf{X})$.

We note that these facts are not self-evident even in the case where R is a field.

The first two will be typical of many questions in this book regarding more encompassing and difficult environments. All three results extend to the case of t-minors, $t \leq m$. In the case of the last two, the extension is proved in a similar way. The corresponding codimension part of the first result reads $(m - t + 1)(n - t + 1)$ instead of $n - m + 1$ and can be shown by similar methods. Instead, the Cohen–Macaulay part is quite a bit more delicate, being a landmark in the post-classical invariant theory. The proof has undergone a long thread by several authors, starting with Northcott [24] until the definite general form by Hochster et al. If R is a field, a proof can be found in [32, Corollary 14] by drawing upon the theory of Stanley–Reisner rings of shellable complexes.

In this book we will proceed as follows: (1) give the proof of the codimension part of the first of these theorems, for arbitrary t, (2) give the full proof of the second theorem assuming known that $I_t(\mathbf{X})$ has pure grade (a consequence of the first theorem), and (3) give the proof of the third theorem in the case where the ground ring is a perfect field (proposing the general case of locally regular Noetherian ground ring as an exercise).

For this, we roughly follow the presentation in [6, Chapter 2]. For those interested in the precise historical roots of each result, we recommend reading [6, Section E. Comments and References] – a curious account of the Eagon–Northcott results is retrievable in [22]. The results have all been recovered later by a mix of combinatorial methods and Gröbner bases. A comprehensive account is to be found in [5].

The proofs look more general – and often easier – if one argues in terms of the grade of an ideal instead of its codimension (height).

Proposition 1.10 (Eagon) *Let $\mathbf{X}$ denote an $m \times n$ matrix of indeterminate entries over a Noetherian ring R, and let $1 \leq t \leq \min\{m, n\}$. Then the grade of the ideal $I_t(\mathbf{X})$ is $(m - t + 1)(n - t + 1)$.*

Proof By Proposition 1.2 and by the preceding observation, it suffices to show that the grade of $I_t(\mathbf{X})$ is at least $(m - t + 1)(n - t + 1)$.

Proceed by induction on t. For $t = 1$, the result is trivial since the generators themselves form an R-sequence. Thus, assume that $t > 1$, and let $\{u_1, \ldots, u_g\} \subset I$ denote an R-sequence of maximal length in $I := I_t(\mathbf{X})$. Set $J := (u_1, \ldots, u_g)$. The elements of I are zerodivisors on R/J; hence $I \subset P$ for certain associated prime P of R/J. Clearly, all three ideals $J \subset I \subset P$ have the same grade g. Since $t > 1$ and $g \leq (m - t + 1)(n - t + 1)$, then $g < mn$. This means that some entry of $\mathbf{X}$ does not belong to P; we may assume that $x_{1,1} \notin P$.

We next move over to the ring of fractions $\widetilde{R} := R[\mathbf{X}][x_{1,1}^{-1}]$, and let $\widetilde{J} \subset \widetilde{I} \subset \widetilde{P}$ denote the corresponding extended ideals. Note that the image of the regular R-sequence is a regular $\widetilde{R}$-sequence, and $\widetilde{P}$ is an associated prime of $\widetilde{R}/\widetilde{J}$; hence all three extended ideals still have grade g.

But now, we perform elementary row and column operations so as to make vanish all entries along the first row and column of the corresponding matrix $\widetilde{\mathbf{X}}$ over $\widetilde{R}$, except $x_{1,1}$. It follows that $\widetilde{I}$ is generated by the $(t - 1) \times (t - 1)$ minors of the submatrix of $\widetilde{\mathbf{X}}$ obtained by omitting its first row and column. It remains to show that we are in a position as to apply the inductive hypothesis, in which case one has

$$\mathrm{grade}(I) = \mathrm{grade}(\widetilde{I}) = (m-1-(t-1)+1)(n-1-(t-1)+1) = (m-t+1)(n-t+1),$$

as required.

Note that we are in a more delicate situation than the localization at the end of the proof of Proposition 1.2, as we have to make sure that the localization is still a polynomial ring over a Noetherian ring. That is, we need that $\widetilde{R}$ be a ring of polynomials over a Noetherian subring. For this, we observe that $\widetilde{R} = \overline{R}[\mathbf{Y}]$, where

$$\overline{R} := R[x_{1,1}, \ldots, x_{1,n}; x_{2,1}, \ldots, x_{m,1}, x_{1,1}^{-1}] \quad \text{and} \quad \mathbf{Y} = \left\{ x_{i,j} - x_{i,1}x_{1,j}x_{1,1}^{-1} \right\}_{\substack{2 \le i \le m \\ 2 \le j \le n}}.$$

Now, since $\{x_{i,j}, 2 \le i \le m, 2 \le j \le n\}$ is an algebraically independent set over $\overline{R}$ and $x_{i,1}x_{1,i}x_{1,1}^{-1} \in \overline{R}$, then so is the set $\mathbf{Y}$. Therefore, up to a trivial isomorphism, $\widetilde{R}$ is of the desired form. $\square$

It is quite surprising that the singular locus of the generic determinantal variety defined by $I_t(\mathbf{X})$ can be determined by a similar inductive procedure, since typically spotting singularities depends on the Jacobian criterion. There is a more encompassing version as above over a coefficient ring which is locally regular everywhere – and even over an arbitrary Noetherian coefficient ring, but we will for simplicity assume that the latter is a field.

Informally, the singular locus of a variety is the set of its singular points. Since there is a long history about it in algebraic geometry, we will play on the safe side and assume throughout that the base field k is algebraically closed (perfect will do). The singular locus will be a closed subset of the variety; hence it is defined by a certain ideal sheaf – in our simple setup of projective varieties, an ideal in the homogeneous coordinate ring of the variety.

Proposition 1.11 *Let $\mathbf{X}$ denote an $m \times n$ generic matrix over a perfect field k, and let $1 \le t \le \min\{m, n\}$. Then the singular locus of the projective variety defined by $I_t(\mathbf{X}) \subset R := k[\mathbf{X}]$ is defined by the ideal $I_{t-1}(\mathbf{X})/I_t(\mathbf{X})$.*

Proof The argument will be purely algebraic. For that we assume known that a point of the variety is non-singular if and only if the local ring of the variety at that point is regular. In this language one proves a bit more, namely, that for a prime ideal $P \subset k[\mathbf{X}]/I_t(\mathbf{X})$ the local ring $(k[\mathbf{X}]/I_t(\mathbf{X}))_P$ is regular if and only if $I_{t-1}(\mathbf{X})/I_t(\mathbf{X}) \not\subset P$.

We induct on t. The result is obviously valid for $t = 1$ since $k[\mathbf{X}]/I_1(\mathbf{X}) = k$ and $\{0\}$ is the only prime ideal of k. Thus, assume that $t \ge 2$.

We may further assume that the given prime ideal P does not contain $I_1(\mathbf{X})/I_t(\mathbf{X})$. Namely, we show that otherwise $(k[\mathbf{X}]/I_t(\mathbf{X}))_P$ is not regular (and clearly $P \supset I_{t-1}(\mathbf{X})/I_t(\mathbf{X})$ as $I_1(\mathbf{X}) \supset I_{t-1}(\mathbf{X})$ for $t \ge 2$). In fact, since $I_1(\mathbf{X})$ is a maximal ideal, we would have $P = I_1(\mathbf{X})$, and it is clear that the localization cannot be regular as $I_t(\mathbf{X}) \subset I_1(\mathbf{X})^2$ (recall that a residue ring A/J of a regular local ring is regular only if J is a subset of a regular system of parameters).

Since $I_1(\mathbf{X})/I_t(\mathbf{X}) \not\subset P$, we may assume (by reordering variables) that the residue $\overline{x_{1,1}}$ of $x_{1,1}$ modulo $I_t(\mathbf{X})$ does not belong to P. Localizing at the powers of this element, we find pretty much as in the proof of the previous proposition

$$(k[\mathbf{X}]/I_t(\mathbf{X}))_{\overline{x_{1,1}}} = k[\mathbf{X}]_{x_{1,1}}/I_t(\mathbf{X})_{x_{1,1}}$$

$$= k[x_{1,1}, \ldots, x_{1,n}; x_{2,1}, \ldots, x_{m,1}, x_{1,1}^{-1}][\widetilde{\mathbf{X}}]/I_{t-1}(\widetilde{\mathbf{X}}), \quad (1.5)$$

where, as usual here, $\widetilde{\mathbf{X}}$ denotes both the matrix and the set $\left\{ x_{i,j} - x_{i,1}x_{1,j}x_{1,1}^{-1} \right\}_{\substack{2 \le i \le m \\ 2 \le j \le n}}$ of algebraically independent elements over k.

We next consider the k-subalgebra $k[\widetilde{\mathbf{X}}]/I_{t-1}(\widetilde{\mathbf{X}})$ of the k-algebra in (1.5).

Note that this ring inclusion is flat (polynomial). Let $\widetilde{P}$ denote the contraction to $k[\widetilde{\mathbf{X}}]/I_{t-1}(\widetilde{\mathbf{X}})$ of the extended ideal of P to the above localization (1.5). Because of flatness, $(k[\mathbf{X}]/I_t(\mathbf{X}))_P$ is regular if and only if $(k[\widetilde{\mathbf{X}}]/I_{t-1}(\widetilde{\mathbf{X}}))_{\widetilde{P}}$ is regular. On the other hand, by the inductive hypothesis, the latter holds if and only if $\widetilde{P}$ does not contain the ideal $I_{t-2}(\widetilde{\mathbf{X}})/I_{t-1}(\widetilde{\mathbf{X}})$, and this happens if and only if P does not contain $I_{t-1}(\mathbf{X})/I_t(\mathbf{X})$. $\qquad\square$

We leave as an exercise proving Proposition 1.11 in the case the coefficient field k is replaced with a locally everywhere regular Noetherian ring. We remark that a similar result as Proposition 1.11 holds as well in the case of an $m \times m$ symmetric generic matrix [1, Theorem 2.2].

For the next result we assume known the Cohen–Macaulay part of the first theorem – in fact, all we need is that $I_t(\mathbf{X})$ has pure grade.

Proposition 1.12 *Let* $\mathbf{X}$ *denote an* $m \times n$ *generic matrix over a Noetherian integral domain* D. *Then* $I_t(\mathbf{X})$ *is a prime ideal for every* $1 \le t \le \min\{m, n\}$.

Proof We induct on t.

The result is trivial for $t = 1$ since $I_1(\mathbf{X}) = (\mathbf{X})$. The proof is in the same spirit of the argument in the previous propositions; hence we only explain the initial step anew in this situation.

Let $t \ge 2$. Since we are assuming that $I_t(\mathbf{X})$ is perfect, it is of pure grade, and the latter has value $(m - t + 1)(n - t + 1)$ by Proposition 1.10. But

$$\text{grade} I_1(\mathbf{X}) = mn > (m - t + 1)(n - t + 1) = mn - (t - 1)(m + n - (t - 1))$$

as $t \ge 2$. Therefore, $I_1(\mathbf{X})$ is not contained in any associated prime of $D[\mathbf{X}]/I_t(\mathbf{X})$, and hence, by prime avoidance, there exists some element in $I_1(\mathbf{X})$ which is a nonzerodivisor on $D[\mathbf{X}]/I_t(\mathbf{X})$.

Alas, we do not know at this stage whether we can assume such an element to be one of the entries of the matrix; otherwise we would be done by localization at the powers of this nonzerodivisor by the inductive hypothesis. But let us be brave and localize at the powers of $x_{1,1}$ and prove it is after all a nonzerodivisor on $D[\mathbf{X}]/I_t(\mathbf{X})$.

At any rate, as before, we get that the localization $D[\mathbf{X}]_{x_{1,1}}/I_t(\mathbf{X})_{x_{1,1}}$ is a polynomial ring over the subring $D[\widetilde{\mathbf{X}}]/I_{t-1}(\widetilde{\mathbf{X}})$, which is a domain by the inductive hypothesis. Therefore, $D[\mathbf{X}]_{x_{1,1}}/I_t(\mathbf{X})_{x_{1,1}}$ is also a domain. Let $P \subset D[\mathbf{X}]$ denote the unique associated prime ideal of $D[\mathbf{X}]/I_t(\mathbf{X})$ not containing $x_{1,1}$ (the inverse image of the null ideal).

Suppose that there exists yet another associated prime $Q \neq P$ of $D[\mathbf{X}]/I_t(\mathbf{X})$. Necessarily, $x_{1,1} \in Q$. On the other hand, $I_1(\mathbf{X})$ is not contained in Q, so there must be some $x_{i,j} \notin Q$, with $(i, j) \neq (1, 1)$. Exchanging the roles of P and Q, upon use of the inductive hypothesis after localizing at $x_{i,j}$, we have $x_{i,j} \in P$. Since the extended prime $P D[\mathbf{X}]_{x_{1,1}}/I_t(\mathbf{X})_{x_{1,1}}$ is the null ideal, the image of $x_{i,j}$ in this extended prime is zero. This gives that $x_{1,1}^r x_{i,j} \in I_t(\mathbf{X})$, for some $r \geq 0$. But this is absurd since no monomial belongs to $I_t(\mathbf{X})$, for $t > 1$ – e.g., because every t-minor, for $t > 1$, vanishes at the point with all coordinates equal to 1, while a monomial does not (We thank W. Bruns for this simple argument).

This contradiction shows that such an associated prime Q does not exist, and hence $x_{1,1}$ is in fact a nonzerodivisor on $D[\mathbf{X}]/I_t(\mathbf{X})$. $\square$

1.5 Linear Sections

Let $k[\mathbf{z}] = k[z_1, \ldots, z_d]$ stand for a polynomial ring over the field k. To proceed, we will adopt the following terminology:

Definition 1.13 Let $\mathbb{L}$ denote an $m \times n$ matrix whose entries are linear forms in a polynomial ring R over a field k, and let $V(\mathbb{L}) \subset R_1$ denote the k-vector space spanned by its entries. A *linear section* of $\mathbb{L}$ is an $m \times n$ matrix $\mathbf{L}$ of linear entries in R such that $V(\mathbf{L}) \subset V(\mathbb{L})$.

By an abuse, we also refer to matrices as above simply as linear matrices for short. Note that any linear matrix can be thought of as a linear section of the fully generic matrix of the same size.

Often, for the sake of a more convenient matrix display, one is lead to view the linear section $\mathbf{L}$ of the linear matrix $\mathbb{L}$ as a specialization – perhaps, a more geometrically inclined way. This is especially useful when $\mathbb{L}$ is the fully generic matrix.

In this manner, one may proceed as follows. Let $\mathbf{L} = (\ell_{i,j})$ be any $m \times n$ matrix of linear forms in the polynomial ring $k[\mathbf{z}] := k[z_1, \ldots, z_d]$, such that $k[\ell_{i,j}] = k[\mathbf{z}]$ (clearly, $d \leq mn$).

Letting $\mathbf{X} = (x_{i,j})$ denote the $m \times n$ generic matrix over k, consider the surjective k-algebra homomorphism $k[\mathbf{X}] \to k[\mathbf{z}]$ mapping $x_{i,j}$ to $\ell_{i,j}$. The kernel of this map is the ideal generated by the linear forms in $\mathbf{X}$ spanning the kernel $\mathcal{L}$ of the underlying k-vector space map $\sum_{i,j} k x_{i,j} \rightarrow \sum_{i,j} k \ell_{i,j}$. Thus, $k[\mathbf{X}]/(\mathcal{L}) \simeq k[\mathbf{z}]$, and since determinants commute with ring homomorphisms, one has

$$k[\mathbf{X}]/(I_t(\mathbf{X}), \mathcal{L}) \simeq k[\mathbf{z}]/I_t(\mathbf{L}), \forall t \geq 1. \tag{1.6}$$

We leave to the reader the task of translating this setup in a more geometric minded language in terms of a linear embedding $\mathbb{P}^{d-1} \hookrightarrow \mathbb{P}^{mn-1}$ and pullback of determinantal varieties.

Note that $\mathcal{L}$ is spanned by a regular sequence of $mn - d$ linear forms.

Example 1.14 Let

$$\mathbf{L} = \begin{pmatrix} z_1 & z_2 & z_3 \\ z_2 & z_3 & z_4 \\ z_3 & z_4 & z_5 \end{pmatrix}$$

stand for the 3×3 generic Hankel matrix. Taking the 3×3 generic matrix $\mathbf{X} = (x_{i,j})$, the map described above has kernel generated by the regular sequence of 1-forms

$$x_{1,2} - x_{2,1}, x_{1,3} - x_{2,2}, x_{2,2} - x_{3,1}, x_{2,3} - x_{3,2}.$$

Many authors introduce Hankel matrices by means of these relations.

1.5.1 General Linear Forms

A distinctive case is that in which the entries $\ell_{i,j} (1 \leq i \leq m, 1 \leq j \leq n)$ of $\mathbf{L}$ are general forms, in the usual sense that the set of their coefficients lies on a dense open set of the parameter space of the coefficients.

For our purpose, the basic result that affords some control of the underlying commutative algebra is the following:

Theorem 1.15 *Let $\mathbf{L}$ denote an $m \times n$ matrix of general linear forms in the polynomial ring $R = k[z_1, \ldots, z_d]$ as above. Then, for every $1 \leq t \leq \min\{m, n\}$, one has*

$$\mathrm{ht}(I_t(\mathbf{L})) = \min\{d, (m - t + 1)(n - t + 1)\}.$$

Proof By Proposition 1.2, the right-hand side is an upper bound. We have to show that this is attained under the present hypotheses. First, observe that the result is trivial for $t = 1$ as the entries, being mn general forms with $mn \geq d$, generate R. Thus, assume that $t \geq 2$.

Also, due to (1.6), it is equivalent to show that

$$\dim k[\mathbf{X}]/(I_t(\mathbf{X}), \mathcal{L}) = \max\{0, d - (m - t + 1)(n - t + 1)\}, \tag{1.7}$$

where $\mathcal{L} = \{L_1, \ldots, L_s\}$ as in the above discussion. Note that, since the entries of φ are general forms, $\mathcal{L}$ is itself a sufficiently general sequence of forms.

We will accomplish this equality stepwise, showing that for $2 \leq t \leq \min\{m, n\}$ and $1 \leq r \leq s := mn - d$, one has

$$\dim k[\mathbf{X}]/(I_t(\mathbf{X}), L_1, \ldots, L_r) = \begin{cases} 0 & \text{if } D \leq r \\ D - r & \text{if } D > r, \end{cases}$$

where $D := \dim k[\mathbf{X}]/I_t(\mathbf{X}) = mn - (m - t + 1)(n - t + 1) = (m + n - t + 1)(t - 1)$.

We induct on r.

Since $t \geq 2$, then $L_1 \notin I_t(\mathbf{X})$ is a nonzerodivisor on $k[\mathbf{X}]/I_t(\mathbf{X})$ because $I_t(\mathbf{X})$ is a prime ideal generated in degree ≥ 2. Therefore,

$$\dim k[\mathbf{X}]/(I_t(\mathbf{X}), L_1) = \dim k[\mathbf{X}]/I_t(\mathbf{X}) - 1 = D - 1 = D - r,$$

while clearly $D \geq 2$ as $t \geq 2$.

Thus, assume $r \geq 2$. By the inductive hypothesis,

$$\dim k[\mathbf{X}]/(I_t(\mathbf{X}), L_1, \ldots, L_{r-1}) = \begin{cases} 0 & \text{if } D \leq r - 1 \\ D - (r - 1) & \text{if } D > r - 1 \end{cases}$$

in the range $2 \leq t \leq \min\{m, n\}$ and $2 \leq r \leq s := mn - d$.

Now consider the sets $\mathrm{Ass}(k[\mathbf{X}]/(I_t(\mathbf{X}), L_1, \ldots, L_{r-1})$, for every t in the range for which $D > r - 1$ (as a slight control, note that since $r \geq 2$ and $D > r - 1$, then $t \geq 2$). This is a finite set of primes.

Now, in this range of D, the ring $k[\mathbf{X}]/(I_t(\mathbf{X}), L_1, \ldots, L_{r-1})$ has positive dimension, and since it is Cohen–Macaulay, then $I_1(\mathbf{X})$ is not an associated prime of $k[\mathbf{X}]/(I_t(\mathbf{X}), L_1, \ldots, L_{r-1})$ for every $2 \leq t \leq \min\{m, n\}$.

Thus, as $L_1, \ldots, L_{r-1}, L_r$ is a sufficiently general sequence of forms, then $L_r \notin P$ (i.e., $L_r \notin P_1$, the degree 1 part of P) for every prime $P \in \bigcup_t \mathrm{Ass}(k[\mathbf{X}]/(I_t(\mathbf{X}), L_1, \ldots, L_{r-1})$.

Then the dimension again drops by 1, i.e., we get

$$\dim k[\mathbf{X}]/(I_t(\mathbf{X}), L_1, \ldots, L_r) = \begin{cases} 0 & \text{if } D \leq r \\ D - r & \text{if } D > r, \end{cases}$$

as claimed.

Applying with $r = mn - d$ yields

$$\dim k[\mathbf{X}]/(I_t(\mathbf{X}), \mathcal{L}) = \begin{cases} 0 & \text{if } D \leq mn - d \\ D - mn + d & \text{if } D > mn - d. \end{cases}$$

Substituting for D yields the required result. $\square$

Remark 1.16 We note that, under the hypothesis of Theorem 1.15, for every $1 \leq t \leq \min\{m, n\}$, the ring $R/I_t(\mathbf{L})$ is Cohen–Macaulay. Indeed, if $d \leq (m - t + 1)(n - t + 1)$, then $R/I_t(\mathbf{L})$ has finite length and hence is trivially Cohen–Macaulay, while if $I_t(\mathbf{L})$ has the generic codimension, then $R/I_t(\mathbf{L})$ is Cohen–Macaulay by the result of Eagon–Hochster (see Section 1.4).

Proposition 1.17 (char$(k) = 0$) *Let* $\mathbf{L}$ *denote an* $m \times n (m \leq n)$ *matrix of general linear forms in the polynomial ring* $R = k[z_1, \ldots, z_d]$. *Then the ring* $R/I_m(\mathbf{L})$ *satisfies the condition* R_r *of Serre, with* $r = \min\{d - (n - m + 2), n - m + 2\}$.

Proof With the standing notation, set $S := k[\mathbf{X}]$. Since $\mathbf{X}$ is the $m \times n$ generic matrix, the Jacobian ideal of $S/I_m(\mathbf{X})$ is $I_{m-1}(\mathbf{X})/I_m(\mathbf{X})$. Let $\mathcal{L} = \{L_1, \ldots, L_s\}$ as in the previous discussion, where $s = mn - d$. By Bertini's theorem [14, Chapter 3], the singular scheme of the scheme-theoretic general hyperplane section $S/(I_m(\mathbf{X}), L_1)$ is the scheme associated to $S/(I_{m-1}(\mathbf{X}), L_1)$. Inducting on the number s of general hyperplane sections yields that the singular scheme of the scheme-theoretic linear section $S/(I_m(\mathbf{X}), \mathcal{L}) \simeq R/I_m(\mathbf{L})$ is the scheme associated to $S/(I_{m-1}(\mathbf{X}), \mathcal{L}) \simeq R/I_{m-1}(\mathbf{L})$. By Theorem 1.15, the latter has codimension at least $\min\{d, 2(n - m + 1)\}$. Since $I_{m-1}(\mathbf{L})$ has codimension $m - n + 1$, $R/I_m(\mathbf{L})$ satisfies Serre's condition $R_{\min\{d-(n-m+2),n-m+2\}}$). $\square$

1.5.2 1-Generic Matrices

The terminology *sparse* is currently used for matrices having a certain number of null entries. It is of course difficult to establish general algebraic results when one has a random distribution of such zeros. Taking instead a suitable strategic distribution will be considered later in this book. For the moment, we wish to consider matrices that are in the opposite extreme; namely, we look for a notion that generalizes the idea that a matrix is not sparse even under the triple action of the general linear group. The notion to follow was studied in [12].

Definition 1.18 Let $\mathbf{L}$ denote an $m \times n$ matrix of linear forms in a polynomial ring R of dimension d over a field k, and let $t \leq \min\{m, n\}$. One says that $\mathbf{L}$ is *t-generic* if, even after arbitrary elementary row and column operations, every set of t entries is linearly independent.

Note that it is equivalent to require that upon arbitrary elementary row and column operations every $t \times t$ submatrix has linearly independent entries (see [18, Linear Determinantal Varieties in General]).

In particular, a 1-generic matrix has no null entry even after elementary row and column operations. The notion is equally defined in terms of conjugation under the usual triple action of $GL(m, k) \times GL(n, k) \times GL(d, k)$, by requiring that every set of t entries in any such conjugate of $\mathbf{L}$ is independent.

Example 1.19 The basic example of a 1-generic matrix is a generic matrix. In fact, the latter is t-generic for every $t \leq \min\{m, n\}$, but the proof of this fact is not totally obvious though it is clearly enough to show that this matrix is t-generic for $t = \min\{m, n\}$ (see [12, Examples. (1) p. 548]).

We assume throughout that the ground field is algebraically closed, as it will be required in parts of the theory as developed in [12].

The following notion of (trace) orthogonality will be useful.

For convenience of later reference, we fix the ground polynomial ring $R = k[x_1, \ldots, x_d]$, where we have traded z_i for x_i. To avoid overcrowding, we write the $m \times n$ generic matrix as $\mathbf{Y} = (y_{i,j})_{1 \leq i \leq m, 1 \leq j \leq n}$.

Let $\mathbf{L}$ denote an $m \times n$ matrix with linear entries in a polynomial ring $R = k[x_1, \ldots, x_d]$. We assume without loss of generality that the entries span the entire k-vector space $\sum_{i=1}^{d} k x_i$. Let f denote the trace of the product $\mathbf{Y} \mathbf{L}^t$, a bihomogeneous form of bidegree $(1, 1)$ in the natural bigrading of the polynomial ring $k[x_1, \ldots, x_d, y_{i,j} \mid 1 \leq i \leq m, 1 \leq j \leq n]$.

Let $\partial f / \partial x_s$ denote the partial derivative of f with respect to the variable x_s, and consider the relations $\{\partial f / \partial x_s = 0, 1 \leq s \leq d\}$ as a k-linear system of equations in the $y_{i,j}$'s. Let $V \subset k[y_{i,j} \mid 1 \leq i \leq m, 1 \leq j \leq n]$ be the k-vector space spanned by a k-linearly independent solution thereof; necessarily, $\dim_k V = mn - d$. Finally specialize the entries of $\mathbf{Y}$ accordingly applying the given dependency relations.

Definition 1.20 The resulting $m \times n$ matrix that has entries in the k-subalgebra $k[V] \subset k[y_{i,j} \mid 1 \leq i \leq m, 1 \leq j \leq n]$ is called the *(trace) orthogonal* matrix to $\mathbf{L}$ and denoted $\mathbf{L}^{\perp}$.

Clearly, $\mathbf{L}^{\perp}$ is only well-defined up to conjugation.

The following criterion was established in [12, Proposition-Definition 1.1]:

Proposition 1.21 (k algebraically closed) *Let $\mathbf{L}$ denote an $m \times n$ matrix of linear forms in a polynomial ring $R = k[x_1, \ldots, x_d]$ and let $t \leq \min\{m, n\}$. Then $\mathbf{L}$ is t-generic if and only if the ideal of $(t + 1)$-minors of $\mathbf{L}^{\perp}$ has height $mn - d$ as an ideal of the k-subalgebra $k[V]$.*

In other words, the criterion requires that the $(t + 1)$-minors generate an ideal which is primary to the irrelevant maximal ideal of $k[V]$. Note that while the definition of being t-generic gives way to no algorithmic procedure, the above proposition gives a largely effective criterion.

We next digress on lower bounds for both the number of ground variables and the codimension of an ideal of minors under the assumption of the t-generic property. As for the number of ground variables, it has already been said that it cannot drop too much; alas, it is not so clear that there might be a similar expectation for the codimension of an ideal of minors. Nevertheless, one has the following theorem:

Theorem 1.22 (k algebraically closed) *Let $\mathbf{L}$ denote an $m \times n$ $(m \le n)$ matrix of linear forms in a polynomial ring $R = k[x_1, \ldots, x_d]$ over a field, and let $1 \le t \le m$ be such that $\mathbf{L}$ is t-generic. Then:*

(a) *The entries of $\mathbf{L}$ span a k-vector space of dimension at least $t(m + n - t)$.*
(b) *More generally, for every $s \le m - t$, the codimension of $I_{s+1}(\mathbf{L}) \subset R$ is at least $t(m + n - t - 2s)$.*

Proof

(a) Recall our convention that the entries of $\mathbf{L}$ span the whole of the ambient k-vector space, i.e., we are assuming that the space of the entries has dimension d. Then the statement is an immediate consequence of the criterion in Proposition 1.21, which tells us that cod $I_{t+1}(\mathbf{L}^{\perp}) = mn - d$. From this, using the by now well-known upper bound for cod $I_{t+1}(\mathbf{L}^{\perp})$ yields

$$d \ge mn - (m - t)(n - t) = t(m + n - t).$$

(b) (Possibly it requires null characteristic) We refer to [12, Corollary 3.3] for the details of an argument using a mixture of linear algebra and the technology of tangent spaces at points on a variety. $\square$

Remark 1.23

(1) We know of no (commutative) algebraic proof of (b) above. Since there are not many obvious classes of linear sections which are t-generic for some intermediate $1 < t < m$, but not $(t + 1)$-generic, one gets slightly astray looking for the main algebraic properties of such classes. Also, it leads to wonder whether there is a clear characteristic-free proof of the result itself.
(2) As a particular case of the theorem, a 1-generic $m \times n$ matrix $\mathbf{L}$ requires at least $m + n - 1$ ground variables. This bound is attained by the generic Hankel matrix (which is shown to be 1-generic by a simple argument in [13, Proposition 6.3]).

1.5.3 Applications

The following result of D. Eisenbud shows that the maximal minors of a 1-generic matrix behave similarly to the case of generic matrices.

Theorem 1.24 *Let $\mathbf{L}$ denote an $m \times n$ $(m \le n)$ matrix of linear forms in $R = k[x_1, \ldots, x_d]$. If $\mathbf{L}$ is 1-generic, then $I_m(\mathbf{L}) \subset R$ is a Cohen–Macaulay prime ideal of codimension $n - m + 1$.*

Proof We provide a sketchy argument. The Cohen–Macaulayness assertion follows the standard procedure by showing that the codimension of $I_m(\mathbf{L})$ has the expected value $n - m + 1$ (by which even the Eagon–Northcott free resolution specializes from the generic case). By Theorem 1.22 (b), the codimension of $I_m(\mathbf{L})$ is at least $m + n - 1 - 2(m - 1) = n - m + 1$, so one is done for this part.

For the primality assertion, one proceeds by induction on m. For $m = 1$, the result is trivial even without the assumption that $\mathbf{L}$ is 1-generic. Thus, one may assume that $m \geq 2$ and consider the submatrix of the first $m - 1$ rows, which is clearly 1-generic as well. For a detailed argument – which resembles many of the prior arguments in this work – see [13, Theorem 6.4], where the author first gives a geometric reasoning to show that the radical of $I_m(\mathbf{L})$ is a prime ideal (does that sound familiar?). Since $R/I_m(\mathbf{L})$ is unmixed, it follows that $I_m(\mathbf{L})$ is a P-primary ideal for some prime ideal $P \subset R$. The rest of the proof consists in localizing at P and using discerningly the inductive hypothesis. $\qquad\square$

The next result is in J. Harris's book [18].

Proposition 1.25 ([18, Proposition 9.4 and Proposition 9.12]) *Let $R = k[x_1, \ldots, x_d]$ stand for a polynomial ring over a field k, and let $\mathbf{L}$ denote a 1-generic $2 \times n$ matrix of linear forms.*

(i) *(Mnemonic: curve) If $n = d - 1$, then $\mathbf{L}$ is conjugate to the $2 \times (d - 1)$ generic Hankel matrix.*

(ii) *(Mnemonic: surface) If $n = d - 2$, then $\mathbf{L}$ is conjugate to the following generic scroll matrix*

$$\begin{pmatrix} x_1 & x_2 & \ldots & x_l & x_{l+2} & x_{l+3} & \ldots & x_{d-1} \\ x_2 & x_3 & \ldots & x_{l+1} & x_{l+3} & x_{l+4} & \ldots & x_d \end{pmatrix},$$

for suitable l.

We observe that the above mnemonic reminders refer to embeddings in $\mathbb{P}^{d-1}$.

The proof of (ii) is quite involved. A slightly different proof is given in [loc.cit.], as due to F. Schreyer, that moreover generalizes the result of (ii) to include rational normal scrolls of any dimension as in Proposition 1.9, depending on the given value of $n \leq d - 1$.

Other examples of 1-generic matrices obtained in [12], besides arbitrary generic Hankel matrices, are the defining matrices of rational normal scrolls (the converse of Harris–Schreyer result above) and generic symmetric matrices (for the latter see Exercise 1.37). Unfortunately, for a given size $m \times n (m \leq n)$, apart from the generic matrix, not many other structured examples are known to be t-generic with $2 \leq t < m$. In Chapter 7 we will discuss yet another class of 1-generic matrices.

To proceed, let $\mathbf{L} \subset \mathbb{L}$ denote an $m \times n$ linear section over a polynomial ring R over a field k, and let $V(\mathbf{L}) \subset V(\mathbb{L})$ stand for the corresponding k-vector spaces spanned by the entries.

Borrowing from the notation in [12], we set $\mathrm{codim}_{\mathbb{L}}\mathbf{L} := \dim_k(V(\mathbb{L})/V(\mathbf{L}))$.

The next theorem is among the main results of [12].

Theorem 1.26 ([12, Theorem 2.1]) *Let $\mathbb{L}$ be an $m \times n$ linear matrix, with $m \leq n$, and let $\mathbf{L}$ be one of its linear sections. Assume that $\mathbb{L}$ is $(m - s)$-generic, for some $0 \leq s \leq m - 1$.*

(i) *If* $\operatorname{codim}_{\mathbb{L}} \mathbf{L} \leq s$, *then* $I_{s+1}(\mathbf{L})$ *has the expected codimension and, hence, is Cohen–Macaulay.*

(ii) *If* $\operatorname{codim}_{\mathbb{L}} \mathbf{L} \leq s - 1$, *then* $I_{s+1}(\mathbf{L})$ *is a prime ideal.*

(iii) $(m > s + 1)$ *If* $\operatorname{codim}_{\mathbb{L}} \mathbf{L} \leq s - 2$, *then* $R/I_{s+1}(\mathbf{L})$ *is a normal domain.*

Remark 1.27 In particular, for $s = m - 1$, one retrieves Theorem 1.24 by taking $\mathbf{L} = \mathbb{L}$.

The proofs of the above theorem are quite technical and mostly geometric. It could be interesting to obtain purely algebraic proofs.

Here is an example (for more encompassing results, see Chapter 6, Section 6.3.3).

Proposition 1.28 *Let* $\mathcal{H}_m$ *denote the* $m \times m$ *generic Hankel matrix. Then the ideal* $I_{m-1}(\mathcal{H}_m)$ *of its submaximal minors is a Cohen–Macaulay prime ideal.*

Proof By [16] (see also Proposition 6.1), $I_{m-1}(\mathcal{H}_m)$ is the ideal of maximal minors of a suitable $(m - 1) \times (m + 1)$ generic Hankel matrix. Therefore, the result follows from the above theorem (or, more directly, from Theorem 1.24). $\qquad\square$

As will be shown in Proposition 6.3 of Chapter 6, the special fiber of $I_{m-1}(\mathcal{H}_m)$ is the special fiber of a Grassmannian in its Plücker embedding. This is a very distinctive facet of a square Hankel matrix as compared with the other Hankel like catalecticants to be discussed in Chapter 7.

Example 1.29 Consider the 2-catalecticant

$$
C_{3,2} = \begin{pmatrix} x_1 & x_2 & x_3 \\ x_3 & x_4 & x_5 \\ x_5 & x_6 & x_7 \end{pmatrix}.
$$

The codimension 4 ideal $I_2(C_{3,2})$ is reduced, but not prime: the ideal (x_1, x_3, x_5, x_7) is a minimal prime thereof. Further aspects of this and other matters will be considered in Chapter 7, e.g., Proposition 7.6.

Exercises

1.30 Given an integer $d \geq 1$ and a plane curve $C \in \mathbb{P}^2$ of degree c, how does one get the homogeneous defining ideal of $v_d(C) \subset \mathbb{P}^N$ under the Veronesean embedding $v_d : \mathbb{P}^2 \to \mathbb{P}^N$?
(HINT: Try $C := \{x_0^c + x_1^c + x_2^c = 0\}$ as a starter; also, use computer assistance to guess a few initial cases.)

1.31 Show that, for $d = 3$, the Veronese embedding of $\mathbb{P}^2$ is arithmetically Gorenstein. Clarify whether this is an accident or comes out of some relation to Grassmannians.

1.32 Describe a similar condition as in Proposition 1.17 in order that $I_t(\mathbf{L})$ be normal for a fixed $1 \leq t \leq m$ (thus, for example, for $t = 2, d \geq mn - (m + n) + 3$ is the required bound).

1.33 Taking for granted that the generic Hankel matrix is 1-generic, show that the bound in part (b) of Theorem 1.22 is attained by the generic Hankel matrix for any value of s as stipulated.
(HINT: Use [16, Lemme 2.3] (see also Chapter 7) to reduce to the case of maximal minors and use Theorem 1.24)

1.34 Prove that the generic matrix of size $m \times n$ ($m \leq n$) is t-generic for any $t \leq m$.
(HINT: Show that the trace orthogonal is the zero matrix.)

1.35 Prove that the trace orthogonal to a $2 \times n$ generic Hankel matrix $\mathcal{H}_{2,n}$ is a degenerate $2 \times n$ Hankel matrix with zero at the right-upper and left-lower corners (up to a sign). Deduce that $\mathcal{H}_{2,n}$ is 1-generic.

1.36 Take for granted that for any $m \geq 1$ the $m \times m$ generic Hankel matrix is 1-generic (which is a fact). If $m = 3$, deduce that its trace orthogonal is not even a symmetric matrix.
(HINT: What is the upper bound of the codimension of the submaximal minors of a symmetric matrix?)

1.37 Prove that the generic symmetric matrix $\mathbf{L}$ is 1-generic.
(HINT: Describe explicitly the partial derivatives of the corresponding trace biform, and show that the trace orthogonal of $\mathbf{L}$ can be expressed as a generic skew-symmetric matrix; then argue that the ideal of 2-minors of the latter has the maximal codimension.)

References

1. M. Barile, Arithmetical ranks of ideals associated to symmetric and alternating matrices. J. Algebra **176**, 59–82 (1995) 17
2. J. Barshay, Graded algebras of powers of ideals generated by A-sequences. J. Algebra **25**, 90–99 (1973) 9
3. A. Bernardi, Ideals of varieties parameterized by certain symmetric tensors. J. Pure Appl. Algebra **212**, 1542–1559 (2008) 9
4. A. Bernardi, J. Brachat, P. Comon, B. Mourrain, General tensor decomposition, moment matrices and applications. Journal of Symbolic Computation, in *International Symposium on Symbolic and Algebraic Computation*, vol. 52, eds. by I. Emiris, E. Schost (2013), pp. 51–71 9
5. W. Bruns, A. Conca, Gröbner bases and determinantal ideals, in *Commutative Algebra, Singularities and Computer Algebra (Sinaia, 2002)*, pp. 9–66. NATO Science Series, II: Mathematics, Physics and Chemistry, vol. 115 (Kluwer Academic Publishers, Dordrecht, 2003) 15
6. W. Bruns, U. Vetter, *Determinantal Rings*, LNM No. 1327 (Springer, Berlin, 1988) 15
7. C. Ciliberto, F. Russo, A. Simis, Homaloidal hypersurfaces and hypersurfaces with vanishing Hessian. Adv. Math. **218**, 1759–1805 (2008) 11
8. P. Corsini, Sulle potenze di un ideale generato da una A-successione. Rend. Sem. Mat. Univ. Padova **39**, 190–204 (1967) 9

9. J.A. Eagon, *Ideals Generated by the Subdeterminants of a Matrix*. Ph.D. Thesis (University of Chicago, Chicago, 1961) 6

10. J.A. Eagon, D.G. Northcott, Ideals defined by matrices and a certain complex associated to them. Proc. R. Soc. Lond. A **269**, 188–204 (1962) 6

11. D. Eisenbud, J. Harris, On varieties of minimal degree, in *Algebraic Geometry, Bowdoin, 1985*. Proceedings of Symposia in Pure Mathematics, vol. 46 (1987), pp. 3–13 11

12. D. Eisenbud, Linear sections of determinantal varieties. Amer. J. Math. **110**, 541–575 (1988) 21, 22, 23, 24

13. D. Eisenbud, *The Geometry of Syzygies: A Second Course in Algebraic Geometry and Commutative Algebra*, GTM No. (Springer, Berlin, 2005) 23, 24

14. H. Flenner, L. O' Carroll, W. Vogel, *Joins and Intersections*. Springer Monographs in Mathematics (Springer, Berlin, 1999) 21

15. L. Goddard, Bases for the prime ideals associated with certain classes of algebraic varieties. Proc. Cambridge Philos. Soc. **39**, 35–48 (1943) 9

16. L. Gruson, C. Peskine, Courbes de L'Espace Projectif: Variétés de Sécantes 1–32, in *Enumerative Geometry and Classical Algebraic Geometry*, ed. by P. Le Barz, Y. Hervier. Progress in Mathematics, vol 24 (Birkhäuser, Boston, 1982) 25, 26

17. M. Hazewinkel, C.F. Martin, A short elementary proof of Grothendieck1s theorem on algebraic vector bundles over the projective line. J. Pure Applied Algebra **25**, 207–221 (1982) 12

18. J. Harris, *Algebraic Geometry*, GTM No. 133 (Springer, Berlin, 1992) 11, 21, 24

19. M. Hochster, Rings of invariants of tori, Cohen-Macaulay rings generated by monomials, and polytopes. Ann. Math. **96**, 318–337 (1972) 10

20. V. Kodiyalam, T. Lam, R. Swan, Determinantal Ideals, Pfaffian Ideals, and the Principal Minor Theorem, in *Noncommutative Rings, Group Rings, Diagram Algebras and Their Applications*, eds. by S.K. Jain. Contemporary Mathematics, vol. 456 (2008) 4

21. J.M. Landsberg, G. Ottaviani, Equations for secant varieties of Veronese and other varieties. Ann. Mat. Pura Appl. **92**, 569–606 (2013) 9

22. A. Micali, Sur des idéaux engendrés par des déterminants, in *Séminaire Dubreil–Pisot, Algèbre et Théorie des Nombres*, Anée 17, Nos. 1 et 18, pp. 1963-1964 15

23. C. Meadows, Linear systems cut out by quadrics on projections of varieties. J. Algebra **90**, 198–207 (1984) 9

24. D.G. Northcott, Semi-regular rings and semi-regular ideals. Quart. J. Math. Oxford **11**, 8–104 (1960) 15

25. A. Parolin, Varietà Secanti alle Varietà di Segre e di Veronese e Loro Applicazioni, Tesi di dottorato, Università di Bologna, A.A. 2003/2004 9

26. M. Pucci, The Veronese variety and catalecticant matrices. J. Algebra **202**, 72–95 (1998) 9

27. Z. Ramos, A. Simis, *Graded Algebras in Algebraic Geometry*. Expositions in Mathematics, vol. 70 (De Gruyter, Berlin, 2022) 7

28. T.G. Room, *The Geometry of Determinantal Loci* (Cambridge University of Press, Cambridge, 1938) 9

29. F. Russo, *On the Geometry of Some Special Projective Varieties*. Lecture Notes of the Unione Matematica Italiana (Springer, Berlin, 2015) 8

30. A. Simis, *Commutative Algebra*, 2nd edn. (De Gruyter Graduate, Berlin-Boston, 2023) 4

31. A. Simis, W. Vasconcelos, R. Villarreal, On the ideal theory of graphs. J. Algebra **167**, 389–416 (1994) 10

32. B. Sturmfels, Gröbner bases and Stanley decompositions of determinantal rings. Math. Z. **205**, 137–144 (1990) 15

33. J. Watanabe, Hankel matrices and Hankel ideals, in *Queen's Papers in Pure and Applied Mathematics X*, vol. 102 (1996), pp. 351–363 11

Chapter 2
Algebraic Preliminaries

Abstract This chapter brings in the ground algebraic material employed in the book. Based on elementary facts from matrices, deeper results are extracted from the point of view of commutative algebra and geometry, with an emphasis on the role of the adjugate matrix, cofactors, and other determinantal minors. The latter are a tool to approach various questions concerning matrices with homogeneous linear entries, such as the structure of the linear syzygies of ideals of cofactors, features of Hessian matrices, and inverse maps to birational maps. Additional emphasis is on linear sections over polynomial rings and the ideal theoretic notions intertwining with matrix theory along other parts of the book. For the reader's convenience, definitions are given ab initio whenever possible. In particular, we develop the role of the partial derivatives of a determinant f and explain the tendency of 2×2-minors to belong to the homogeneous defining ideal of the dual variety to f. In that thread, from the geometric side, some deeper results are collected that will be of importance throughout subsequent chapters, such as those concerning the algebraic properties of rational maps, the dual variety, and symbolic blowup theory. As before, the standing reference for basic commutative algebra is Simis (2023. Commutative algebra, 2nd edn. De Gruyter Graduate, Berlin).

2.1 Minors and Cofactors

In this section we elaborate on selected properties of the ideals of minors and the cofactors of a square matrix.

2.1.1 The Adjugate of a Square Matrix

The notion of the adjugate of a square matrix goes back to the origins of the well-established theory of determinants, having been introduced by Cauchy in the early nineteenth century. Its usefulness goes well beyond the traditional role in linear

Z. Ramos, A. Simis, *Determinantal Ideals of Square Linear Matrices*,
https://doi.org/10.1007/978-3-031-55284-7_2

algebra. In particular, in this book it will play an important role in a variety of theorems and their proofs.

Let $\mathcal{M}$ denote an $m \times m$ matrix with entries in a fixed commutative ring – henceforth called the *ground ring*.

The *adjugate matrix* of $\mathcal{M}$ is the transpose of the $m \times m$ matrix whose entries are the cofactors of $\mathcal{M}$. We denote it by $\mathrm{adj}(\mathcal{M})$. Its basic structure property is the relation established by Cauchy:

$$\mathcal{M}\,\mathrm{adj}(\mathcal{M}) = \mathrm{adj}(\mathcal{M})\,\mathcal{M} = (\det \mathcal{M})\mathbb{I}_m, \tag{2.1}$$

where $\mathbb{I}_m$ denotes the $m \times m$ identity matrix.

The adjugate is often called the *classical adjoint* or the *adjunct*, not to be confused with the adjoint operator.

One has in addition the following well-known relation:

Proposition 2.1 *Suppose that the determinant of an $m \times m$ matrix $\mathcal{M}$ is a regular element in the ground ring. Then*

$$\mathrm{adj}(\mathrm{adj}(\mathcal{M})) = (\det \mathcal{M})^{m-2}\,\mathcal{M}. \tag{2.2}$$

Proof From the basic adjugate relation, since $\det \mathcal{M}$ is regular, $\det \mathrm{adj}(\mathcal{M}) = (\det \mathcal{M})^{m-1}$. Applying to $\mathrm{adj}(\mathcal{M})$ yields

$$\mathrm{adj}(\mathcal{M})\,\mathrm{adj}(\mathrm{adj}(\mathcal{M})) = (\det \mathcal{M})^{m-1}\mathbb{I}_m.$$

Now multiply both members of the last relation by $\mathcal{M}$, then apply the basic adjugate relation once more to $\mathcal{M}$, and finally cancel one copy of $\det \mathcal{M}$ everywhere. □

Although the adjugate concerns cofactors – hence, minors of submaximal size – it often interacts with minors of smaller sizes. The following result gives a useful relationship between the determinant of a square matrix and the 2-minors of its adjugate.

Theorem 2.2 *Let L denote an $m \times m$ ($m \geq 2$) matrix over a Noetherian ground ring R containing an infinite field k. Suppose that L satisfies the following conditions:*

(1) *$f := \det L$ is a regular element in R.*
(2) *The determinant of some $(m-1) \times (m-1)$ submatrix of L is a regular element over $R/(f)$.*

Then, for any 2×2 submatrix T of the adjugate $\mathrm{adj}(L)$, f is a factor of $\det T$.

Proof Say,

$$T = \begin{pmatrix} t_{i,j} & t_{i,j'} \\ t_{i',j} & t_{i',j'} \end{pmatrix}.$$

To prove the claim, start with the full $2 \times m$ submatrix $M_{i,i'}$ of $\mathrm{adj}(L)$, with ith and i'th rows, so that T is a submatrix thereof.

As guaranteed by assumption (2), let $\tilde{N}$ be an $(m-1) \times (m-1)$ submatrix of L such that $\det \tilde{N} \in I_{m-1}(L)$ is regular over $R/(f)$. Let N denote the unique $m \times (m-1)$ submatrix of L containing $\tilde{N}$ as a submatrix.

The following relation

$$M_{i,i'} N \equiv \mathbf{0} \,(\mathrm{mod}\, f)$$

stems from the adjugate formula (2.1)

$$\mathrm{adj}(L)L = f\, \mathbb{I}_m \equiv \mathbf{0} \,(\mathrm{mod}\, f).$$

Here, $\mathbf{0}$ denotes the null matrix.

For even more reason,

$$M_{i,i'}\, N \,\mathrm{adj}(\tilde{N}) \equiv \mathbf{0} \,(\mathrm{mod}\, f). \tag{2.3}$$

Up to a row permutation we can assume that $\tilde{N}$ is the upper $(m-1) \times (m-1)$ submatrix of N. Indeed, for an arbitrary matrix $A \in \mathrm{GL}(m,k)$, we have $M_{i,i'}\, N = M_{i,i'}\, A^{-1} A\, N$ and $I_2(M_{i,i'}) = I_2(M_{i,i'} A^{-1})$.

Then one has

$$N \,\mathrm{adj}(\tilde{N}) = \left(\begin{array}{cccc}
\det \tilde{N} & & & \\
& \det \tilde{N} & & \\
& & \ddots & \\
& & & \det \tilde{N} \\
\hline
p_1 & p_2 & \cdots & p_{m-1}
\end{array}\right),$$

for suitable elements $p_i \in R$, where the empty slots have null entries.

Therefore, writing

$$M_{i,i'} = \begin{pmatrix} t_{i,1} & \cdots & t_{i,j} & \cdots & t_{i,j'} & \cdots & t_{i,m} \\ t_{i',1} & \cdots & t_{i',j} & \cdots & t_{i',j'} & \cdots & t_{i',m} \end{pmatrix},$$

where

$$T = \begin{pmatrix} t_{i,j} & t_{i,j'} \\ t_{i',j} & t_{i',j'} \end{pmatrix},$$

yields

$$M_{i,i'}\, N \,\mathrm{adj}(\tilde{N}) = \begin{pmatrix} \cdots & t_{i,j} \det \tilde{N} + t_{i,j'}\gamma & \cdots \\ \cdots & t_{i',j} \det \tilde{N} + t_{i',j'}\gamma & \cdots \end{pmatrix},$$

where $\gamma = 0$ if $j' \neq m$, and $\gamma = p_j$ if $j' = m$.

Then, by (2.3),

$$T \begin{pmatrix} \det \tilde{N} \\ \gamma \end{pmatrix} \equiv \mathbf{0} \, (\mathrm{mod} \, f).$$

In particular,

$$\mathrm{adj}(T) T \begin{pmatrix} \det \tilde{N} \\ \gamma \end{pmatrix} \equiv \mathbf{0} \, (\mathrm{mod} \, f),$$

hence,

$$\begin{pmatrix} \det T & 0 \\ 0 & \det T \end{pmatrix} \begin{pmatrix} \det \tilde{N} \\ \gamma \end{pmatrix} \equiv \mathbf{0} \, (\mathrm{mod} \, f).$$

Thus,

$$\det T \det \tilde{N} \in (f).$$

Since $\det \tilde{N}$ is a regular element over $R/(f)$, then $\det T \in (f)$. $\square$

2.1.2   Matrices Over Polynomial Rings

Most of the results in this book concern the case where the matrices have entries in a polynomial ring over a field. When geometric translation is needed, we assume that the corresponding ground field is algebraically closed. However, we do not wish to bluntly assume characteristic zero, and in fact, there will often be some effort to state characteristic-free results.

A great deal of the underlying material tosses around a generic matrix as largely considered in Chapter 1. Thus, we mean an $m \times n$ matrix whose entries $\{x_{i,j} | 1 \leq i \leq m, 1 \leq j \leq n\}$ are indeterminates over a ground ring A (mostly, a field k). The polynomial ring $A[\mathbf{X}] := A[x_{i,j} | 1 \leq i \leq m, 1 \leq j \leq n]$ is referred to as the ground polynomial ring of the generic matrix. If no confusion arises, we often denote both the matrix and its set of entries by $\mathbf{X}$.

We will need two propositions that have been stated elsewhere. Proofs are given for the reader's convenience.

The first is a suitable generalization of [3, Theorem 10.16 (b)]. The short proof below has been given in [6, Lemma 2.7].

Proposition 2.3 *Let $R = k[x_1, \ldots, x_n]$ be a polynomial ring over a field k, and let $M = (p_{i,j})$ be an $m \times m$ matrix whose entries are homogeneous polynomials in R of the same degree (zeros allowed). Consider the inclusion of k-algebras $k[\Delta_{i,j} | 1 \leq$*

$i, j \leq m] \subset k[p_{i,j} | 1 \leq i, j \leq m]$, *where* $\Delta_{i,j}$ *denotes the* (i, j)*th cofactor of* M. *If* $\det M \neq 0$, *then the extended inclusion to the respective fields of fractions is algebraic.*

Proof Using the relation $\det \mathrm{adj}(M) = (\det M)^{m-1}$, as coming from (2.1) and (2.2), one has the field extensions

$$k((\det M)^{m-1}, \{(\det M)^{m-2} p_{i,j} | 1 \leq i, j \leq m\}) \subset k(\Delta_{i,j} | 1 \leq i, j \leq m)$$
$$\subset k(p_{i,j} | 1 \leq i, j \leq m).$$

Since the inclusion

$$k((\det M)^{m-1}, \{(\det M)^{m-2} p_{ij} | 1 \leq i, j \leq m\}) \subset k(p_{i,j} | 1 \leq i, j \leq m)$$

is trivially algebraic, so is $k(\Delta_{i,j} | 1 \leq i, j \leq m) \subset k(p_{i,j} | 1 \leq i, j \leq m)$ too. $\qquad\square$

For the next proposition, we need the notion of the initial ideal of a polynomial ideal over a field, in the sense of Gröbner bases theory. There are many excellent sources for the general theory of monomial ideals and Gröbner bases; we refer to the recent book [9].

Proposition 2.4 *Let* $\mathbf{X} = (x_{i,j})_{1 \leq i, j \leq m}$ *denote the* $m \times m$ *generic matrix over a field* k, *and let* X *denote its set of entries away from the main anti-diagonal of every submaximal minor. Then* X *is a regular sequence modulo the ideal* $I_{m-1}(\mathbf{X}) \subset k[x_{i,j} | 1 \leq i, j \leq m]$ *generated by the submaximal minors of* $\mathbf{X}$.

Proof For easy visualization, X is the set of bulleted entries below (for $m \geq 6$):

$$\begin{pmatrix}
\bullet & \bullet & \bullet & \cdots & \bullet & \bullet & \bullet & x_{1,m-1} & x_{1,m} \\
\bullet & \bullet & \bullet & \cdots & \bullet & \bullet & x_{2,m-2} & x_{2,m-1} & x_{2,m} \\
\bullet & \bullet & \bullet & \cdots & \bullet & x_{3,m-3} & x_{3,m-2} & x_{3,m-1} & \bullet \\
\bullet & \bullet & \bullet & \cdots & x_{4,m-4} & x_{4,m-3} & x_{4,m-2} & \bullet & \bullet \\
\vdots & \vdots & \vdots & & \vdots & \vdots & \vdots & \vdots & \vdots \\
\bullet & x_{m-2,2} & x_{m-2,3} & \cdots & x_{m-2,m-4} & \bullet & \bullet & \bullet & \bullet \\
x_{m-1,1} & x_{m-1,2} & x_{m-1,3} & \cdots & \bullet & \bullet & \bullet & \bullet & \bullet \\
x_{m,1} & x_{m,2} & \bullet & \cdots & \bullet & \bullet & \bullet & \bullet & \bullet
\end{pmatrix}$$

(A similar picture can be depicted for $m \leq 5$.) Consider the lexicographic order of monomials in $\mathbf{X}$ upon an ordering of the entries respecting rows and columns as in [22]. Clearly, the cardinality of X is $2\binom{m-1}{2} = (m-1)(m-2)$. In particular, the set $\{a_2, \ldots, a_{(m-1)(m-2)}\}$ of entries in X is thus ordered as well. Then, [22, Theorem 1] and the assumption that no a_i is on the support of the initial term of any submaximal minor imply that, for any $i \geq 2$, the initial ideal of the ideal $(a_1, \ldots, a_i, I_{m-1}(\mathbf{X}))$ is $(a_1, \ldots, a_i, \mathrm{in}(I_{m-1}(\mathbf{X})))$. Clearly, a_{i+1} is not a zerodivisor modulo the latter ideal, and hence, by a well-known preliminary of initial ideal theory, it is not a zerodivisor modulo $(a_1, \ldots, a_i, I_{m-1}(\mathbf{X}))$ either. $\square$

The above proposition made strong us of the characterization of the initial ideal of a minor in the generic matrix. In the case of a symmetric matrix, a similar result is possibly available. We give the following special case, which is all we will use in the book.

Lemma 2.5 *Let $X = (x_{i,j})_{1 \leq i \leq j \leq m}$ denote the $m \times m$ symmetric generic matrix over a ground field k. Consider the lexicographic order of the monomials in $k[X]$ induced from the ordering of the variables respecting rows*

$$x_{1,m} > x_{1,m-1} > \cdots > x_{1,1} > x_{2,m} > x_{2,m-1} > \cdots > x_{2,2} > \cdots$$

$$\cdots > x_{m-1,m} > x_{m-1,m-1} > x_{m,m}.$$

For any set of indices $\mathbf{i} = \{1 \leq i_1 < \cdots < i_u \leq m\}$, one has $\mathrm{in}(\delta_{\mathbf{i},\mathbf{i}}) = \prod\limits_{s+t=u+1} x_{i_s,j_t}$, where $\delta_{\mathbf{i},\mathbf{i}}$ denotes the corresponding central minor.

Proof The proof is by induction on u. For $u = 1, 2$ the result is obvious. Now, suppose $u \geq 2$. Laplace block expansion along the first and the last rows of the $u \times u$ matrix $(x_{i_s,i_t})_{1 \leq s \leq t \leq u}$ yields

$$\delta_{\mathbf{i},\mathbf{i}} = \delta_{\{i_1,i_u\},\{i_1,i_u\}}\delta_{\mathbf{i}',\mathbf{i}'} + T = \left(x_{i_1,i_1}x_{i_u,i_u} - x_{i_1,i_u}^2\right)\delta_{\mathbf{i}',\mathbf{i}'} + T,$$

where $\mathbf{i}' = \{i_2, \ldots, i_{u-1}\}$ and the monomial terms involved in T are of degree at most 1 in the variable x_{i_1,i_u}. Thus, since x_{i_1,i_u} is the first variable, we have that

$$\mathrm{in}(\delta_{\mathbf{i},\mathbf{i}}) = x_{i_1,i_u}^2 \, \mathrm{in}(\delta_{\mathbf{i}',\mathbf{i}'}).$$

From this equality and induction, we have the desired result. $\square$

The following symmetric analog of Proposition 2.4 is a useful consequence.

Proposition 2.6 *Let $\mathcal{X} \subset X$ be the set of variables $x_{i,j}$ of the $m \times m$ generic symmetric matrix X such that $i + j \leq m - 1$ or $i + j \geq m + 3$. Then, $\mathcal{X}$ is a regular sequence modulo the ideal $I_{m-1}(X)$.*

Proof Rewriting, say, $\mathcal{X} = \{a_1, \ldots, a_v\}$. Since $R/I_{m-1}(X)$ is Cohen–Macaulay of codimension 3, in order to prove the assertion it is enough to show that $\mathrm{ht}\,(a_1, \ldots, a_v, I_{m-1}(X)) = v + 3$.

By Lemma 2.5,

$$\mathrm{in}(f) = \prod_{i+j=m+1} x_{i,j}, \quad \mathrm{in}(\Delta_{m,m}) = \prod_{i+j=m} x_{i,j} \quad \text{and} \quad \mathrm{in}(\Delta_{1,1}) = \prod_{i+j=m+2} x_{i,j}.$$

Since the variables of X are not on the support of the initial term of f, $\Delta_{m,m}$, $\Delta_{1,1}$, we have

$$\mathrm{ht}\,(a_1, \ldots, a_v, \mathrm{in}(f), \mathrm{in}(\Delta_{m,m}), \mathrm{in}(\Delta_{1,1})) = v + 3.$$

But, $f, \Delta_{m,m}, \Delta_{1,1} \in I_{m-1}(X)$, thus

$$(a_1, \ldots, a_v, \mathrm{in}(f), \mathrm{in}(\Delta_{m,m}), \mathrm{in}(\Delta_{1,1})) \subset \mathrm{in}(a_1, \ldots, a_v, I_{m-1}(X)).$$

Hence, $v + 3 \le \mathrm{ht}\,(a_1, \ldots, a_v, I_{m-1}(X)) \le v + 3$. $\qquad\square$

2.1.2.1 Bringing in Partial Derivatives

Quite generally, let $R = k[x_1, \ldots, x_r]$ be a polynomial ring over a field k, and let $f \in R$ be a homogeneous polynomial. Let $S_f \subset R/(f)$ stand for the k-subalgebra generated by the residues of the partial derivatives $\partial f / \partial x_i$, $1 \le i \le r$. Consider the k-algebra map

$$\partial : k[y_1, \ldots, y_r] \twoheadrightarrow S_f, \quad y_i \mapsto \partial f / \partial x_i (\mathrm{mod}\, f), \tag{2.4}$$

where the y_i's are new indeterminates. From this definition follows immediately that if f is irreducible (respectively, reduced), then $\ker (\partial)$ is a prime ideal (respectively, a radical ideal).

Our interest in $\ker (\partial)$ resides on the fact that, when f is reduced, it can be identified with the homogeneous defining ideal of the dual variety $V(f)^* \subset (\mathbb{P}^{r-1})^*$ to the variety $V(f) \subset \mathbb{P}^{r-1}$ – in other words, the homogeneous defining ideal of the image of the Gauss map of $V(f)$ under the Plücker embedding [16, Chapter 9, Section 9.1].

Parts of the book will deal with this notion and its relation to the *Hessian matrix* of f, defined as the $r \times r$ Jacobian matrix of its partial derivatives

$$H(f) = \left(\frac{\partial^2 f}{\partial x_i \partial x_j} \right)_{i,j=1,\ldots,r}.$$

Its determinant $h(f) \in k[x_1, \ldots, x_r]$ is the *Hessian* of f. A more detailed treatment will be given in Chapter 3 and Chapter 8. A useful tool to examine such relation is a well-known theorem of B. Segre [18, Theorem 1]:

Theorem 2.7 $(\mathrm{char}(k) \ne 2)$ *Let* $f \in k[x_1, \ldots, x_r]$ *be a reduced form. Let* $V(f)^*$ *and* $H(f)$ *denote, respectively, the dual variety and the Hessian matrix of* f. *Then*

$$\dim(V(f)^*) = \mathrm{rank}\,H(f)\,(\mathrm{mod}\, f) - 2, \tag{2.5}$$

where rank $H(f)$ $(\mathrm{mod}\ f)$ *means the largest order of a minor of* $H(f)$ *not a multiple of* f.

Segre makes a remark to the purpose that the exceptional characteristic is due to a singular behavior of quadrics. However, it is plausible that this hypothesis may be lifted in many situations. Note that f is allowed to have (reduced) factors.

Remark 2.8 It is important to observe that the notation $(\mathrm{mod}\ f)$ means here what is said, not to be confused with the slightly tighter notion of the rank of $H(f)$ as a matrix over the residue ring $S = k[x_1, \ldots, x_r]/(f)$ (the latter meaning the largest $t \geq 0$ such that the ideal of minors $I_t(H(f))$ has a regular element over S). Throughout the book one adheres to this meaning.

2.1.3   Linear Sections Endowed with Decomposition

Let $\mathbf{X} := \{x_{i,j}\}_{1 \leq i,j \leq m}$ denote the set of entries of the $m \times m$ generic matrix over a field k. As before, $\mathbf{X}$ will denote both the matrix itself and the set of its entries, hoping no confusion will be caused.

In the sequel, let $L = (\ell_{i,j})_{1 \leq i,j \leq m}$ denote an $m \times m$ matrix whose entries $\ell_{i,j}$ are 1-forms of the polynomial ring $k[\mathbf{X}] := k[x_{i,j} | 1 \leq i, j \leq m]$. Such a matrix is called a linear section of the $m \times m$ generic matrix $\mathbf{X} = (x_{i,j})_{1 \leq i,j \leq m}$, as discussed in Section 1.5. A linear section of $\mathbf{X}$ carries not only an "internal" aspect in that its entries span a k-vector subspace of $k[\mathbf{X}]_1$ but also an "external" one in the sense that it specializes the generic matrix through the obvious k-algebra endomorphism $x_{i,j} \mapsto \ell_{i,j}$ of the polynomial ring $k[\mathbf{X}]$. Both may be useful.

Let now $X := \{x_{i,j}\}_{1 \leq i \leq j \leq m}$ denote the set of entries of the $m \times m$ generic symmetric matrix over k. By analogy, a symmetric matrix $L = (\ell_{i,j})_{1 \leq i \leq j \leq m}$ is a *symmetric linear section* of X if its entries are 1-forms in the polynomial ring $k[X]$.

Among the ways of studying special square linear sections, the following one seems to be fairly structural.

Definition 2.9 One is given a decomposition of the set $\mathbf{X}$ as the disjoint union $\mathbf{X} = \mathbf{X}_1 \cup \mathbf{X}_2$ of two subsets. A linear section $L = (\ell_{i,j})_{1 \leq i,j \leq m}$ of the generic matrix $\mathbf{X}$ is said to be *endowed with this decomposition* if $\ell_{i,j} \in \mathbf{X}_1$ or $\ell_{i,j}$ is a linear form with entries in the polynomial ring $k[\mathbf{X}_2]$. Note that some linear forms $\ell_{i,j} \in k[\mathbf{X}_2]$ may be zero. If all of them vanish, so as to allow for so-called sparse sections, we refer to a linear section *endowed with a sparse decomposition*.

One defines similarly when a symmetric linear section of the generic symmetric matrix is said to be endowed with a decomposition of the set X.

It is important to stress that one allows for an extreme case such as when $\mathbf{X}_2 = \emptyset$, as well as for entry repetition in the set of entries of L with indices (i, j) such that $x_{i,j} \in \mathbf{X}_1$. Clearly, a linear section may be endowed with many different decompositions and is often the case where one is more convenient than any other, in particular because most of the invariants attached to L are stable under elementary row and column operations. The applications to be stated in the subsequent subsections usually assume that $\mathbf{X}_1 \neq \emptyset$ and is often reasonably large. Matrices whose entries are sufficiently general linear forms or zeros may require an entirely different approach.

A typical case of the above decomposition is that of a so-called *coordinate linear section* $L = (\ell_{i,j})_{1 \leq i,j \leq m}$, by which one means that either $\ell_{i,j} \in \mathbf{X}$ (i.e., is a variable of $k[\mathbf{X}]$) or else $\ell_{i,j} = 0$. Here, the head decomposition of $\mathbf{X}$ is quite arbitrary, while the forms with entries in $k[X_2]$ are all zeros. For example, this includes the generic case of symmetric, catalecticant (Hankel), alternating matrices (in which, extremely $X_2 = \emptyset$), and some of their sparse versions.

The following is a generalization of a result usually referred to as a consequence of [8]. An independent proof of the latter appeared in [14, Proposition 5.3.1], in which the rough assumption was that of a coordinate linear section as above. We restate it in larger scope and give a short proof for the reader's convenience.

Proposition 2.10 *Let* $L = (\ell_{i,j})_{1 \leq i,j \leq m}$ *stand for a linear section of the* $m \times m$ *generic matrix* $\mathbf{X}$ *endowed with a decomposition* $\mathbf{X} = \mathbf{X}_1 \cup \mathbf{X}_2$*. Then, for every* $1 \leq i, j \leq m$ *such that* $\ell_{i,j} = x_{u,v} \in \mathbf{X}_1$*, the partial derivative of* $f = \det L$ *with respect to* $x_{u,v}$ *is the sum of the cofactors of* $x_{u,v}$ *in all its slots as an entry of* L*.*

Proof Quite generally, let $\mathcal{L} = (l_{i,j})$ denote any linear section of $\mathbf{X}$ without a decomposition prescription. Write $f := \det \mathcal{L} \subset R$ and $g := \det \mathbf{X}$. Let σ denote the k-endomorphism of $k[\mathbf{X}]$ given by $x_{i,j} \mapsto l_{i,j}$. For any $1 \leq u, v \leq m$, the ordinary chain rule yields

$$\partial f / \partial x_{u,v} = \sum_{1 \leq i,j \leq m} (\partial l_{i,j} / \partial x_{u,v}) \, \sigma(\partial g / \partial x_{i,j})$$

$$= \sum_{1 \leq i,j \leq m} c_r^{i,j} \, \sigma(\partial g / \partial x_{i,j}), \tag{2.6}$$

where $c_r^{i,j}$ is the coefficient of $x_{u,v}$ in $l_{i,j}$.

Now, when $\mathcal{L} = L$, the only non-vanishing terms on the right-hand side of (2.6) correspond to the entries $\ell_{i,j} = x_{u,v}$. On the other hand, it is well-known or easy to see that $\partial g / \partial x_{i,j}$ is the cofactor of $x_{i,j}$ in the generic matrix $\mathbf{X}$. Therefore, whenever an entry $\ell_{i,j}$ equals some $x_{u,v}$, the summand $c_r^{i,j} \sigma(\partial g / \partial x_{i,j}) = \sigma(\partial g / \partial x_{i,j})$ is the cofactor of the entry $x_{u,v}$ in slot (i, j) of L. $\qquad\qquad\square$

2.1.4 A Role of the 2 × 2 Minors

Let there be given a linear section L of the generic $m \times m$ generic matrix or of the generic symmetric matrix over an infinite field k of coefficients. Denote the set of entries of any of them by the same $\mathbf{X}$, hoping this will cause no confusion. Let $\tilde{\mathbf{X}}$

denote the subset of $\mathbf{X}$ whose elements effectively appear as variables in the entries of L. If $\mathbf{Y}$ stands for the $m \times m$ generic matrix in the dual variables and $L_{\mathbf{Y}}$ denotes the mutadis–mutandis corresponding linear section, then $\tilde{\mathbf{Y}}$ stands for the analogous set in the "dual" variables.

If L is moreover endowed with a decomposition $\mathbf{X} = \mathbf{X}_1 \cup \mathbf{X}_2$, then $L_{\mathbf{Y}}$ is endowed with a corresponding decomposition $\mathbf{Y} = \mathbf{Y}_1 \cup \mathbf{Y}_2$. Any 2×2 submatrix of $L_{\mathbf{Y}}$ whose entries belong to $\mathbf{Y}_2$ will be called *resilient*, and the corresponding determinant will be called a *resilient minor*.

Theorem 2.11 *Let $L = (\ell_{i,j})$ denote a linear section of the $m \times m$ generic matrix or a symmetric linear section of the $m \times m$ generic symmetric matrix ($char(k) \neq 2$). Assume that L is endowed with a decomposition $\mathbf{X} = \mathbf{X}_1 \cup \mathbf{X}_2$ and satisfies the following conditions*:

(a) $f := \det L \neq 0$.
(b) *Some $(m-1) \times (m-1)$ minor of L is regular over $k[\tilde{\mathbf{X}}]/(f)$.*
(c) *Let $\ell_{i,j}, \ell_{i',j'}$ belong to $\mathbf{X}_1$. Suppose that $(i, j) \neq (i', j')$ in the generic case (respectively, $\{i, j\}$ and $\{i', j'\}$ are distinct as sets in the symmetric generic case). Then $\ell_{i,j} \neq \ell_{i',j'}$.*

Then, up to a change of variables, the ideal $\mathfrak{I}_2(\tilde{\mathbf{Y}})$ generated by the resilient minors is contained in the kernel of the map ∂ as in (2.4), with $\{x_1, \ldots, x_r\}$ (respectively, $\{y_1, \ldots, y_r\}$) replaced by $\tilde{\mathbf{X}}$ (respectively, $\tilde{\mathbf{Y}}$).

Proof Since ∂ commutes with determinants, up to a change of variables the determinant of a resilient 2×2 submatrix of $L_{\mathbf{Y}}$ maps to

$$\begin{pmatrix} \partial f/\partial x_{i,j} & \partial f/\partial x_{i,j'} \\ \partial f/\partial x_{i',j} & \partial f/\partial x_{i',j'} \end{pmatrix} \ (\mathrm{mod}\ f).$$

In the generic case, by Proposition 2.10 and the assumption in item (c), the entries of the above matrix are cofactors of L.

In the symmetric case, item (c) warrants that an entry in $\mathbf{X}_1$ is only repeated if the two are symmetrically seated. In this case, again by Proposition 2.10, any of the entries is either a cofactor or else a cofactor affected with a coefficient 2. Such nonzero coefficients could have been omitted before in the definition of the resilient minor or the map ∂.

Thus, in any of the two cases, the matrix is a submatrix of the adjugate of L. Therefore, the result follows from Theorem 2.2. $\square$

Remark 2.12

(1) The moral of Theorem 2.11 is that, when f is reduced, the ideal $\mathfrak{I}_2(\tilde{\mathbf{Y}})$ generated by the resilient 2×2 minors in the dual variables is contained in the homogeneous defining ideal of the dual variety $V(f)^*$. This has a strong impact to the nature of $V(f)^*$ since, e.g., it typically implies that the latter is deficient (i.e., not a hypersurface), such as is the case of generic and generic symmetric matrices (Section 3.3 and Chapter 8).

(2) Note that, since the theorem applies to symmetric linear sections of the $m \times m$ generic symmetric matrix, the determinant of a resilient 2×2 matrix, such as

$$\begin{pmatrix} y_{i,j} & y_{i,j'} \\ y_{i,j'} & y_{i',j'} \end{pmatrix},$$

will belong to the kernel of the Gaussian map ∂.

An easy consequence of Theorem 2.11 is the following sufficiently known result:

Proposition 2.13 *Let* $g := \det \mathbf{X}$ *denote the determinant of the* $m \times m$ *generic matrix or of the* $m \times m$ *generic symmetric matrix. Then*

$$\operatorname{rank} H(g) \ (\mathrm{mod}\ g) \leq \begin{cases} 2m, & \text{if } \mathbf{X} \text{ is the } m \times m \text{ generic matrix} \\ m + 1, & \text{if } \mathbf{X} \text{ is the } m \times m \text{ generic symmetric matrix.} \end{cases}$$

Proof Let $V(g)^*$ denote the dual variety of $V(g)$ in its natural embedding. With $\mathbf{Y}$ standing for the dual variables in any of the two cases, Theorem 2.11 implies

$$\dim V(g)^* = \dim k[\mathbf{Y}]/\ker(\partial) - 1 \leq \dim k[\mathbf{Y}]/I_2(\mathbf{Y}) - 1$$

$$= \begin{cases} m^2 - (m-1)^2 - 1, & \text{if } \mathbf{X} \text{ is the } m \times m \text{ generic matrix} \\ \binom{m+1}{2} - \binom{m}{2} - 1, & \text{if } \mathbf{X} \text{ is the } m \times m \text{ generic symmetric matrix} \end{cases}$$

$$= \begin{cases} 2m, & \text{if } \mathbf{X} \text{ is the } m \times m \text{ generic matrix} \\ m + 1, & \text{if } \mathbf{X} \text{ is the } m \times m \text{ generic symmetric matrix.} \end{cases}$$

Finally, use Theorem 2.7, by which $\dim V(g)^* = \operatorname{rank} H(g) \ (\mathrm{mod}\ g) - 2$. $\square$

It carries along a similar bound for the sparse decomposition case:

Proposition 2.14 *Let* $L = (\ell_{i,j})$ *denote a linear section of the* $m \times m$ *generic matrix or a symmetric linear section of the generic symmetric matrix, with determinant* f, *endowed with a sparse decomposition* $\mathbf{X} = \mathbf{X}_1 \cup \mathbf{X}_2$. *Then*

$$\operatorname{rank} H(f) \ (\mathrm{mod}\ f) \leq \begin{cases} 2m, & \text{if } \mathbf{X} \text{ is the } m \times m \text{ generic matrix} \\ m + 1, & \text{if } \mathbf{X} \text{ is the } m \times m \text{ generic symmetric matrix.} \end{cases}$$

Proof This is really an outcome of specialization via Proposition 2.13.

Since L is endowed with a sparse decomposition $\mathbf{X} = \mathbf{X}_1 \cup \mathbf{X}_2$, then $\tilde{\mathbf{X}} \subset \mathbf{X}_1$, where, as before, $\tilde{\mathbf{X}}$ denotes the set of nonzero entries of L. Let $g := \det \mathbf{X}$ denote the generic determinant. Write

$$g = f + \sum_{x_{i,j} \notin \tilde{\mathbf{X}}} x_{i,j} h_{i,j}$$

for certain $h_{i,j} \in k[\mathbf{X}]$.

Given an integer

$$t \geq \begin{cases} 2m + 1, & \text{if } \mathbf{X} \text{ is the } m \times m \text{ generic matrix} \\ m + 2, & \text{if } \mathbf{X} \text{ is the } m \times m \text{ generic symmetric matrix,} \end{cases}$$

consider a subset S of t variables of $\tilde{\mathbf{X}}$. The Hessian matrix $H(g)$ of g with respect to the variables in S has the following shape:

$$B := \left(\frac{\partial^2 f}{\partial x_{i,j} \partial x_{i',j'}} + \sum_{x_{i,j} \notin \tilde{\mathbf{X}}} x_{i,j} \frac{\partial^2 h_{i,j}}{\partial x_{i,j} \partial x_{i',j'}} \right)_{x_{i,j}, x_{i',j'} \in S} .$$

As a consequence of Proposition 2.13,

$$\det B = pg$$

for some $p \in k[\mathbf{X}]$. Consider the specialization k-algebra map $\zeta : k[\mathbf{X}] \to k[\tilde{\mathbf{X}}]$ such that $\zeta(x_{i,j}) = \ell_{i,j}$ if $x_{i,j} \in \mathbf{X}_1$ and $\zeta(x_{i,j}) = 0$ otherwise. Evaluating $\det B$ gives

$$\zeta(\det B) = \zeta(p) f.$$

But $\zeta(\det B)$ is the t-minor of the Hessian of f with respect to the variables in S. This proves the statement. $\qquad\qquad\qquad\qquad\qquad\qquad\qquad\qquad\qquad\qquad\qquad\square$

The main application of Theorem 2.11 and the above proposition is toward the structure of the dual variety of determinantal hypersurfaces of various kinds (Chapter 8).

2.2 Ideal Theory Basics

We revise a few tools from commutative ideal theory. For the basic terminology, we refer to the book [20]. We note in particular the extensive use throughout of the notion of *grade*, *height*, and *codimension*, the last two often used interchangeably even if the ground ring is arbitrary. Recall that the *grade* of a proper ideal $I \subset R$ is the maximal length of an R-sequence (regular sequence) contained in I. We assume known (see [20, Theorem 5.4.3]) that the grade of an ideal does not exceed its height and that, by a reformulation of the so-called Macaulay's theorem, a Noetherian ring R is Cohen–Macaulay if and only if the grade and the height of any of its ideals coincide.

2.2.1 Special Fiber, Analytic Spread, and Reduction Number

For any given ideal $I \subset R$ of a ring, one defines the *Rees algebra* $\mathcal{R}_R(I) :=$ $\bigoplus_{i \geq 0} I^i$. It can be accommodated as the R-subalgebra $R[It] \subset R[t]$, where t is an indeterminate over R. In particular, it inherits the standard grading of the polynomial ring $R[t]$ over R and is a torsion-free R-algebra.

Now, let $(R, \mathfrak{m})$ denote a Noetherian local ring and its maximal ideal (respectively, a standard graded ring over a field and its irrelevant maximal ideal). Let $I \subset \mathfrak{m}$ be an ideal (respectively, a homogeneous ideal $I \subset \mathfrak{m}$). The *special fiber* of I is the $(R/\mathfrak{m})$-algebra $\mathcal{R}_R(I)/\mathfrak{m}\mathcal{R}_R(I)$. It is often designated as the *fiber cone* of I and will be denoted $\mathcal{F}_R(I)$. Note that this is an algebra over the residue field of R. The (Krull) dimension of this algebra is called the *analytic spread* of I and is denoted $\ell(I)$.

If the ground ring R is sufficiently clear or fixed in the context, we frequently omit it as a script in the notation.

The analytic spread admits lower and upper bounds. If $R/\mathfrak{m}$ is infinite, one has $\mathrm{ht}(I) \leq \ell(I)$. Of an easier substance is the inequality

$$\ell(I) \leq \min\{\mu(I), \dim(R)\},$$

where $\mu(I)$ stands for the minimal number of generators of I. If the latter inequality turns out to be an equality, one says that I has maximal analytic spread. Typically, though not always, the ideals discussed in this work will have $\dim R \leq \mu(I)$; hence being of maximal analytic spread means in this case that $\ell(I) = \dim R$.

Quite generally, given ideals $J \subset I$ in a ring R, J is said to be a *reduction* of I if there exists an integer $n \geq 0$ such that $I^{n+1} = JI^n$. An ideal shares the same radical with all its reductions. Therefore, they share the same set of minimal primes and have the same codimension. A reduction J of I is called *minimal* if no ideal strictly contained in J is a reduction of I. The *reduction number* of I with respect to a reduction J is the minimum integer n such that $JI^n = I^{n+1}$. It is denoted by $\mathrm{red}_J(I)$. The (absolute) *reduction number* of I is defined as $r(I) = \min\{\mathrm{red}_J(I) \mid J \subset I \text{ is a minimal reduction of } I\}$.

If $(R, \mathfrak{m})$ is Noetherian local with $R/\mathfrak{m}$ infinite, then every minimal reduction of I is minimally generated by exactly $\ell(I)$ elements. In particular, in this case, every reduction of I contains a reduction generated by $\ell(I)$ elements.

We next digress on some standard properties of the special fiber of a homogeneous ideal I in the case where it is Cohen–Macaulay. In particular, the reduction number $r(I)$ will have a distinctive role. For this, recall the definition of regularity in the case of a homogeneous ideal I in a standard graded polynomial ring R over an infinite field k. Let

$$\cdots \longrightarrow \bigoplus_j R(-d_{i,j}) \longrightarrow \cdots \longrightarrow \bigoplus_j R(-d_{1,j}) \longrightarrow R \longrightarrow R/I \to 0$$

stand for the minimal free graded resolution of R/I over R. The (Castelnuovo–Mumford) *regularity* of R/I is $\mathrm{reg}(R/I) := \max_{i,j}\{d_{i,j} - i\}$.

Now, assume that $k := R/\mathfrak{m}$ is infinite, and consider a graded presentation $R[It] \simeq R[\mathbf{Y}]/\mathcal{I}$, with $R[\mathbf{Y}]$ a polynomial ring. It induces a presentation $\mathcal{F}(I) \simeq k[\mathbf{Y}]/J$, for a suitable ideal J. This way one can talk about the regularity of $\mathcal{F}(I)$ as coming from its minimal free graded resolution over $k[\mathbf{Y}]$.

A great more deal can be obtained if the homogeneous ideal I is *equigenerated*, to mean that it is generated by forms of the same degree.

The following result has been observed in the literature (see, e.g., [7, Proposition 1.2]):

Proposition 2.15 *With the above notation, assume that the ideal I is moreover equigenerated. If the special fiber $\mathcal{F}(I)$ is Cohen–Macaulay, then the reduction number $r(I)$ of I coincides with the regularity of $\mathcal{F}(I)$.*

The gimmick of the proof rests on [24, Proposition 1.85], by which, when the special fiber is Cohen–Macaulay, one can read $r(I)$ off the Hilbert series of $\mathcal{F}(I)$ as the degree of the polynomial in the numerator in its fractional form (the so-called *h-polynomial*).

It affords the following:

Corollary 2.16 *Let $I \subset R$ denote an equigenerated homogeneous ideal such that its special fiber $\mathcal{F}(I)$ is Cohen–Macaulay. Setting $n := \mu(I)$, the following conditions are equivalent:*

 (i) *The reduction number $r(I)$ is at most $\ell(I) - 1$.*
 (ii) *The largest shift in the minimal graded resolution of $\mathcal{F}(I)$ over R is at most $n - 1$.*
 (iii) *The Hilbert function $H_{\mathcal{F}(I)}(t)$ of $\mathcal{F}(I)$ coincides with its Hilbert polynomial for all $t \geq 1$.*

Proof The equivalence of (i) and (ii) follows immediately from Proposition 2.15 since the homological dimension of $\mathcal{F}(I)$ is $n - \ell(I)$. On the other hand, again by Vasconcelos [24, Proposition 1.85], $r(I)$ is the degree of the h-polynomial in the Hilbert series of $\mathcal{F}(I)$. Therefore, since $\dim \mathcal{F}(I) = \ell(I)$, the equivalence of (i) and (iii) follows from [2, Proposition 4.3.5 (c)].   □

Remark 2.17 In the above corollary, $r(I)$ is frequently said to have the expected value if equality takes place in item (i). The extreme lowest value $r(I) = 1$ corresponds of course to the event that $\mathcal{F}(I)$ has minimal multiplicity (degree); hence the corresponding projective variety is either a rational normal scroll or a cone over a Veronese.

The following question was posed by the senior author and Wolmer Vasconcelos:

Conjecture 2.18 Let I be an equigenerated homogeneous ideal in a standard graded algebra R over an infinite field. If the Rees algebra of I is Cohen–Macaulay, then so is the special fiber of I.

Here, of course, the equigeneration assumption is essential. The conjecture is true in the equimultiple case (i.e., $\operatorname{ht} I = \ell(I)$) [17, Corollary 2.6], provided R is Cohen–Macaulay. It also holds in the very special landscape of a codimension 2 perfect ideal $I \subset R = k[x, y, z]$ such that $\mu(I) \geq 4$, $\ell(I) = 3$ and the defining ideal of $\mathcal{F}(I)$ is generated in degrees ≥ 3 [17, Theorem 3.3]. Another case where the conjecture holds is when I is a monomial ideal and $\mathcal{R}_R(I)$ is in addition locally regular in codimension 1 (so $\mathcal{R}_R(I)$ is actually normal).

2.2.2 Linear Rank

Suppose now that R is a standard graded ring over an infinite field k. Then all the above statements in the Noetherian local case hold for a homogeneous ideal $I \subset \mathfrak{m}$. Moreover, suppose that I is generated by forms of the same degree $d \geq 1$. Set $r = \mu(I) = \dim_k [I]_d$, and consider a graded free presentation [20, Chapter 7] of I:

$$R(-(d+1))^\ell \oplus \sum_{j \geq 2} R(-(d+j)) \xrightarrow{\varphi} R(-d)^r \longrightarrow I \longrightarrow 0.$$

Assuming, as will typically be the case, that I contains a regular element, then φ has (well-defined) rank $r - 1$.

Of much interest in this work is the value of ℓ, which can often be zero. The image of $R(-(d+1))^\ell$ by φ is the *linear part of* φ – often denoted φ_1. One can see that ℓ does not depend on the particular minimal system of homogeneous generators of I. If φ_1 has a well-defined rank, we call it the *linear rank* of I. One says that I has *maximal linear rank*, provided its linear rank is $r - 1 (= \operatorname{rank}(\varphi))$. Clearly, the latter condition is trivially satisfied if $\varphi = \varphi_1$, in which case I is said to have *linear presentation* (or is *linearly presented*).

Note that φ is a graded matrix whose columns generate the (first) *syzygy module of* I (corresponding to the given choice of generators) and a *syzyzy* of I is an element of this module – that is, a relation of degree 1 with coefficients in R of the chosen generators. Due to obvious degree reasons, φ_1 can be taken as the submatrix of φ whose entries are forms of degree 1 in the standard graded ring R. Thus, the linear rank is the rank of the matrix of the linear syzygies.

Remark 2.19 The existence at all of linear syzygies is a reflection of the degeneracy of the ideal generators, in the sense that ideals generated by general forms will typically have few linear syzygies or no linear syzygies at all, such as is the case of a complete intersection in degree ≥ 2. The size of this degeneracy is actually what makes it possible to understand particular ideals from the point of view of their minimal resolutions.

2.2.3 Ideals of Linear Type

Our basic reference for this topic is [20, 7.2.2 and 7.3.4].

Given an ideal $I \subset R$ in any commutative ring, there is a natural surjective homomorphism $\mathcal{S}_R(I) \twoheadrightarrow \mathcal{R}_R(I)$ of graded R-algebras of the symmetric algebra of I onto its Rees algebra. The ideal I is said to be of *linear type* if this surjection is injective as well. By looking at the respective defining ideals of the two algebras, with a slight speech abuse, this happens if and only if the R-relations of I come from its linear R-relations (i.e., its syzygies).

Given an integer $s \geq 2$, an ideal $I \subset R$ in a Noetherian ring satisfies the *condition* G_s if, locally at any prime ideal $P \supset I$ such that ht $P \leq s - 1$, the minimal number of generators of I is at most ht P. Clearly, if $t \leq s$, then $G_s \Rightarrow G_t$. If this property is satisfied for $s \geq \dim R + 1$ (hence, for every $s \geq 1$), then the condition has been variously called G_∞ or F_1.

The condition was originally introduced in [1] and has since been explored by many authors. As is well-known, it can be characterized in terms of the codimension of the Fitting ideals of I:

Proposition 2.20 *Let R be a Noetherian ring and $I \subset R$ an ideal with a finite free presentation*

$$R^r \xrightarrow{\varphi} R^m \longrightarrow I \to 0.$$

Then I satisfies the condition G_s if and only if

$$\text{ht } I_j(\varphi) \geq m - j + 1, \quad \text{for } m - s + 1 \leq j \leq m - 1. \tag{2.7}$$

For a proof, see, e.g., [12, Corollary 2.2]. We note that, in particular, G_2 (hence, G_s for any $s \geq 2$) implies that I has a rank. Also, G_∞ is characterized by the condition

$$\text{ht } I_j(\varphi) \geq \text{rank } \varphi - j + 2, \quad \text{for } 1 \leq j \leq \text{rank } \varphi. \tag{2.8}$$

A weaker condition is obtained by replacing 2 by 1 in the above inequality. Because of its similarity to F_1, it has been dubbed F_0. Its impact is on the dimension of the symmetric algebra of I.

As is easy to verify, if $I \subset R$ is an ideal of linear type in a Noetherian ring, then it satisfies G_∞. A second useful property implied by the linear type property in a Noetherian ring is that I has no proper reductions locally everywhere, which means that I has maximal analytic spread locally everywhere or still that I is generated by analytically independent elements locally everywhere.

While the second of these conditions has a geometric flavor in various directions, the first is more algebraic. Unfortunately, the two properties are not mutually complementary toward achieving the linear type property, so often one has to delve into it quite a bit more.

In this book a special case is prominent, namely, when I is generated by the partial derivatives of a homogeneous polynomial $f \in R = k[\mathbf{X}]$, where k is a field. We recall that I is called the *Jacobian ideal* of f or the *gradient ideal* of f and will be denoted J_f.

An important feature of such an ideal J_f is that the linear type property implies the algebraic independence of the partial derivatives of f over k – equivalently, J_f has maximal analytic spread, or at least if $\mathrm{char}(k) = 0$, the Hessian determinant of f does not vanish (Chapter 3). Although a difficult property to come around, it is quite common in the case where f is the determinant of a square linear matrix.

Remark 2.21 Estimating the codimension of a Fitting ideal is usually tricky and, depending on the sizes involved, can discourage anyone to resort to a computer. There is a slight grip on this if the matrix has linear entries and is 1-generic or more (Section 1.5.2), but otherwise it can be a hard knuckle. The problem is that most matrices of syzygies of ideals are seldom linear and 1-generic unless they belong to a couple of well-known classes. Much less so, syzygies of the gradient ideal $J_f \subset R$ of a form $f \in R$. Yet, in the case where f is the determinant of a linear square matrix, partial knowledge about the linear syzygies may come out of Proposition 2.10, as the cofactors have many linear syzygies thanks to the adjugate formula. Alas, most common cases of such specializations of the generic square matrix dash our hopes.

For some recent account on the linear type property regarding J_f, we refer to [19, Chapter 5].

2.2.4 Perfectness in Height 2

One of the earliest results relating the condition G_∞ and ideals of linear type takes place in the realm of perfectness. We briefly recall some related definitions.

Pretty generally, a finitely generated module E over a Noetherian ring A is *perfect* if the grade of E (i.e., the grade of its annihilator ideal) coincides with the homological (projective) dimension of E over A. This is quite a strong requirement since E must at least be of finite homological dimension. By abuse, we say that an ideal $I \subset A$ is perfect if A/I is a perfect A-module.

This notion is often disguised behind the notion of Cohen–Macaulayness due to the following principle: If A is a Cohen–Macaulay ring and $I \subset A$ is an ideal such that A/I has no nontrivial idempotents and has finite homological dimension over A, then A/I is Cohen–Macaulay if and only if A/I is perfect.

One important property of a perfect module E (as of a Cohen–Macaulay E module over an equicodimensional Cohen–Macaulay ring A) is that of being of pure grade, a condition already encountered in Chapter 1, Section 1.4. Namely, $\mathrm{grade}\, E = \mathrm{grade}\, P$ for every associated prime P of E. In the case where A is itself a Cohen–Macaulay ring, this property is the same as the commonly used *height unmixedness*.

Proposition 2.22 ([10, Theorem 1.1], [2, Proposition 5], [21, Theorem 3.4]) *Let R be a local domain, and let $I \subset R$ be a perfect ideal of height 2. Then:*

(i) *I is of linear type if and only if it satisfies the condition G_∞.*
(ii) *If I is of linear type, then the symmetric algebra $S(I)$ is a complete intersection and hence is Cohen–Macaulay when R is Cohen–Macaulay.*

We now discuss the impact of condition G_s on perfect ideals of height 2.

We look more closely at the graded components of the canonical R-algebra surjection $\mathcal{S}_R(I) \twoheadrightarrow \mathcal{R}_R(I)$. Let I denote a perfect ideal of height 2 over a Cohen–Macaulay ring, with an $m \times (m-1)$ linear presentation matrix φ. Then, for every integer t, the following complex was studied in [2]:

$$\mathcal{K}_t \; : \; 0 \to F_{m-1} \to F_{m-2} \to \cdots \to F_1 \to F_0 \to 0, \tag{2.9}$$

where

$$F_i := \bigwedge^i R^{m-1} \otimes_R S_{t-i}(R^m) \qquad (\text{with } S_u(R^m) = 0 \text{ for } u < 0)$$

and $\delta_i : F_i \to F_{i-1}$ is given by

$$\delta_i(e_1 \wedge \ldots \wedge e_i \otimes g) := \sum_{l=1}^{i} e_1 \wedge \ldots \wedge \widehat{e_l} \wedge \ldots \wedge e_i \otimes \varphi(e_l)g,$$

with $\{e_1, \ldots, e_{m-1}\}$ denoting a basis of R^{m-1}.

Theorem 2.23 [2] *Let R be a Cohen–Macaulay ring, and let I be a perfect ideal of R of height 2. If I satisfies the G_s condition, then*

(a) *The complex $\mathcal{K}_t$ is a free resolution for $S_t(I)$ for every $t < s$.*
(b) *$S_t(I) \simeq I^t$ for every $t < s$.*

So far, we have dealt with the symmetric algebra. We now bring up the Rees algebra to make a case for a strong relation between the two. For this, let $R = k[x_1, \ldots, x_d]$ be a polynomial ring over an infinite field k, and let $I \subset R$ be a linearly presented perfect ideal of height 2, with an $m \times (m-1)$ linear presentation matrix φ as above. Choosing new variables $y_1, \ldots, y_m$ over k, a defining ideal of $S(I)$ is generated by $I_1(\mathbf{y}\varphi)$, where $\mathbf{y} := (y_1 \ldots y_m)$. Now, there exists a unique $d \times (m-1)$ matrix B with linear entries in $k[y_1, \ldots, y_m]$ such that

$$\mathbf{y}\varphi = \mathbf{x}B, \tag{2.10}$$

where $\mathbf{x} := (x_1 \ldots x_d)$.

CLAIM: If $m > d$, then $(I_1(\mathbf{x}B), I_d(B))$ belongs to the defining ideal $\mathcal{J}$ of $\mathcal{R}_R(I)$ over $k[y_1, \ldots, y_m]$.

In fact, for any $d \times d$ submatrix $\tilde{B}$ of B, we have $I_1(\mathbf{x}\tilde{B}) \subset I_1(\mathbf{x}B) \subset \mathcal{J}$.

In particular,

$$(x_1 \det \tilde{B}, \ldots, x_d \det \tilde{B}) = I_1(\mathbf{x}\tilde{B} \operatorname{adj}(\tilde{B})) \subset \mathcal{J},$$

where $\operatorname{adj}(\tilde{B})$ is the adjugate. Thus, since $\mathcal{J}$ is a prime ideal and $x_i \notin \mathcal{J}$ for every $1 \leq i \leq d$, then $\det \tilde{B} \in \mathcal{J}$.

The inclusion in the claim is an equality under the G_d assumption:

Theorem 2.24 ([13, Theorem 1.3]) *Let $R = k[x_1, \ldots, x_d]$ be a polynomial ring over an infinite field k, and let $I \subset R$ be a perfect ideal of height 2 with an $m \times (m-1)$ linear presentation matrix φ. Assume that $m > d$. If I satisfies the G_d condition, then $\ell(I) = d$, $r(I) = d$, $\mathcal{J} = (I_1(\mathbf{x}B), I_d(B))$ and $\mathcal{R}(I)$ is Cohen–Macaulay.*

Remark 2.25 In particular, in the above theorem the presentation ideal of the special fiber $\mathcal{F}(I) = \mathcal{R}(I)/(\mathbf{x})\mathcal{R}(I)$ is $I_d(B)$. Since B is a $d \times (m-1)$ matrix and $\operatorname{ht} I_d(B) = (m-1) - d + 1$, then the ideal $I_d(B) \subset k[\mathbf{y}]$ is resolved by the Eagon–Northcott complex [20, Section 6.4]. Hence, the d-minors of B are k-linearly independent.

2.2.5 On the Symbolic Rees Algebra

We discuss certain preliminaries on the subject having in mind later applications (Chapter 9).

Given an ideal $I \subset R$ in a Noetherian ring and an integer $r \geq 1$, the rth *symbolic power* $I^{(r)}$ of I is the contraction of $U^{-1}I^r$ under the natural homomorphism $R \to U^{-1}R$ of fractions, where U is the complementary set of the union of the associated primes of R/I. In this part I will be a codimension 2 perfect ideal in a polynomial ring over a field; hence R/I is Cohen–Macaulay and so I is a pure (unmixed) ideal. In this setup then $I^{(r)}$ is precisely the intersection of the primary components of the ordinary power I^r relative to the associated primes of R/I, i.e., the unmixed part of I^r.

A slightly different way to envisage symbolic powers is by noting that, given an integer $r \geq 1$, the R/I-torsion of the conormal module I^{r-1}/I^r of order $r - 1$ is $(I^{(r)} \cap I^{r-1})/I^r$. Taking the direct sum over all $r \geq 0$ yields the R/I-torsion of the associated graded ring of I; hence the nontriviality of symbolic powers gives a measure of the torsion of the latter. In particular, there is no nonzero torsion if and only if $I^{(r)} = I^r$ for every $r \geq 0$ – in which case one says that the ideal I is *normally torsion-free*. However, this information is most of the times pretty useless once it holds. What matters for a substantial class of ideals – codimension 2 perfect ideals included – is to guess some sort of asymptotic behavior for the equality of the two powers, more like an "inf-asymptotic" such behavior in the sense that one has equality throughout up to a certain exponent order, thereafter comparison gets disorganized or even chaotic.

We observe that, like the ordinary powers, the symbolic powers constitute a decreasing multiplicative filtration, so one can consider the corresponding *symbolic*

Rees algebra $\mathcal{R}_R^{(I)} = \bigoplus_{r \geq 0} I^{(r)} t^r \subset R[t]$. However, unlike the ordinary Rees algebra, this algebra may not be finitely generated over R. Alas, there are no definite effective ways to check when $\mathcal{R}_R^{(I)}$ is Noetherian. The necessary and sufficient conditions of Huneke [11, Theorems 3.1 and 3.25] obtained in dimension 3 are not effective, and neither is the necessary condition of Cowsik–Vasconcelos [5], [19, Proposition 3.5]. Nevertheless, the latter becomes quite effective, provided one has a good guess about what finitely generated subalgebra looks like a strong candidate. In a precise way, one has the following strategy.

First recall that, given an ideal $I \subset R$, where R is a Noetherian domain with field of fractions K, the *ideal transform* of R relative to I is the R-subalgebra $T_R(I) := R :_K I^\infty \subset K$. We will draw on the following two fundamental facts:

- [23, Proposition 7.1.4] If $C \subset T_R(I)$ is a finitely generated R-subalgebra such that $\mathrm{depth}_{IC}(C) \geq 2$, then $C = T_R(I)$.
- [23, Proposition 7.2.6] If R moreover satisfies the condition (S_2) of Serre, then

$$\mathcal{R}_R^{(I)} \simeq T_{\mathcal{R}(I)}(J) \subset R[t]$$

as R-subalgebras of $R[t]$ for a suitable choice of the ideal $J \subset R$.

Our idea of applying these principles is summarized in the following result, of immediate verification:

Proposition 2.26 *Let $R = k[x_1, \ldots, x_n]$ denote a standard graded polynomial ring over an infinite field k, with irrelevant maximal ideal $\mathfrak{m} := (x_1, \ldots, x_n)$. Let $I \subset R$ stand for a homogeneous ideal satisfying the following properties:*

(i) *For every $r \geq 1$, the R-module $I^{(r)}/I^r$ is either zero or $\mathfrak{m}$-primary.*
(ii) *$\mathrm{depth}_{\mathfrak{m}C}(C) \geq 2$ for some finitely generated graded R-subalgebra $C \subset \mathcal{R}_R^{(I)}$ containing the Rees algebra $\mathcal{R}_R(I)$.*

Then $C = \mathcal{R}_R^{(I)}$.

Note that if the subalgebra C itself satisfies Serre's property (S_2), then condition (ii) above is granted.

We observe that the typical graded R-subalgebra $C \subset \mathcal{R}_R^{(I)}$ containing the Rees algebra $\mathcal{R}_R(I)$ as above has the form $C = R[It, I^{(2)}t^2, \ldots, I^{(s)}t^s] \subset R[t]$, for suitable $s \geq 1$. Although the non-vanishing of certain of the R-modules $I^{(r)}/I^r$ gives a measure of how far one has to go (provided the symbolic Rees algebra is finitely generated), it is really the R-modules

$$\frac{I^{(r)}}{\sum_{1 \leq j \leq r-1} I^{(r-j)} \cdot I^{(j)}}$$

that count for the search of *fresh* (or *genuine*) generators of the algebra. Although this is a well-known simple observation, it often encrypts some subtleties in a particular case.

2.3 Algebraic Aspects of Rational Maps

The overall reference for this part and its language is [16].

Throughout, $\mathbb{P}^n = \mathbb{P}^n_k$ denotes the nth projective space over an infinite field k.

The following is a characteristic-free generalized version of [4, Proposition 4.5 (iii)].

Theorem 2.27 ([16, Theorem 3.2.26]) *Let* $\mathfrak{F}\colon \mathbb{P}^n \dashrightarrow \mathbb{P}^m$ *be a rational map, defined by* $m + 1$ *forms* $\mathbf{f} = \{f_0, \ldots, f_m\}$ *of a fixed degree. If the image of* $\mathfrak{F}$ *has dimension* n *and if the ideal* $(\mathbf{f}) \subset R$ *has maximal linear rank* $(= m)$, *then* $\mathfrak{F}$ *is birational onto its image.*

We consider the notion of the *inversion factor* of a birational map [16, Section 4.2]. Inversion factors of *Cremona maps* in characteristic zero have a curious property.

Theorem 2.28 $((\mathrm{char}(k) = 0)$ [16, Proposition 4.2.2]) *Let* $\mathfrak{F}$ *denote a Cremona map of* $\mathbb{P}^n$ *defined by forms* $\mathbf{f} : \{f_0, \ldots, f_n\}$ *in* R *without nontrivial common factor, and let* $\Theta(\mathbf{f})$ *denote the Jacobian matrix of* $\mathbf{f}$. *Then* $\det \Theta(\mathbf{f})$ *divides a power of the source inversion factor* F *of* $\mathfrak{F}$. *In particular, if* $\det \Theta(\mathbf{f})$ *is reduced, then it divides* F.

The classical theory of plane Cremona maps in characteristic zero relates the Jacobian of a homaloidal net to the principal curves of the corresponding Cremona map. The next largely algebraic proposition has close analogy to this result.

Theorem 2.29 ([15, Proposition 2.11]) *Let* $R = k[x_1, \ldots, x_d]$ *be a standard graded polynomial ring over a field* k *of characteristic zero, and let* $\mathcal{L} = (\ell_{ij})$ *be a* $d \times (d-1)$ *matrix whose entries are linear forms in* R. *For every* $i = 1, \ldots, d$, *write* Δ_i *for the signed* $(d-1)$-*minor of* $\mathcal{L}$ *obtained by omitting the* i*th row, and let* $\Theta = \Theta(\mathbf{\Delta})$ *denote the Jacobian matrix of* $\mathbf{\Delta} := \{\Delta_1, \ldots, \Delta_d\}$.

If the ideal $I_{d-1}(\mathcal{L}) = (\mathbf{\Delta}) \subset R$ *is of linear type, then the rational map* $\mathfrak{F} : \mathbb{P}^{d-1} \dashrightarrow \mathbb{P}^{d-1}$ *defined by* $\mathbf{\Delta}$ *is a Cremona map, and the associated source inversion factor is* $\frac{1}{d-1} \det \Theta$.

There is a very basic and somewhat surprising result relating inversion factors to symbolic powers which seems to have gone unnoticed in the classical birational theory.

Theorem 2.30 ([16, Proposition 4.2.4]) *Let* $\mathfrak{F} : \mathbb{P}^n \dashrightarrow \mathbb{P}^m$ *be a birational map onto the image, with base ideal* $I \subset R = k[\mathbf{X}]$ *generated in degree* $d \geq 2$. *Let* $F \in R$ *denote the source inversion factor relative to a given representative of the inverse map. If* $\mathrm{depth}\,(R/I) > 0$, *then:*

(a) F *is an element of the symbolic power* $I^{(d')}$, *where* d' *is the degree of the coordinates of the representative. In particular,* $I^{(d')} \neq I^{d'}$.

(b) *If, moreover,* $I^{(\ell)} = I^{\ell}$, $\ell \leq d' - 1$, *then* F *is not iterated from lower order powers.*

(c) *Moreover, if $I^{(d')}$ is generated in standard degree $\geq dd' - 1$, where d is the coordinate degree of $\mathfrak{F}$, then F is a homogeneous minimal generator of the symbolic Rees algebra.*

Exercises

2.31 Consider the matrix

$$
M = \begin{pmatrix}
0 & y & 0 \\
0 & -z & -y^2 \\
y & 0 & 0 \\
-z & 0 & x^2
\end{pmatrix}
$$

with entries in $k[x, y, z]$ (k, a field). Set $I := I_3(M)$.

1. Determine ht I.
2. Determine the analytic spread of I.
3. Show that the special fiber $\mathcal{F}(I)$ is a hypersurface ring.
4. Determine the homological dimension of I.

2.32 Let $I \subset R = k[x_1, x_2, x_3, x_4]$ be the polynomial ideal over the field k generated by the maximal minors of the 6×5 matrix

$$
\varphi = \begin{bmatrix}
x_1 & 0 & 0 & 0 & h_1 \\
x_2 & x_1 & 0 & 0 & h_2 \\
x_3 & x_2 & x_1 & 0 & h_3 \\
0 & x_3 & x_2 & 0 & h_4 \\
0 & 0 & x_3 & x_1 & h_5 \\
0 & 0 & 0 & x_4 & x_3^d
\end{bmatrix},
$$

where $h_1, \ldots, h_5$ are forms of a given degree $d \geq 1$ such that I has height 2 (and hence is a perfect ideal).

1. Prove that I satisfies condition G_3, but not condition G_4.
   (HINT: Use the criterion of Proposition 2.20.)
2. Deduce that I is generically a complete intersection. Digress on the difficulty of obtaining this result from the mere definitions.

2.33 Let $I \subset R = k[x, y, z]$ be the polynomial ideal over the field k generated by the maximal minors of the 6×5 matrix

$$\varphi = \begin{bmatrix} y & 0 & 0 & 0 & 0 \\ -z & y & 0 & 0 & 0 \\ 0 & -z & x & 0 & 0 \\ 0 & 0 & -z & x & 0 \\ 0 & 0 & 0 & -z & xy \\ 0 & 0 & 0 & 0 & -z^2 \end{bmatrix}.$$

1. Find all associated (i.e., minimal) primes of I.
2. Prove that I does not satisfy condition G_3 (without computing Fitting ideals).
3. Prove that the special fiber of I is not Cohen–Macaulay.

   (HINT: Derive the Hilbert series of $k[x, y, z]/I$ from the free resolution of I.)

2.34 Let $I \subset k[x, y, z]$ be the ideal generated by the maximal minors of the following matrix:

$$\begin{bmatrix} z^2 & 0 & 0 & 0 \\ x^2 & z^2 & 0 & 0 \\ 0 & y^2 & z & 0 \\ 0 & 0 & x & z \\ 0 & 0 & 0 & y \end{bmatrix}.$$

1. Prove that I is not generically a complete intersection.
2. Writing $\mathcal{F}(I) \simeq k[y_1, \ldots, y_5]/Q$, show that the maximal minors of the matrix

$$\begin{bmatrix} y_3 & y_1^2 \\ -y_4 & -y_2^2 \\ y_5 & y_3^2 \end{bmatrix}$$

   belong to Q.
3. Deduce that $\mathcal{F}(I)$ is Cohen–Macaulay.

2.35 This problem may require computer assistance. Consider the following rational maps $\mathfrak{F} : \mathbb{P}^n \dashrightarrow \mathbb{P}^n$, defined by

1. $(n = 2)\ (x_1 x_2 : x_0 x_2 : x_0 x_1)$
2. $(n = 2)\ (x_0^3 : x_1 x_2^2 : x_0^2 x_2)$
3. $(n = 2)\ (x_0^3 : x_0 x_1 x_2 : x_1 x_2^2)$
4. $(n = 3)$ the maximal minors of the 4×3 matrix

$$\begin{pmatrix} 0 & 0 & -x_1 \\ -x_0 & x_0 - x_1 & x_1 \\ x_0 & 0 & 0 \\ x_2 & -x_3 & x_3 \end{pmatrix}.$$

For each of these maps:

- Prove that it is a Cremona map, and compute its inverse map.
- Compute the corresponding source inversion factor, and confirm Theorem 2.28.

References

1. M. Artin, M. Nagata, Residual intersections in Cohen-Macaulay rings. J. Math. Kyoto Univ. **12**, 307–323 (1972) 44
2. L. Avramov, Complete intersections and symmetric algebras. J. Algebra **73**, 248–263 (1981) 46
3. W. Bruns, U. Vetter, *Determinantal Rings*. Lecture Notes in Mathematics, vol. 1327 (Springer, Berlin, 1988) 32
4. C. Ciliberto, F. Russo, A. Simis, Homaloidal hypersurfaces and hypersurfaces with vanishing Hessian. Advances Math. **218**, 1759–1805 (2008) 49
5. R.C. Cowsik, Symbolic powers and number of defining equations, in *Algebra and its Applications*. Lecture Notes in Pure and Applied Mathematics, vol. 91 (Marcel Dekker, New York, 1984), pp. 13–14 48
6. R. Cunha, M. Mostafazadehfard, Z. Ramos, A. Simis, Coordinate sections of generic Hankel matrices. J. Algebra **611**, 285–319 (2022) 32
7. M. Garrousian, A. Simis, Ş. Tohăneanu, A blowup algebra for hyperplane arrangements. Algebra & Number Theory **12**, 1401–14210 (2018) 42
8. M.A. Golberg, The derivative of a determinant. Amer. Math. Mon. **79**, 1124–1126 (1972) 37
9. J. Herzog, T. Hibi, Monomial ideals, in *Graduate Texts in Mathematics*, vol. 260 (Springer, Berlin, 2011) 33
10. C. Huneke, On the symmetric algebra of a module. J. Algebra **69**, 113–119 (1981) 46
11. C. Huneke, Hilbert functions and symbolic powers. Michigan Math. J. **34**, 293–318 (1987) 48
12. N.P.H. Lan, On Rees algebras of linearly presented ideals. J. Algebra **420**, 186–200 (2014) 44
13. S. Morey, B. Ulrich, Rees algebras of ideals with low codimension. Proc. Amer. Math. Soc. **124**, 3653–3661 (1996) 47
14. M. Mostafazadehfard, *Hankel and sub-Hankel Determinants—a Detailed Study of their Polar Ideals*, PhD Thesis (Universidade Federal de Pernambuco, Brazil, 2014) 37
15. Z. Ramos, A. Simis, Symbolic powers of perfect ideals of codimension 2 and birational maps. J. Algebra, **413**, 153–197 (2014) 49
16. Z. Ramos, A. Simis, *Graded Algebras in Algebraic Geometry*. Expositions in Mathematics, vol. 70 (De Gruyter, Berlin, 2022) 35, 49
17. Z. Ramos, A. Simis, Tight and adjusted equigenerated sequences, and the special fiber. São Paulo J. Math. Sci. **17**(1), 320–344 (2023). Special Issue in Honor of Rafael H. Villarreal on the Occasion of his 70th Birthday, eds. by I. Gitler, C. Renteria, A. Simis 43
18. B. Segre, Bertini forms and Hessian matrices. J. London Math. Soc. **26**, 164–176 (1951) 35
19. A. Simis, B. Ulrich, W. Vasconcelos, Jacobian dual fibrations. Amer. J. Math. **115**, 47–75 (1993) 48
20. A. Simis, *Commutative Algebra*, 2nd edn. (De Gruyter Graduate, Berlin, 2023) 40, 43, 44, 47
21. A. Simis, W.V. Vasconcelos, On the dimension and integrality of symmetric algebras Math. Z. **177**, 341–358 (1981) 46
22. B. Sturmfels, Gröbner bases and Stanley decompositions of determinantal rings. Math. Z. **205**, 137–144 (1990) 33
23. W. Vasconcelos, *Arithmetic of Blowup Algebras*. London Mathematical Society, Lecture Notes Series, vol. 195 (Cambridge University Press, Cambridge, 1994) 48
24. W. Vasconcelos, *Integral closure*. Springer Monographs in Mathematics (Springer, Berlin, 2005) 42

Chapter 3
Geometric Oversight

Abstract In this chapter we give an overview of the main geometric players related to a projective hypersurface, among them the polar map and image, the gradient ideal and the Hessian matrix and its determinant. A discussion is enticed about the rank of the Jacobian matrix of a set of forms in a polynomial ring over a field, and a characteristic-free proof is given of the rank of the Grassmann Jacobian (i.e., the Jacobian matrix of the maximal minors of a generic matrix over a field). Complete coverage delivers the main properties of the determinant of the generic square matrix and the generic symmetric matrix over a field of characteristic $\neq 2$. Some consideration is given to the question as to when a projective hypersurface is defined by the determinant of a matrix of linear entries and how the algebraic features of this matrix as the ones in the book may reflect back into nontrivial traits of the hypersurface.

3.1 The Polar Map

Let $f \in R = k[\mathbf{X}] := k[x_0, \ldots, x_n]$ be a nonzero homogeneous polynomial, where k is an infinite field. We will use this notation for the variables in order to comply with the usual geometric convention of dealing with $\mathbb{P}^n$ (instead of the more awkward $\mathbb{P}^{n-1}$) but subsequently will not resist changing back to variables $x_1, \ldots, x_n$. This part is a second, more detailed, drilling into the role of the partial derivatives of f (see Section 2.1.2.1 for a previous insight).

Again, since the notions below have a geometric flare, we assume whenever needed that k is algebraically closed. In addition, playing with partial derivatives, especially when invoking the classical Euler relation, leads to restriction on the field characteristic. To play safely, one may bluntly assume that $\text{char}(k) = 0$. Yet, throughout the subsequent material we will be tempted to assume a more flexible characteristic, for which reinstated proofs have to be clearly established.

Let $V(f) \subset \mathbb{P}^n$ denote the corresponding algebraic hypersurface. Set $d := \deg f$ throughout.

We will often denote by f_i the partial derivative $\frac{\partial f}{\partial x_i}$, $i = 0, \ldots, n$.

Z. Ramos, A. Simis, *Determinantal Ideals of Square Linear Matrices*,
https://doi.org/10.1007/978-3-031-55284-7_3

Let $p = (p_0 : \cdots : p_n) \in \mathbb{P}^n$. For every positive integer $s < \deg f$, consider the polynomial

$$\Delta_{\mathbf{p}}^s f = \left(p_0 \frac{\partial}{\partial x_0} + \cdots + p_n \frac{\partial}{\partial x_n} \right)^{(s)} f(\mathbf{x}) \in R,$$

where the symbol (s) means, as usual, the operations involving products and derivatives up to degree s. The polynomial $\Delta_{\mathbf{p}}^s f$ of degree $\deg f - s$ defines a hypersurface $V_p^s(f) := V(\Delta_{\mathbf{p}}^s f) \subset \mathbb{P}^n$, which depends only on p and on $V(f)$; it is called the *s-th polar* of $V(f)$ with respect to p.

For $s = 1$ the polar $V_p^s(f)$ is often called *the first polar* or simply the *polar* of $V(f)$ with respect to p and denoted simply $V_p(f)$. It is a certain k-linear combination of the partial derivatives $f_0, \ldots, f_n$, hence an element of the k-vector space $k[f_0, \ldots, f_n]_{d-1}$. One often expresses this in the geometric jargon by saying that it varies in a linear system with base locus scheme the singular locus $\mathrm{Sing}(V(f))$ of $V(f)$, defined by the Jacobian (or gradient) ideal generated by the partial derivatives $f_0, \ldots, f_n$.

Although higher polars are useful in treating certain varieties of classical algebraic geometry, we will in this book only consider first polars. Closely associated to the polar of $V(f)$, and playing a substantial role in the book, is the notion of the polar map of f (or of $V(f)$).

Definition 3.1 The rational map $\varphi_f : \mathbb{P}^n \dashrightarrow (\mathbb{P}^n)^{\check{}}$ defined by the partial derivatives of f is called the *polar map* of f. If this map is birational – i.e., a Cremona transformation – then f is said to be *homaloidal*. The (closed) image of φ_f is the projective variety whose homogeneous coordinate ring is $k[\partial f/\partial x_0, \ldots, \partial f/\partial x_n]$ up to a degree normalization – we let $\mathfrak{P}(f)$ denote this image and call it the *polar image* of f (or of the hypersurface $V(f)$).

3.2   The Hessian of a Projective Hypersurface

Recall that the Hessian matrix of f is the $(n + 1) \times (n + 1)$ Jacobian matrix of its partial derivatives

$$H(f) = \left(\frac{\partial^2 f}{\partial x_i \partial x_j} \right)_{i,j=0,\ldots,n}.$$

Its determinant $h(f) \in k[x_0, \ldots, x_n]$ is the *Hessian* of f.

The associated variety $V(h(f)) \subset \mathbb{P}^n$ is called the *Hessian hypersurface*.

3.2.1 Parabolism and the Dual Variety

GEOMETRIC MOTIVATION Classically, a point $p \in \mathbb{P}^n$ lies on $V(f) \cap V(h(f))$ if and only if either $p \in \mathrm{Sing}(V(f))$ or p is a *parabolic point* of $V(f)$ – the latter meaning that the tangent cone $C_p(f)$ at p of the intersection of $V(f)$ with the tangent hyperplane $\pi_p(f)$ (necessarily singular at p) has a vertex of positive dimension (see [21, p. 71]). More precisely, a point $p \in V(f)$ is said to be $\mathfrak{v}$-*parabolic*, $\mathfrak{v} \geq 0$, if the vertex of $C_p(f)$ has dimension $\mathfrak{v}$. In that case, p is a point of multiplicity $\mathfrak{v}$ for the Hessian hypersurface $V(h(f))$ (see [19]). If f is irreducible and the general point of $V(f)$ is $\mathfrak{v}$-parabolic, for certain $\mathfrak{v} \geq 0$, then $f^{\mathfrak{v}}$ divides $h(f)$; in particular, if $\mathfrak{v} > 0$, then $V(f) \subset V(h(f))$.

Conversely, suppose that f is irreducible and that $V(f) \subset V(h(f))$. Then the dual variety $V(f)^*$ of $V(f)$ (Section 2.1.2.1) is deficient, in the sense that $\dim V(f)^* < n - 1$, and the general point $p \in V(f)$ is $\mathfrak{v}$-parabolic with $\mathfrak{v} := n - 1 - \dim V(f)^* > 0$ (see [18], §§4–5, [19]). In this case, the Hessian hypersurface $V(h(f))$ contains $V(f)$ with multiplicity at least $n - 1 - \dim V(f)^*$ – this value being known as the *expected multiplicity*.

Borrowing the above terminology, we will say that a homogeneous polynomial $f \in R$ is *parabolic* if f is a factor of $h(f)$ with multiplicity ≥ 1 (not necessarily the expected multiplicity). This means, in particular, that $h(f) \neq 0$ and that the (true) multiplicity of f as a factor of $h(f)$ is an upper bound for $n - 1 - \dim V(f)^*$. Using this terminology, even if the multiplicity is not the expected one, seems like a good idea as the two properties can be handled separately. Besides, the concept is much weaker than that of a totally Hessian form, and there are many examples of parabolic determinantal forms which are not totally Hessian – here a form f is called *totally Hessian* if up to a nonzero scalar multiple $h(f)$ is a power of f. Thus, if $f \in k[x_0, \ldots, x_n]$ has degree $d \geq 1$, being totally Hessian means that

$$h(f) = cf^{\frac{(d-2)(n+1)}{d}},$$

for some $c \in k \setminus \{0\}$. The totally Hessian situation is to be considered as an extremal condition, whereby one might expect that, more commonly, a certain power of f with positive exponent divides the Hessian determinant.

3.2.2 The Rank of the Jacobian Matrix

There is a more encompassing principle relating the dimension of the image of the polar map to the rank of the Hessian matrix of f which goes back in essence to E. Noether's days. Since our next dealing is mostly algebraic, we will write the ground polynomial ring in variables $x_1, \ldots, x_n$ instead.

Proposition 3.2 *Let $A = k[\mathbf{g}] = k[g_1, \ldots, g_r] \subset R = k[x_1, \ldots, x_n]$ be a k-subalgebra, and let $\Theta(\mathbf{g})$ stand for the Jacobian matrix of $\mathbf{g} = \{g_1, \ldots, g_r\}$ with respect to the variables of R. Set L and K for the respective fields of fractions of A and R. Then:*

(i) $\operatorname{rank} \Theta(\mathbf{g}) \leq \dim A$.
(ii) *If* $\operatorname{char}(k) = 0$ *then* $\operatorname{rank} \Theta(\mathbf{g}) = \dim A$.
(iii) *If* $\{g_1, \ldots, g_r\}$ *is algebraically independent over k and if the field extension $K|L$ is separable algebraic, then* $\operatorname{rank} \Theta(\mathbf{g}) = \dim A$.

Proof A proof of item (ii) is given in [23, Proposition 1.1], but we will proceed ab initio.

(i) Let $A \simeq k[\mathbf{y}]/P$ be a presentation of A by mapping $y_j \mapsto g_j$. Say, $P = (\mathbf{f}) = (f_1, \ldots, f_m)$, with $f_i \in k[\mathbf{y}]$. Let $\Theta(\mathbf{f})$ denote the Jacobian matrix of $\mathbf{f}$, and let $\Theta(\mathbf{f})(\mathbf{g})$ stand for the matrix obtained by evaluating its entries by $y_j \mapsto g_j$. Then the well-known presentation of the module $\Omega_{A/k}$ of Kähler differentials [24, Proposition 4.2.8] can be written as

$$A^m \xrightarrow{\Theta(\mathbf{f})(\mathbf{g})^*} A^r \longrightarrow \Omega_{A/k} \to 0,$$

where $*$ denotes dual into A. Since R is a polynomial ring over k, the extension $K|k$ is separably generated; hence one has $\operatorname{rank}_A \Omega_{A/k} = \dim A$ ([13, Theorem 5.9 (iii)], using the obvious equality $\operatorname{rank}_A \Omega_{A/k} = \operatorname{rank}_L \Omega_{L/k}$; see also [24, Proposition 4.2.13] when k is perfect).

Therefore, the above exact sequence implies that $r - \operatorname{rank}_R \Theta(\mathbf{f})(\mathbf{g}) = \dim A$.

On the other hand, by the chain rule of derivatives, $\Theta(\mathbf{f})(\mathbf{g})$ fits in the following complex of free R-modules:

$$R^m \xrightarrow{\Theta(\mathbf{f})(\mathbf{g})^*} R^r \xrightarrow{\Theta(\mathbf{g})^*} R^n,$$

from which one gets $\operatorname{rank}_R \Theta(\mathbf{g}) \leq r - \operatorname{rank}_R \Theta(\mathbf{f})(\mathbf{g})$. To conclude it suffices to observe that the rank of $\Theta(\mathbf{g})$ (respectively, of $\Theta(\mathbf{f})(\mathbf{g})$) is the same over A as over R.

(ii) We claim the reverse inequality $\operatorname{rank} \Theta(\mathbf{g}) \geq \dim A$.

Picking a transcendence basis of A over k out of the generators $\mathbf{g}$, the dimension of the corresponding algebra stays the same, while the rank of the corresponding Jacobian matrix can only get smaller. Therefore, we may assume that $\mathbf{g}$ is an algebraically independent set over k.

We induct on the difference $n - r \geq 0$.

First, $n = r$. Fixing $1 \leq i \leq n$, let $\mathcal{P}_i \in L[Y]$ denote the minimal polynomial of x_i over $L = k(\mathbf{g})$. Since $\operatorname{char}(k) = 0$, x_i is separable over L, i.e., $d\mathcal{P}_i/dY \neq 0$. Eliminating "denominators" in $\mathcal{P}_i$ and pulling back to variables $y_1, \ldots, y_n, y_{n+1}$ over k such that $y_j \mapsto g_j$, for $1 \leq j \leq r = n$, and

$y_{n+1} \mapsto x_i$, one gets a polynomial $F_i \in k[y_1, \ldots, y_n, y_{n+1}]$ such that

$$F_i(\mathbf{g}, x_i) = 0, \qquad \frac{\partial F_i}{\partial y_{n+1}}(\mathbf{g}, x_i) \neq 0. \tag{3.1}$$

The chain rule of derivatives affords the relation

$$\left(\frac{\partial F_i}{\partial y_1}(\mathbf{g}, x_i) \; \ldots \; \frac{\partial F_i}{\partial y_{n+1}}(\mathbf{g}, x_i) \right) \Theta(\mathbf{g}, x_i) = 0,$$

for every i, $1 \leq i \leq n$, where $\Theta(\mathbf{g}, x_i)$ denotes the Jacobian matrix of the elements $\{\mathbf{g}, x_i\}$. Since

$$\Theta(\mathbf{g}, x_i) = \begin{pmatrix} \Theta(\mathbf{g}) \\ 0 \ldots \underbrace{1}_{i} \ldots 0 \end{pmatrix},$$

we get a matrix equality

$$\begin{pmatrix} \frac{\partial F_1}{\partial y_1}(\mathbf{g}, x_1) & \ldots & \frac{\partial F_1}{\partial y_n}(\mathbf{g}, x_1) \\ \vdots & \vdots & \vdots \\ \frac{\partial F_n}{\partial y_1}(\mathbf{g}, x_n) & \ldots & \frac{\partial F_n}{\partial y_n}(\mathbf{g}, x_n) \end{pmatrix} \Theta(\mathbf{g}) = - \begin{pmatrix} \frac{\partial F_1}{\partial y_{n+1}}(\mathbf{g}, x_1) & & \\ & \ddots & \mathbf{0} \\ \mathbf{0} & & \frac{\partial F_n}{\partial y_{n+1}}(\mathbf{g}, x_1) \end{pmatrix}.$$

Therefore, (3.1) implies that $\det \Theta(\mathbf{g}) \neq 0$, as was to be shown.

To complete the inductive procedure, let now $n \geq r + 1$. Pick a polynomial $h \in k[x_1, \ldots, x_n]$ such that $\{\mathbf{g}, h\}$ is algebraically independent over k. By the inductive hypothesis, $\Theta(\mathbf{g}, h)$ is of rank $> \dim k[\mathbf{g}, h] = r + 1$; hence the ideal of $(r + 1)$-minors $I_{r+1}(\Theta(\mathbf{g}, h)) \subset k[x_1, \ldots, x_n]$ is nonzero. Since $I_{r+1}(\Theta(\mathbf{g}, h)) \subset I_r(\Theta(\mathbf{g}))$, the latter is nonzero as well; hence $\Theta(\mathbf{g})$ has rank at least $r = \dim A$, as desired.

(iii) This follows from (i) and from the proof of the inequality rank $\Theta(\mathbf{g}) \geq \dim A$ claimed in (ii), starting at the place where $\mathbf{g}$ was assumed to be algebraically independent over k. $\qquad\square$

In the notation of Definition 3.1, one has the following corollary:

Corollary 3.3 *Let $f \in k[x_1, \ldots, x_n]$ be a reduced polynomial.*

(a) *In characteristic zero one has* $\dim \mathfrak{P}(f) = \operatorname{rank} H(f)$.

(b) *If the partial derivatives of f are algebraically independent over k and the algebraic field extension*

$$k(\partial f/\partial x_1, \ldots, \partial f/\partial x_n) \subset k(x_1, \ldots, x_n)$$

is separable, then $\operatorname{rank} H(f) = n$.

The failure of the equality in (b) in certain strategic characteristic, although a source of inconvenience, is totally expected due to Euler's formula. Thus, if $f \in k[x_1, \dots, x_n]$ is a homogeneous polynomial of degree m, then the Euler formula gives

$$(m-1)\frac{\partial f}{\partial x_i} = \sum_{j=1}^{n} x_j \frac{\partial^2 f}{\partial x_i x_j},$$

where the right side vanishes if $\mathrm{char}(k)$ divides $m - 1$. Therefore, if this is the case, the column $(x_1 \ \dots \ x_n)^t$ is a syzygy of $H(f)$; hence rank $H(f) < n$.

Example 3.4 Let f stand for the determinant of the 3×3 generic matrix over a field of characteristic 2. Then the polar image of f has dimension 9, while rank $H(f) = 8$. In addition, one can see where the proof of item (ii) of Proposition 3.2 breaks down: the minimal polynomial of an entry, say, $x_{1,1}$, over the field $L = k(\partial f/\partial x_{i,j} | 1 \leq i, j \leq 3)$ is of the form $t^2 + l$, with $l \in L$, and hence its derivative vanishes in characteristic 2.

Remark 3.5

(1) Thus, in characteristic zero, the hypersurface $V(f)$ has vanishing Hessian if and only if the derivatives $f_1, \dots, f_n$ are algebraically dependent over k. Also, $V(f)$ is smooth if and only if $\{\partial f/\partial x_1, \dots, \partial f/\partial x_n\}$ is a regular sequence; in particular, if $V(f)$ is smooth, then $\det H(f) \neq 0$. Thus, having vanishing Hessian implies at least that $\mathrm{Sing}(V(f)) \neq \emptyset$. How big is this locus is given by a result of Zak [27, Proposition 4.9], telling that the dual variety to the polar image $\mathfrak{P}(f)$ is contained in $\mathrm{Sing}(V(f))$.
(2) Clearly, $V(f)^* \subset \mathfrak{P}(f)$. Moreover, if $V(f)$ is (reduced and) irreducible, then there are geometric restrictions implying that this inclusion is proper. It would be highly interesting to formulate purely algebraic conditions for this to take place. In fact, it will be the case in many situations in this book.

3.2.3   The Grassmann Algebra

A question comes up as to when the equality in Proposition 3.2 (ii) holds in arbitrary characteristic. Although there are some grassroots feeling that most questions about maximal minors of sufficiently generic matrices over a field are not sensitive to the characteristic, yet the following outstanding case seems to be underrated in the literature.

For convenience, we introduce the following terminology and notation.

Definition 3.6 Let $\mathbf{X} = (x_{i,j})$ denote an $m \times n (m \leq n)$ generic matrix over an arbitrary field k. The k-subalgebra $G(n, m)$ of the polynomial ring $k[\mathbf{X}]$, generated by the m-minors of $\mathbf{X}$, is called the *Grassmann algebra* of size (n, m).

We also record the following well-known result.

Proposition 3.7 ([1, Corollary 5.12]) $\dim G(n, m) = m(n - m) + 1$.

In characteristic zero the next result is a consequence of Corollary 3.3 (a). We give a proof in a characteristic-free environment as we were not able to find an explicit argument for it in the literature.

Theorem 3.8 (Arbitrary characteristic) *Let* $\Theta(n, m)$ *denote the Jacobian matrix of the m-minors of* $\mathbf{X}$ *as above over an arbitrary field. Then* $\operatorname{rank} \Theta(n, m) = \dim G(n, m)$.

Proof Throughout we list the m-minors $\Delta_{j_1,\ldots,j_m}$ according to the ordering of their vector of indices in the lexicographic order. Likewise, we assume that $\Theta(n, m)$ is written with respect to this ordering.

By Proposition 3.7, we are to show that $\operatorname{rank} \Theta(n, m) = m(n - m) + 1$.

CLAIM 1 $\operatorname{rank} \Theta(n, m) \geq m(n - m) + 1$.

Decompose the transpose of $\mathbf{X}$ in the following way:

$$
\mathbf{X}^t = \begin{pmatrix} M \\ L_1 \\ \vdots \\ L_{n-m} \end{pmatrix},
$$

where M is an $m \times m$ matrix and L_i $(i = 1, \ldots, n - m)$ are the subsequent row matrices.

For every $1 \leq i \leq n-m$, denote by $f_{i,1}, \ldots, f_{i,m+1}$ the ordered signed m-minors of the $(m + 1) \times m$ matrix $\begin{pmatrix} M \\ L_i \end{pmatrix}$.

Note that for any two indices $1 \leq i, i' \leq n - m$, with $i \neq i'$, the only common m-minor of the respective sets of m-minors is $f := \det M = f_{i,m+1} = f_{i',m+1}$.

SUBCLAIM $L_i \operatorname{adj}(M) = \begin{pmatrix} f_{i,1} & \cdots & -f_{i,m} \end{pmatrix}$, where $\operatorname{adj}(M)$ denotes the adjugate of the matrix M.

In fact,

$$
\begin{pmatrix} M \\ L_i \end{pmatrix} \operatorname{adj}(M) = \left(\begin{array}{ccc} f & \cdots & 0 \\ \vdots & \ddots & \vdots \\ 0 & \cdots & f \\ \hline g_1 & \cdots & g_m \end{array} \right).
$$

Thus,

$$
\mathbf{0} = \begin{pmatrix} f_{i,1} & \cdots & f_{i,m+1} \end{pmatrix} \begin{pmatrix} M \\ L_i \end{pmatrix} \operatorname{adj}(M) = \begin{pmatrix} f_{i,1} f + g_1 f & \cdots & f_{i,m} f + g_m f \end{pmatrix},
$$

and hence $g_j = -f_{i,j}$, for $j = 1, \ldots, m$.

Now, let H denote the $(m(n-m)+1) \times (m(n-m)+1)$ Jacobian matrix of the forms

$$\{f_{1,1}, \ldots, f_{1,m}\} \cup \cdots \cup \{f_{n-m,1}, \ldots, f_{n-m,m}\} \cup \{f\}$$

with respect to the variables of the set $\{x_{i,j} \in L_1\} \cup \cdots \cup \{x_{i,j} \in L_{n-m}\} \cup \{x_{m,m}\}$.

By the Subclaim, for any given $1 \leq i \leq n - m$ and $1 \leq j \leq m$, one has

$$f_{i,j} = -L_i(j\text{th column of adj}(M)).$$

Since L_i and M have no entries in common and neither do L_i and $L_{i'}$ for $i \neq i'$, we conclude that H has the following shape:

$$\left(\begin{array}{cccc|c}
-\mathrm{adj}(M) & \mathbf{0} & \cdots & \mathbf{0} & 0 \\
\mathbf{0} & -\mathrm{adj}(M) & \cdots & \mathbf{0} & 0 \\
\vdots & \vdots & \ddots & \vdots & \vdots \\
\mathbf{0} & \mathbf{0} & \cdots & -\mathrm{adj}(M) & 0 \\
\hline
* & * & \cdots & * & \Delta(M)_{m,m}
\end{array}\right)$$

where $\Delta(M)_{m,m}$ is the cofactor of M with respect to the (m, m)th entry, in particular, a non-vanishing form. Since $\det \mathrm{adj}(M) = (\det M)^{m-1} = f^{m-1} \neq 0$, then $\det H \neq 0$. But H is up to elementary row/column operations a submatrix of the Jacobian matrix of $I_m(\mathbf{X})$, thus wrapping up the proof of Claim 1.

CLAIM 2 rank $\Theta(n, m) \leq m(n - m) + 1$.

Write $R := k[\mathbf{X}]$, and consider the R-homomorphism $R^{nm} \xrightarrow{\Theta(n,m)} R^{\binom{n}{m}}$ defined by $\Theta(n, m)$ in the canonical basis of R^{nm} and $R^{\binom{n}{m}}$ – which we call informally the *Jacobian map*. It suffices to show that the kernel of the Jacobian map has rank at least $m^2 - 1$.

To see this, for every $1 \leq i \leq m$, let ℓ_i denote the ith column of the transpose of $\mathbf{X}$. Now, for every $1 \leq i \leq m - 1$, consider the following $(nm) \times m$ matrix:

$$\mathfrak{s}(i) = \left(\begin{array}{ccccccccc}
\ell_i & \mathbf{0} & \mathbf{0} & \cdots & \mathbf{0} & \cdots & \mathbf{0} & \mathbf{0} \\
\mathbf{0} & \ell_i & \mathbf{0} & \cdots & \mathbf{0} & \cdots & \mathbf{0} & \mathbf{0} \\
\vdots & \vdots & \vdots & \ddots & \vdots & \cdots & \vdots & \vdots \\
\mathbf{0} & \mathbf{0} & \mathbf{0} & \cdots & \ell_i & \cdots & \mathbf{0} & \mathbf{0} \\
\mathbf{0} & \mathbf{0} & \mathbf{0} & \cdots & \mathbf{0} & \cdots & \mathbf{0} & \mathbf{0} \\
\vdots & \vdots & \vdots & \ddots & \vdots & \cdots & \vdots & \vdots \\
\mathbf{0} & \mathbf{0} & \mathbf{0} & \cdots & \mathbf{0} & \cdots & \ell_i & \mathbf{0} \\
\mathbf{0} & \mathbf{0} & \mathbf{0} & \cdots & -\ell_m & \cdots & \mathbf{0} & \ell_i
\end{array}\right),$$

where the column with the entry ℓ_m is the ith column, and in addition consider the $(nm) \times (m-1)$ matrix

$$
\mathfrak{s}(m) =
\begin{pmatrix}
\ell_m & \mathbf{0} & \mathbf{0} & \cdots & \mathbf{0} & \cdots & \mathbf{0} \\
\mathbf{0} & \ell_m & \mathbf{0} & \cdots & \mathbf{0} & \cdots & \mathbf{0} \\
\vdots & \vdots & \vdots & \ddots & \vdots & \cdots & \vdots \\
\mathbf{0} & \mathbf{0} & \mathbf{0} & \cdots & \ell_m & \cdots & \mathbf{0} \\
\mathbf{0} & \mathbf{0} & \mathbf{0} & \cdots & \mathbf{0} & \cdots & \mathbf{0} \\
\vdots & \vdots & \vdots & \ddots & \vdots & \cdots & \vdots \\
\mathbf{0} & \mathbf{0} & \mathbf{0} & \cdots & \mathbf{0} & \cdots & \ell_m \\
\mathbf{0} & \mathbf{0} & \mathbf{0} & \cdots & \mathbf{0} & \cdots & \mathbf{0}
\end{pmatrix}.
$$

SUBCLAIM Any among the $m(m-1)+m-1 = m^2-1$ columns of these matrices belongs to the kernel of the Jacobian map.

To prove it, consider the $m \times m$ submatrix of $\mathbf{X}$ with column indices $j_1, \ldots, j_m$

$$
M_{j_1,\ldots,j_m} =
\begin{pmatrix}
x_{1,j_1} & \cdots & x_{1,j_m} \\
\vdots & \ddots & \vdots \\
x_{m,j_1} & \cdots & x_{m,j_m}
\end{pmatrix},
$$

and set $D_{j_1 \cdots j_m} := \det M_{j_1,\ldots,j_m}$. For $1 \le r \le m$, let $\partial^{(r)}_{j_1,\ldots,j_m}$ denote the vector of the partial derivatives of $D_{j_1 \cdots j_m}$ with respect to the variables which are the ordered entries of ℓ_r. Since $\frac{\partial D_{j_1 \cdots j_m}}{\partial x_{i,j_k}} = \Delta(M_{j_1,\ldots,j_m})_{i,k}$ and $\frac{\partial D_{j_1 \cdots j_m}}{\partial x_{i,j}} = 0$ if $j \notin \{j_1, \ldots, j_m\}$, then, by the adjugate relation,

$$
\mathrm{adj}(M_{j_1,\ldots,j_m})\, M_{j_1,\ldots,j_m} - M_{j_1,\ldots,j_m}\, \mathrm{adj}(M_{j_1,\ldots,j_m}) = D_{j_1 s j_m}\, \mathbb{I}_m,
$$

one gets

$$
\partial^{(r)}_{j_1,\ldots,j_m}\, \ell_i =
\begin{cases}
0, & \text{if } i \ne r \\
D_{j_1 s j_m}, & \text{if } i = r.
\end{cases}
$$

Hence,

$$
\partial^{(r)}_{j_1,\ldots,j_m}\, \ell_i = 0, \quad \text{if } i \ne r \tag{3.2}
$$

and

$$
\partial^{(i)}_{j_1,\ldots,j_m}\, \ell_i - \partial^{(m)}_{j_1,\ldots,j_m}\, \ell_m = 0 \quad \text{for } 1 \le i \le m-1. \tag{3.3}
$$

The relations (3.2) and (3.3) prove the stated subclaim.

To wrap up the proof of Claim 2, let $\Bbbk(n, m)$ stand for the concatenation of the matrices $\mathfrak{s}(i)$ $(1 \leq i \leq m - 1)$ and $\mathfrak{s}(m)$, whose columns as just seen belong to the kernel of the Jacobian map. We now show that $I_{m^2-1}(\Bbbk(n, m))$ has positive height, which implies in particular that $\Bbbk(n, m)$ has rank $m^2 - 1$. For this, note that up to column permutation, we can write the $nm \times (m^2 - 1)$ matrix $\Bbbk(n, m)$ in the form

$$\begin{pmatrix} \mathbf{X}^t & & & & \\ & \mathbf{X}^t & & & \\ & & \ddots & & \\ & & & \mathbf{X}^t & \\ * & * & \cdots & * & \overline{\mathbf{X}}^t \end{pmatrix},$$

where the empty slots are null entries and $\overline{\mathbf{X}}$ denotes the $(m - 1) \times n$ submatrix obtained from $\mathbf{X}$ omitting the mth row. Thus, for every $m \times m$ submatrix A of $\mathbf{X}$ and every $(m-1) \times (m-1)$ submatrix B of $\overline{\mathbf{X}}^t$, we have the following $(m^2-1) \times (m^2-1)$ submatrix of $\Bbbk(n, m)$:

$$\begin{pmatrix} A^t & & & & \\ & A^t & & & \\ & & \ddots & & \\ & & & A^t & \\ * & * & \cdots & * & B^t \end{pmatrix}$$

whose determinant is $\det A^{m-1} \det B$. Thus, a minimal prime of $I_{m^2-1}(\Bbbk(n, m))$ contains $\det A \det B$ for all choices of A and B. It follows that such a prime contains either $I_m(\mathbf{X})$ or $I_{m-1}(\overline{\mathbf{X}})$, in particular, it has height $\geq n - m + 1$ or $\geq n - m + 2$. Thus, ht $I_{m^2-1}(\Bbbk(n, m)) \geq 1$. $\qquad\square$

Corollary 3.9 *If* $m < n$, *the kernel of the Jacobian map* $R^{nm} \overset{\Theta(n,m)}{\longrightarrow} R^{\binom{n}{m}}$ *is generated by the columns of* $\Bbbk(n, m)$, *i.e., the induced R-free complex*

$$0 \to R^{m^2-1} \longrightarrow R^{nm} \longrightarrow R^{\binom{n}{m}} \tag{3.4}$$

is exact.

Proof By the above theorem and its proof, the ranks add up to nm. Still by the above, the argument at the end of the proof of Claim 2 actually gives ht $I_{m^2-1}(\Bbbk(n, m)) \geq 2$ provided $m < n$. Then the Buchsbaum–Eisenbud criterion [2, Theorem, Section 1] applies. $\qquad\square$

Remark 3.10 For $m = n$, the Jacobian map is the homomorphism $R^{m^2} \longrightarrow R$ defined by the gradient of $\det \mathbf{X}$, which is given by the submaximal minors of $\mathbf{X}$ (Proposition 2.10). The latter generate a codimension 4 Gorenstein ideal with minimal free resolution given by the complex of Gulliksen–Negård [8]. In this case the generating syzygies are still linear, but $2(m^2 - 1)$ in numbers.

3.3 Propadeutics: Generic Determinantal Hypersurfaces

The properties established in this section are mostly well-known, but we stress their characteristic-free background. They mean to be phase zero for the more involved theory to come.

3.3.1 The Generic Square Matrix

Throughout, k is a field of arbitrary characteristic $\neq 2$.

Proposition 3.11 (char$(k) \neq 2$) *Let* $\mathbf{X} = (x_{i,j})_{1 \leq i,j \leq m}$ *denote an* $m \times m$ *generic matrix over k, and let $f = \det \mathbf{X}$. Then:*

(a) *f is a homaloidal polynomial, and the polar map of f is an involution up to signs.*

(b) *The source inversion factor of the polar map of f coincides with the $(m - 2)$th power of f.*

(c) *The dual variety $V(f)^*$ is defined by the 2×2-minors of the $m \times m$ generic matrix in the dual variables; in particular,* $\dim V(f)^* = 2m - 2$.

(d) *f is totally Hessian; in particular, the exponent of f as a factor of $h(f)$ has the expected value $m(m - 2)$.*

Proof We first remark that in the generic case the partial derivative $\partial f / \partial x_{i,j}$ is the cofactor of the entry $x_{i,j}$ of the matrix $\mathbf{X}$, which is immediate from the data (or directly, from Proposition 2.10). Thus, the polar map is defined by cofactors of the matrix $\mathbf{X}$.

(a) and (b) These items hold in any characteristic. Namely, apply (2.2) with $M := \mathrm{adj}(\mathbf{X})$ to get $\mathrm{adj}(\mathrm{adj}(\mathbf{X})) = f^{m-2}\,\mathbf{X}$. Since the polar map φ_f is defined by the entries of $\mathbf{X}$, this means that it is an involution and, moreover, the source inversion factor is f^{m-2}. Thus, we are done with items (a) and (b).

(c) Let $\mathfrak{D}$ denote the homogeneous defining ideal of $V(f)^*$ in its embedding in the dual projective space. By Theorem 2.11 and Proposition 2.13, one has $I_2(\mathbf{Y}) \subset \mathfrak{D}$ and rank $H(f) \pmod{f} \leq 2m$, where $H(f)$ denotes the the Hessian matrix of f.

Thus, by Proposition 2.7, it suffices to prove that rank $H(f) \pmod{f} \geq 2m$.

For this, consider the $(m - 1) \times m$ submatrix of $\mathbf{X}$ omitting the first row, and let $\mathfrak{J}$ denote the Jacobian matrix of its ordered $(m - 1)$-minors. By Theorem 3.8, rank $\mathfrak{J} = (m - 1)(m - (m - 1)) + 1 = m$.

Write the Hessian matrix $H(f)$ in the block form

$$H(f) = \begin{pmatrix} \mathbf{0} & \mathfrak{J} \\ \mathfrak{J}^t & * \end{pmatrix},$$

where $\mathfrak{J}^t$ denotes the transpose of $\mathfrak{J}$.

Let $\mathfrak{J}'$ denote an $m \times m$ submatrix of $\mathfrak{J}$ with $\det \mathfrak{J}' \neq 0$. Now, $f \in (x_{1,1}, \ldots, x_{1,m})$, while $\det \mathfrak{J}' \notin (x_{1,1}, \ldots, x_{1,m})$. This means that $\det \mathfrak{J}' \neq 0$ even modulo f. Since $\mathfrak{J}'$ above is a submatrix of $\mathfrak{J}$ of rank m modulo f, then rank $H(f)$ (mod f) $\geq 2m$. This wraps up the proof of item (c).

(d) For the assertion about f being totally Hessian, we have to show that $h(f) = f^{m(m-2)}$. By Proposition 2.28, $h(f)$ divides a power of the inversion factor $D = f^{m-2}$ and hence is itself a power of f as f is irreducible. A degree count gives that $h(f) = f^{m(m-2)}$.

As for the subsumed assertion, by (c) the dual variety $V(f)^*$ has dimension $2m - 2$. Since in the present situation, $n = m^2 - 1$, we deduce that $n - 1 - \dim V(f)^* = m(m-2)$, as required. $\qquad\qquad\square$

Remark 3.12 In characteristic zero, geometric proofs of item (c) above are known drawing upon tangent spaces and, as such, more generally, the dual variety to generic determinantal varieties of arbitrary size is known (see [7, Proposition 4.11]). In his book [9, Chapters 15–17] Harris cautions the reader against the problem of the characteristic. The proofs in Proposition 3.11 are characteristic-free – the exception is the use of Segre's formula which seems to require $\operatorname{char}(k) \neq 2$ (but perhaps not in this particular context).

3.3.2 The Generic Symmetric Matrix

The symmetric case follows the same pattern as the square generic case, with certain detour in the proof.

Proposition 3.13 ($\operatorname{char}(k) \neq 2$) *Let* $\mathbf{X} = (x_{i,j})_{1 \leq i \leq j \leq m}$ *denote a generic symmetric* $m \times m$ *matrix over the field* k, *and let* $f = \det \mathbf{X}$. *Then:*

(a) *f is a homaloidal polynomial, and the polar map of f is an involution up to a projective transformation.*

(b) *The source inversion factor of the polar map of f coincides with the $(m-2)$th power of f.*

(c) *The dual variety $V(f)^*$ is defined by the 2×2-minors of the $m \times m$ generic symmetric matrix in the dual variables; in particular,* $\dim V(f)^* = m - 1$.

(d) *f is totally Hessian; in particular, the exponent of f as a factor of $h(f)$ has the expected value* $\binom{m}{2} - 1$.

Proof (a) and (b) The two items hold in characteristic $\neq 2$. In the case of the generic symmetric matrix, each $(m-1)$-minor appears twice as a cofactor; hence the partial derivative relative to a variable off the main diagonal will be the corresponding cofactor multiplied by 2 (see Proposition 2.10). Now, the same proof as in the previous proposition yields that the rational map defined by the cofactors is an involution with source inversion factor f^{m-2}. Therefore, the result can be transferred to the polar map by adjusting the coefficients accordingly.

(c) We proceed as in the proof of Proposition 3.11 (c), with the needed adjustments.

Since $\mathrm{char}(k) \neq 2$, by Theorem 2.11 and Proposition 2.13, one has $I_2(\mathbf{Y}) \subset \mathfrak{D}$ and rank $H(f) \pmod{f} \leq m+1$, where $H(f)$ is the Hessian matrix of f.

Thus, by Proposition 2.7, it suffices to prove that rank $H(f) \pmod{f} \geq m+1$.

As we cannot make use of Theorem 3.8, a different approach is needed in order to write a convenient $(m+1) \times (m+1)$ submatrix of the Hessian matrix. We proceed as follows.

(Note for the record that since we assume $\mathrm{char}(k) \neq 2$, now the partial derivatives of the main diagonal carry 2 as a multiplier according to Proposition 2.10). Consider the following subset of the partial derivatives:

$$\mathfrak{F} := \left\{ \frac{\partial f}{\partial x_{1,1}}, \frac{\partial f}{\partial x_{1,2}}, \ldots, \frac{\partial f}{\partial x_{1,m-1}}, \frac{\partial f}{\partial x_{1,m}}, \frac{\partial f}{\partial x_{2,m}} \right\},$$

and let Θ denote its Jacobian matrix with respect to these very variables. Up to permutation of rows and columns, Θ is an $(m+1) \times (m+1)$ submatrix of $H(f)$. Thus, it suffices to show that $\det \Theta$ does not vanish modulo f.

Since the partial derivative of f with respect to $x_{i,j}$ coincides with the cofactor of the entry $x_{i,j}$ (possibly, with a coefficient of 2), obviously, none of the first m elements of $\mathfrak{F}$ involves $x_{1,1}$. Consequently,

$$\frac{\partial^2 f}{\partial x_{1,1} \partial x_{1,j}} = \frac{\partial^2 f}{\partial x_{1,j} \partial x_{1,1}} = 0, \text{ for } j = 1, \ldots, m$$

and

$$\frac{\partial^2 f}{\partial x_{1,j} \partial x_{k,l}} = \frac{\partial^2 f}{\partial x_{k,l} \partial x_{1,j}} \tag{3.5}$$

do not involve $x_{1,1}$, for $j = 1, \ldots, m$ and $x_{k,l} \in \mathbf{X}$.

Therefore, up to permutations Θ has the following shape:

$$\Theta = \left(\begin{array}{c|c} \dfrac{\partial^2 f}{\partial x_{1,1} \partial x_{2,m}} & L \\ \hline \mathbf{0} & \Omega \end{array} \right),$$

where $\mathbf{0}$ is the $m \times 1$ zero matrix and Ω is the $m \times m$ Jacobian matrix of the set $\mathfrak{F} \setminus \{\frac{\partial f}{\partial x_{2,m}}\}$ with respect to the variables $\{x_{1,2}, x_{1,3}, \ldots, x_{1,m}, x_{2,m}\}$. Then

$$\det \Theta = \frac{\partial^2 f}{\partial x_{1,1} \partial x_{2,m}} \cdot \det \Omega,$$

which does not involve x_{11} by (3.5). Since f depends on $x_{1,1}$, it follows that $\det \Theta$ does not vanish modulo f, provided it does not vanish on the polynomial ring.

Now, $\Delta_{1,1}$ depends effectively on $x_{2,m}$, and hence the second derivative above is nonzero. Thus, it suffices to see that $\det \Omega \neq 0$. This will follow from the following specialization: map any variable along the subdiagonal $\mathbf{v} := \{i + j = m + 2\}$ to itself and the remaining ones to zero. The specialized matrix $\Omega(\mathbf{v})$ has a non-vanishing determinant as it is an anti-diagonal matrix whose entries are nonzero monomials on the entries of $\mathbf{v}$.

This proves the statement of this item.

(d) As in the generic case, by the same token (since f is still irreducible), $h(f)$ is a power of f. A degree counting gives that the multiplicity is

$$\frac{m-2}{m} \binom{m+1}{2} = \frac{(m-2)(m+1)}{2} = \binom{m}{2} - 1.$$

On the other hand, one has $n - 1 - \dim V(f)^* = \binom{m+1}{2} - 2 - (m-1) = \binom{m}{2} - 1$, showing that the expected multiplicity takes place. $\square$

Remark 3.14

(1) In both Propositions 3.11 and 3.13, the ideal of $k[\mathbf{X}]$ generated by the cofactors is of linear type and linearly presented ([10] and [12], respectively). Therefore, item (a) of both propositions is also a consequence of [15, Corollary 3.2.27].

(2) Item (d) of both propositions has been proved in [20], but no assumption on the field of coefficients is stated.

(3) It is known that the determinant of the generic symmetric matrix coincides with the so-called *discriminant hypersurface of quadrics*. The curious reader is referred to [5] for the details.

3.3.3 The Circulant

The generic *circulant* based on the polynomial ring $k[x_1, \ldots, x_n]$ is the square matrix

$$
C_n := \begin{pmatrix}
x_1 & x_2 \; x_3 \; \ldots \; x_{n-1} & x_n \\
x_2 & x_3 \; x_4 \; \ldots \; x_n & x_1 \\
x_3 & x_4 \; x_5 \; \ldots \; x_1 & x_2 \\
\vdots & \vdots \; \vdots \; \ldots \; \vdots & \vdots \\
x_{n-1} \; x_n \; x_1 & \ldots \; x_{n-3} & x_{n-2} \\
x_n & x_1 \; x_2 \; \ldots \; x_{n-2} & x_{n-1}
\end{pmatrix}.
$$

It is a Hankel matrix (Chapter 1, Section 1.3.1, and also Chapter 6) of a special kind, with many special features. In the classical literature it is often given in a slightly different shape using row operations that force the entries of the main diagonal to be all x_1 (see, e.g., [17, Chapter VIII, 23–24]). Though its determinant $f_n := \det C_n$ expansion according to the classical rules has a slightly complicated polynomial expression, some of its algebraic and homological properties are pretty understood:

1. (Proposition 2.10) For every $1 \leq i \leq n$, the partial derivative of $f = \det C_n$ with respect to x_i is the sum of the cofactors of x_i in all its slots as an entry of C_n.
   Since each variable appears n times as an entry of C_n, if $\mathrm{char}(k)$ divides n, then all partial derivatives of f_n vanish. Therefore, we assume that $\mathrm{char}(k)$ does not divide n. Let Δ_i denote the cofactor of any x_i.
2. $f_n = \sum_{i=1}^{n} x_i \Delta_i$.
   This follows from the Euler relation and item 1, by which $\partial f_n / \partial x_i = n \Delta_i$.
3. f_n is a free divisor.
   The gradient ideal $J_{f_n} = (\Delta_1, \ldots, \Delta_n)$ is the ideal of maximal minors of the $n \times (n-1)$ submatrix φ_n of C_n with the $n-1$ rightmost columns of C_n. It suffices to show that f_n is reduced; hence J_{f_n} has codimension ≥ 2 [15, Proposition 5.2.6]. This follows, e.g., from the following:
4. Over a field extension of k containing an nth root of unity, the form f_n is the product of n mutually non-proportional linear forms.
   In characteristic zero this is classically approached through the eigenvalues of C_n (see again [17, Chapter VIII, 23–24] and, more recently, [26, Corollary 1.11.2]). Letting ζ denote an nth root of unity in an extension of k (for example, in $\mathbb{C}$), one has

$$
f_n = \prod_{j=0}^{n-1} \sum_{i=1}^{n} (\zeta^j)^i \, x_i.
$$

5. J_{f_n} is an ideal of linear type.
   Since J_{f_n} is a codimension 2 perfect ideal, it suffices to show that it satisfies condition G_∞ [22, Theorem 3.4]:

$$
\mathrm{ht}\, I_t(\varphi_n) \geq n - 1 - t + 2 = n - t + 1, \quad \text{for } 1 \leq t \leq n - 1.
$$

For this, it suffices to note that, due to the special feature of φ_n along its antidiagonals, for any t in the above interval, there are $n - t + 1$ t-minors having respective nonzero pure terms $x_1^t, \ldots, x_{n-t+1}^t$. Then, we conclude via, e.g., a standard argument about the initial ideal of $I_t(\varphi_n)$ in a convenient monomial order.

6. f_n is homaloidal.

This follows from the previous item and the linearity of φ_n [15, Corollary 3.2.27].

Some of the above properties and a few additional ones are more easily obtained by assuming that k contains an nth root of unity. In this case, item 4 tells us that f_n is the product of n linear forms. These forms are independent since their Jacobian matrix is a Vandermonde with non-vanishing determinant as the n roots of unity are all distinct – this special Vandermonde is also called the *discrete Fourier transform matrix*. Therefore, by changing coordinates and renaming, we may assume that $f_n(\mathbf{X}) = g_n(\mathbf{Y}) = y_1 \cdots y_n$ over a polynomial ring $k[\mathbf{Y}] = k[y_1, \ldots, y_n]$.

This passage depends strongly on the assumption that $\mathrm{char}(k)$ does not divide n and that k contains an nth root of unity (automatic if $\mathrm{char}(k)$ is prime). And yet, over an arbitrary field k, g_n and its gradient ideal $J_{g_n} = (y_2 \cdots y_n, \ldots, y_1 \cdots y_{n-1})$ appear under many disguises, both in geometry and in combinatorics. Geometrically, the corresponding polar map has been classically designated as the standard involution (also, one finds it as the *Magnus involution*), while in hyperplane arrangement theory it is the standard Boolean arrangement.

The partial derivatives of J_{g_n} are the maximal minors of the sparse matrix

$$
\psi_n := \begin{pmatrix}
y_1 & 0 & 0 & \ldots & 0 \\
0 & y_2 & 0 & \ldots & 0 \\
0 & 0 & y_3 & \ldots & 0 \\
\vdots & \vdots & \vdots & \ldots & \vdots \\
0 & 0 & 0 & \ldots & y_{n-1} \\
-y_n & -y_n & -y_n & \ldots & -y_n
\end{pmatrix}.
$$

As always, being a free divisor, g_n is up to a scalar multiple of the determinant of the $n \times n$ matrix obtained by concatenating the above matrix with the transpose of $(y_1 \cdots y_n)$, perhaps a matrix not so inspiring as the original circulant.

Proposition 3.15 *With the above notation, assume that k is an arbitrary field. Then:*

(i) *J_{g_n} is an ideal of linear type.*
(ii) *g_n is homaloidal.*
(iii) *The Hessian determinant of g_n is $h(g_n) := (-1)^{n-1}(n - 1)g_n^{n-2}$; hence it vanishes if and only if $\mathrm{char}(k)$ divides $n - 1$.*

Proof

(i) We could actually deduce this from the result of item 5 above, but we choose to give an independent argument. By the same token it suffices to prove the property (F_1) (Section 2.2.3). For this, one notes that, for $1 \leq t \leq n - 1$, the ideal $I_t(\psi_n)$ is generated by the squarefree t-products of $y_1, \ldots, y_n$. This ideal is well-known to be a specialization of the ideal of t-minors of a $t \times n$ generic matrix and hence has height $n - t + 1$.

(ii) As in item 6 above.

(iii) The polar map of g_n is an involution whose inversion factor is g_n^{n-2}. On the other hand, as explained above the polar map is defined by the maximal minors of an $n \times (n - 1)$ matrix of linear entries. Hence, Proposition 2.29 implies that $h(g_n) = (n - 1)g_n^{n-2}$. $\qquad\qquad\square$

The reason to introduce this example is that it encloses many of the desirable properties we will be looking at when dealing with other sort of determinantal hypersurfaces. Typically, properties such as homaloidness and freeness become quite apart, since the first has to do with the nature of the polynomial relations of the gradient ideal, while the second depends on its homological behavior.

3.3.4   A Glimpse of the Inverse Problem: When Is a Homogeneous Polynomial the Determinant of a Linear Matrix?

This part is more algebraic; hence we change notation to $R = k[x_1, \ldots, x_n]$. Given $f \in R$, one may wonder when f is the determinant of a square matrix with entries in R. As such, this is too broad a question. Thus, e.g., every $f \in R$ such that the affine hypersurface $V(f) \subset \mathbb{A}^n$ is smooth turns out to be the determinant of a suitable $n \times n$ matrix with polynomial entries in R. This is a consequence of the fact that such hypersurface is a free divisor (see, e.g., [15, Proposition 5.2.6 and Example 5.2.5]).

A similar result is false in the projective category, where we assume that f is homogeneous in the standard graded structure of R and $V(f) \subset \mathbb{P}^{n-1}$ is smooth as projective hypersurface. On the other hand, homogeneous free divisors still preserve the same property of being the determinant of a suitable $n \times n$ matrix with polynomial entries in R (see [16] and also [15, Subsection 5.2.2]).

A modified question asks when is f the determinant of an $r \times r$ matrix, where r is the highest degree of f. Restricting to a homogeneous $f \in R$, one is asking whether $f = \det L(f)$, where $L(f)$ is a $\deg(f) \times \deg(f)$ matrix of linear forms. As such, the problem has caught the attention of many authors far back in time. The related literature is long – for a feeling see [5, Historical Notes, p. 186 ff]. An obstruction was found by L. E. Dickson quite late back in this streaming. For that, recall the natural triple action of the general linear group on the left (respectively, right) of a matrix and on the variables of R.

Theorem 3.16 ([3, Theorem 1 and Theorem 6]) *Let $f \in R = k[x_1, \ldots, x_n]$ ($n \geq 3$), where k is an algebraically closed field. Suppose that $f = \det L(f)$, where $L(f)$ is an $r \times r$ matrix of linear forms. Then up to the triple action of* $\mathrm{GL}(r, k) \times \mathrm{GL}(n, k) \times \mathrm{GL}(r, k)$, *the number of terms of f is at most $(n - 2)r^2 + 2$.*

One should note the subtlety of allowing for the above action, as otherwise one could, for any given r, write an $r \times r$ matrix whose entries are general linear forms (e.g., with random coefficients throughout) in the given variables $x_1, \ldots, x_n$. Then the number of nonzero terms of the determinant is maximum, i.e., $\binom{n-1+r}{r}$, which, for given r and n sufficiently large, easily goes over the stated bound.

The proof is quite involved and perhaps uninspiring in the present book evolving. Dickson also gave a positive result for the reverse question, namely, he showed that any plane projective curve admits a determinantal equation with linear entries. In this regard he had been preceded by Dixon [4], who showed that the curve admits an equation which is the determinant of a symmetric matrix with linear entries. While Dixon's proof appeals to a classical argument on bitangents, that of Dickson's is more algebraic – actually it has been noticed [14] that Dixon's should be slightly corrected by adding an argument about the existence of a non-vanishing even theta characteristic for the given curve. The advantage of Dickson's representation is that, up to a change of variables, the explicit forms of the entries of the matrix are solved by a linear system, while Dixon's procedure is less computable.

The computational difficulty of the latter motivated Edge, quite some years later, to search for the practical finding of the entries of the matrix. He then solved completely the case of the Fermat quartic $x^4 + y^4 + z^4$ and some of its deformations [6]. Here is an easy example of his findings:

$$x^4+y^4+z^4-x^2y^2-x^2z^2-y^2z^2 = \frac{1}{4}\det\begin{pmatrix} 0 & x-y & -x+z & y+z \\ x-y & 0 & y-z & x+z \\ -x+z & y-z & 0 & x+y \\ y+z & x+z & x+y & 0 \end{pmatrix}. \tag{3.6}$$

Note that since this equation is irreducible, the only possible representation as a Pfaffian would have to be a block diagonal of double size with skew-symmetric blocks. And, indeed such a study has been pursued as well.

The present state of the determinantal representation of polynomials is highly multifaceted, being strongly related from one side to convexity theory and semidefinite programming (SDP) (over a normed ground field) and from the other side to parts of algebraic geometry. The recent literature is too extensive to be mentioned here. For the state of the art the reader is referred to [11, 14, 25], mainly for a symmetric representation.

Both classical and recent proofs are quite involved, falling beyond the scope of this book. From the prevalent point of view here, even in dimension 3 it would be interesting to investigate if such an explicit representation could bring out some insight into the following questions about the corresponding form $f \in k[x, y, z]$:

1. When is f irreducible?
2. When is f smooth?
3. When is depth $k[x, y, z]/J_f = 0$, where J_f is the gradient ideal of f?
4. When is f a free divisor (i.e., when is J_f a perfect ideal of codimension 2)?
5. Can one guess aspects of the nature of the dual variety to $V(f)$?
6. When is J_f an ideal of linear type?
7. Help understanding, if not reproving, Dolgachev's homaloidal theorem.

Exercises

3.17 Let $\mathbf{X}$ denote the 2×3 generic matrix

$$\begin{pmatrix} x_1 & x_2 & x_3 \\ x_4 & x_5 & x_6 \end{pmatrix}$$

over a field k, and $P = I_2(\mathbf{X})$. Extend the ground polynomial ring $R := k[x_1, \ldots, x_6]$ with two additional variables, to get $S := k[x_1, \ldots, x_7, x_8]$ and set $I = (P, f) \subset S$, with $f = x_1 x_8 - x_2 x_7$.

1. Justify that I is a height 3 Cohen–Macaulay ideal.
2. Determine the minimal prime ideals of I, and show that they are Gorenstein ideals.
3. Argue that I is a radical ideal.
   (HINT: Either compute the intersection of the ideals found in item 1, or else, justify that R/I satisfies Serre's property R_0 by taking the Jacobian matrix $\mathfrak{J}$ of the generators of I and proving that the enlarged Fitting $(I_3(\mathfrak{J}), I)$ has height ≥ 4.)
4. Write the minimal free resolution of S/I.
   (HINT: Show it is the mapping cone of the chain map determined by multiplication by f on the minimal free resolution of R/P shifted and extended over S.)
5. Prove that I is an ideal of linear type.
   (HINT: Since I is an almost complete intersection, it suffices to show that it satisfies condition G_∞.)

3.18 Let $n \geq 3$ be an integer.

1. Show that the $n \times n$ circulant is not 1-generic by simple elementary column operations that expose null entries.
2. Confirm the result via both the criterion of Proposition 1.21 and item (a) of Theorem 1.22.
3. Is there a relation between the number of possible null entries by item 1 and the large gaps found in both uses in item 2?

3.19 (This problem may require computer assistance.)

Let f denote the quartic in (3.6), and let $I \subset R = k[x, y, z]$ denote the ideal of 3-minors of its symmetric matrix representation.

1. Show that both R/J_f and R/I have depth 0.
2. Show that the respective unmixed components of R/J_f and R/I coincide, and give their expression in terms of generators.
3. Find the singular points of $V(f)$ and its respective multiplicities.
   (HINT: First get the defining linear equations of these points from the previous item.)

3.20 A determinantal representation of some homogeneous polynomials can be obtained by the method of the Jacobian dual in the theory of rational maps [15, Chapter 3]. This exercise concerns the case of the equation of a smooth cubic surface f in $\mathbb{P}^3$. For this assume, as is well-known, that f is obtained from the blowup of six points in $\mathbb{P}^2$ in linear general position and not lying on a conic.
   Translate the above geometric method in algebraic language:

1. Argue that the defining ideal $I \subset k[x, y, z]$ of six points as above is a height 2 perfect ideal generated by four cubics.
2. Prove that the eliminated polynomial of the four cubics in the first item is a smooth space cubic.
3. Show that the rational map $\mathbb{P}^2 \dashrightarrow \mathbb{P}^3$ defined by the above four cubics is birational onto its image.
4. Deduce that the determinant of the associated Jacobian dual matrix retrieves the smooth space cubic.

3.21 Consider the matrix

$$\varphi = \begin{pmatrix} x & y & z \\ y & z & 0 \\ z & 0 & x \\ 0 & x & y \end{pmatrix}$$

with entries in $k[x, y, z]$, and let $\mathfrak{F} : \mathbb{P}^2 \dashrightarrow \mathbb{P}^3$ be the rational map defined by the 3-minors of φ.

1. Argue that $\mathfrak{F}$ is birational onto its image.
2. Show that the image is defined by a cubic polynomial $f \in k[t_0, t_1, t_2, t_3]$ represented by the determinant of the matrix

$$\begin{pmatrix} t_0 & t_1 & t_2 \\ t_3 & t_0 & t_1 \\ t_2 & t_3 & t_0 \end{pmatrix}.$$

3. Determine the singular points of f.

References

1. W. Bruns, U. Vetter, *Determinantal Rings*. Lecture Notes in Mathematics, vol. 1327 (Springer, Berlin, 1988) 59
2. D. Buchsbaum, D. Eisenbud, What makes a complex exact?. J. Algebra **25**, 259–268 (1973) 62
3. L.E. Dickson, Determination of all general homogeneous polynomials expressible as determinants with linear elements. Trans. Amer. Math. Soc. **22**, 167–179 (1921) 70
4. A.C. Dixon, Note on the reduction of a ternary quantic to asymmetrical determinant. Proc. Cambridge Phil. Soc. **11**, 350–351 (1902) 70
5. I. Dolgachev, *Classical Algebraic Geometry—A Modern View* (Cambridge University Press, New York, 2012) 66, 69
6. W.L. Edge, Determinantal representations of $x^4 + y^4 + z^4$. Math. Proc. Cambridge Phil. Soc. **34**, 6–21 (1938) 70
7. I. Gelfand, M. Kapranov, A. Zelevinsky, *Discriminants, Resultants, and Multidimensional Determinants* (Birkhäuser Boston, Inc., Boston, 1994) 64
8. T.H. Gulliksen, O. Negård, Un complexe résolvant pour certains idéaux déterminantiels. C. R. Acad. Sci. Paris **274**, 16–18 (1972) 63
9. J. Harris, *Algebraic Geometry*, GTM No. 133 (Springer, Berlin, 1992) 64
10. C. Huneke, Determinantal ideals of linear type. Arch. Math. **47**, 324–329 (1986) 66
11. D. Kerner, V. Vinnikov, Determinantal representations of singular hypersurfaces in $\mathbb{P}^n$. Adv. Math. **231**, 1619–1654 (2012) 70
12. B.V. Kotsev, Determinantal ideals of linear type of a generic symmetric matrix. J. Algebra **139**, 488–504 (1991) 66
13. H. Matsumura, *Commutative Algebra* (W. A. Benjamin Co., New York, 1970) 56
14. R. Quarez, Symmetric Determinantal Representation of Polynomials. Linear Algebra Appl. **436**, 3642–3660 (2012) 70
15. Z. Ramos, A. Simis, *Graded Algebras in Algebraic Geometry*. Expositions in Mathematics, vol. 70 (De Gruyter, Berlin, 2022) 66, 67, 68, 69, 72
16. K. Saito, Theory of logarithmic differential forms and logarithmic vector fields. J. Fac. Sci. Univ. Tokyo Sect. 1A Math. **27**, 265–291 (1980) 69
17. R.F. Scott, *The Theory of Determinants and Their Applications*, 2nd edn. (Cambridge University Press, Cambridge, 1904) 67
18. C. Segre, Preliminari di una teoria delle varietà luogi di spazî. Rend. Circ. Mat. Palermo **30**, 87–121 (1910) 55
19. B. Segre, Bertini forms and Hessian matrices. J. London Math. Soc. **26**, 164–176 (1951) 55
20. B. Segre, Sull' hessiano di taluni polinomi (determinanti, pfaffiani, discriminanti, risultanti, hessiani) I, II. Atti Acc. Lincei **37**, 109–117 e 215–221 (1964) 66
21. B. Segre, *Prodromi di Geometria Algebrica* (Cremonese, Roma, 1971) 55
22. A. Simis, W. Vasconcelos, On the dimension and integrality of symmetric algebras. Math. Z. **177**, 341–358 (1981) 67
23. A. Simis, Two differential themes in characteristic zero, in *Topics in Algebraic and Noncommutative Geometry*. Proceedings in Memory of Ruth Michler, eds. by C. Melles, J.-P. Brasselet, G. Kennedy, K. Lauter, L. McEwan. Contemporary Mathematics, vol. 324 (American Mathematical Society, Providence, RI, 2003), pp. 195–204 56
24. A. Simis, *Commutative Algebra*, 2nd edn. (De Gruyter Graduate, Berlin, 2023) 56
25. V. Vinnikov, LMI representations of convex semialgebraic sets and determinantal representations of algebraic hypersurfaces: past, present, and future, in *Mathematical Methods in Systems, Optimization, and Control. Operator Theory: Advances and Applications*, eds. by H. Dym, M. de Oliveira, M. Putinar (Birkhäuser, Basel, 2012), pp. 325–349 70
26. A. Wyn-jones, *Circulants* (2013); available on http://www.circulants.org/circ/index.html 67
27. F.L. Zak, Determinants of projective varieties and their degrees, in *Algebraic Transformation Groups and Algebraic Varieties*. Proceedings of the Conference "Interesting Algebraic Varieties Arising in Algebraic Transformation Group Theory". Encyclopaedia of Mathematical Sciences, vol. 132 (Springer, Berlin, 2004), pp. 207–238 58

Part II

Linear Sections of Notable Structured Square Matrices

Chapter 4
Linear Sections of the Generic Square Matrix

Abstract This chapter captures quite a bit of the spirit of the book in that one views the generic square matrix as a primeval model for many other matrices. One deals with ladder ideals that are Gorenstein, and in the codimension three case, a full discussion of its properties is given, including an explicit alternating matrix for the corresponding maximal Pfaffians and the structure of its lower Pfaffians. In the sequel, based on such Gorenstein ideals, a family of codimension four Gorenstein ideals is discussed that does not seem to be immediately available as such in the known literature. Structural properties of this family are surveyed, some of which are still conjectured. The next topic concerns homological and differential minded discussions of various types of linear sections of the generic square matrix, some tightly related to the generic model, while others more qualified as sparse sections of certain strategic shape. For all these, the corresponding gradient ideal, polar map, and ideals of minors are thoroughly discussed. Throughout the chapter quite a heavy use is made of previous chapters.

4.1 Ta(i)l(e)s of Ladder Ideals

Given an integer $m \geq 2$, let $\mathbf{X} = (x_{l,l})_{1 \leq l, l \leq m}$ denote the $m \times m$ generic matrix over k and let $k[\mathbf{X}] := k[x_{i,j} \mid 1 \leq i, j \leq m]$ stand for the polynomial ring on its entries. As we go along, like in previous parts, and whenever no confusion arises, $\mathbf{X}$ will denote both the matrix itself and its set of entries.

Versions of ladder ideals will come up in the structure of polar maps and dual varieties to be considered in this and later chapters. Typically, they rise as one-sided such ideals; hence, we will develop some related questions thereof.

With the above notation, let $I_j(\mathbf{X}) \subset k[\mathbf{X}]$ stand, as in previous chapters, for the ideal of $j \times j$ minors of the $m \times m$ generic matrix $\mathbf{X}$, for $1 \leq j \leq m - 1$.

In this part we have in mind the following sort of one-sided ladder:

© The Author(s), under exclusive license to Springer Nature Switzerland AG 2024
Z. Ramos, A. Simis, *Determinantal Ideals of Square Linear Matrices*,
https://doi.org/10.1007/978-3-031-55284-7_4

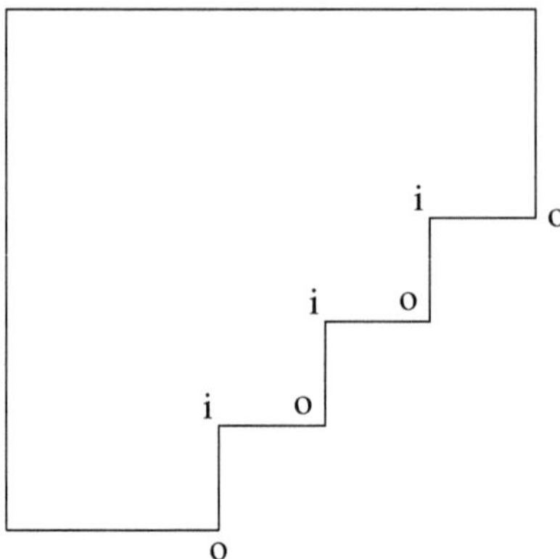

We note that here ladders will be depicted as in [16] – i.e., rotated around the horizontal axis with respect to the ladder depicted in [7].

More precisely, we are looking at the one-sided ladder $\mathcal{L}_c(m)$ defined by the following data, where c denotes the number of inside corners:

- Inside corners: $\{(i, j)|i + j = 2m - c - 1\} = \{(m - c, m - 1), (m - c + 1, m - 2), \ldots, (m - 1, m - c)\}$.
- Outside corners: $\{(i, j)|i + j = 2m - c\} = \{(m - c, m), (m - c + 1, m - 1), \ldots, (m, m - c)\}$.

4.1.1   Gorenstein Plethora

If m is fixed in the discussion, we denote $\mathcal{L}_c(m) = \mathcal{L}_c$. If no confusion arises, we denote by $k[\mathcal{L}_c]$ the polynomial subring of $k[\mathbf{X}]$ spanned by the entries inside $\mathcal{L}_c$. Accordingly, denote by $I_j(\mathcal{L}_c) \subset k[\mathcal{L}_c]$ the ideal generated by the determinants of the $j \times j$ minors confined to $\mathcal{L}_c$. Quite generally, one knows that $k[\mathcal{L}_c]/I_j(\mathcal{L}_c)$ is a normal Cohen–Macaulay domain [7, 12, 16].

In addition, $I_{m-c}(\mathcal{L}_c)$ is a Gorenstein ideal [7, Theorem 4.9], so for mnemonic, set $\mathfrak{g}_{m-c} := I_{m-c}(\mathcal{L}_c) \subset k[\mathcal{L}_c]$. The $(m - c)$-minors generating $\mathfrak{g}_{m-c}$ will be referred to as *canonical generators*.

If Φ is an alternating matrix, we denote by $\mathrm{Pf}_{2j}(\Phi)$ the ideal of $2j$-Pfaffians of Φ.

The main theorem of this section is as follows:

Theorem 4.1 *Keeping the above notation, fixing m, one has*:

(i) *For a given $c \geq 1$, the ideal $\mathfrak{g}_{m-c}$ is minimally generated by*

$$\sum_{i=0}^{c} \binom{m - i}{m - c} \binom{m - c + i - 1}{m - c - 1}$$

$(m - c)$-minors. In addition, its codimension in $k[\mathcal{L}_c]$ is $(c + 1)^2 - \binom{c+1}{2} = \binom{c+2}{2}$, which does not depend on m.

(ii) $I_j(\mathcal{L}_c)$ is Gorenstein if and only if $j = m - c$.

(iii) If $c = 1$ (hence, $\mathrm{ht}\, \mathfrak{g}_{m-1} = 3$), then $\mathfrak{g}_{m-1}$ is the ideal of maximal Pfaffians of the $(2m - 1) \times (2m - 1)$ alternating matrix

$$\Phi_m = \begin{pmatrix} 0 & \cdots & 0 & x_{m,1} & x_{m-1,1} & \cdots & x_{1,1} \\ \vdots & & \vdots & \vdots & \vdots & & \vdots \\ 0 & \cdots & 0 & x_{m,m-1} & x_{m-1,m-1} & \cdots & x_{1,m-1} \\ -x_{m,1} & \cdots & -x_{m,m-1} & 0 & -x_{m-1,m} & \cdots & -x_{1,m} \\ -x_{m-1,1} & \cdots & -x_{m-1,m-1} & x_{m-1,m} & 0 & \cdots & 0 \\ \vdots & & \vdots & \vdots & \vdots & & \vdots \\ -x_{1,1} & \cdots & -x_{1,m-1} & x_{1,m} & 0 & \cdots & 0 \end{pmatrix}.$$

(iv) $I_j(\mathcal{L}_1) = \mathrm{Pf}_{2j}(\Phi_m)$, for every $1 \leq j \leq m - 1$.

In particular, $\mathrm{Pf}_{2j}(\Phi_m)$ is a prime ideal of height $(m - j + 1)^2 - 1$.

(v) $\sqrt{I_{2j}(\Phi_m)} = \sqrt{I_{2j-1}(\Phi_m)} = I_j(\mathcal{L}_1)$, for every $1 \leq j \leq m - 1$.

Proof

(i) For given $c \geq 1$ and $1 \leq j \leq m - 1$, one has

$$\mathrm{ht}\, I_j(\mathcal{L}_c) = (m - j + 1)^2 - \binom{c + 1}{2} \tag{4.1}$$

as an ideal of $k[\mathcal{L}_c]$. This formula has been repeatedly found by several authors [22, Theorem 5.3 and remarks on p. 243)] and [12, Theorem 4.6 and Corollary 4.7]. In the present format it follows from [10, Proposition 8] by counting the number of solid j-minors inside $\mathcal{L}_c$ in the sense established there.

Taking $j = m - c$, the stated value of $\mathrm{ht}\, \mathfrak{g}_{m-c}$ is obvious.

As for the minimal number of generators, we proceed as follows.

For every $0 \leq i \leq c$, let M_i denote the $(m - c + i) \times (m - i)$ submatrix of $\mathcal{L}_c$ whose vertex entries have coordinates: $(1, 1)$, $(1, m - i)$, $(m - c + i, 1)$ and $(m - c + i, m - i)$. Set

$$\mathcal{N} = \{(m - c)\text{-minors of } \mathcal{L}_c\}$$

$$\mathcal{N}_0 = \{(m - c)\text{-minors of } M_0\}$$

$$\mathcal{N}_i = \{(m - c)\text{-minors of } M_i \text{ fixing the last row}\}, \quad 1 \leq i \leq c.$$

Inspection shows that $\mathcal{N}_i \cap \mathcal{N}_j = \emptyset$ for $i \neq j$ and $\mathcal{N} = \mathcal{N}_0 \cup \cdots \cup \mathcal{N}_c$. Since $\mathcal{N}_i$ is a generic matrix, one knows that

$$\mathcal{N}_i = \binom{m-i}{m-c}\binom{m-c+i-1}{m-c-1}.$$

This proves the statement.

(ii) This is [7, Theorem 4.9 (c)].

(iii) Let $\Delta_{i,j}$ denote the (i,j)th entry of the adjugate matrix $\mathrm{adj}(\mathbf{X})$. Then the canonical generators of $\mathfrak{g}_{m-1}$ are identified with the $2m-1$ cofactors

$$\Delta_{i,m}(1 \leq i \leq m),\ \Delta_{m,j}(1 \leq j \leq m-1).$$

Among the relations obtained from the adjugate formula (2.1) of $\mathbf{X}$, we single out the following ones:

$$\sum_{j=1}^{m} x_{i,j}\Delta_{j,m} = 0 \quad (1 \leq i \leq m-1),$$

$$\sum_{j=1}^{m-1} x_{m,j}\Delta_{j,m} - \sum_{i=1}^{m-1} x_{i,m}\Delta_{m,i} = 0,$$

$$\sum_{i=1}^{m} x_{i,j}\Delta_{m,i} = 0 \quad (1 \leq j \leq m-1).$$

Rewriting the canonical generators as

$$\{\Delta_{1,m}, \Delta_{2,m}, \cdots, \Delta_{m-1,m}, -\Delta_{m,m}, -\Delta_{m,m-1}, \cdots, -\Delta_{m,2}, -\Delta_{m,1}\}, \qquad (4.2)$$

these relations give the columns of Φ_m; hence, the latter becomes a matrix of syzygies of the ideal $\mathfrak{g}_{m-1}$.

In order to show that Φ_m is the full matrix of syzygies of $\mathfrak{g}_{m-1}$, one may resort to the following complex of free R-modules:

$$0 \to R(-(2m-1)) \xrightarrow{\Delta^t} R(-(m))^{2m-1} \xrightarrow{\Phi} R(-(m-1))^{2m-1} \xrightarrow{\Delta} R \qquad (4.3)$$

where Φ, Δ, and Δ^t denote the maps, in the canonical bases, associated, respectively, to Φ_m, the $1 \times (2m-1)$ matrix whose entries are the cofactors in (4.2), and its transpose. We now apply the Buchsbaum–Eisenbud criterion [3, Section 1, Theorem].

Obviously, the ranks add up correctly, while grade $I_1(\Delta) = \mathrm{ht}\, \mathfrak{g}_{m-1} = 3$ and grade $I_{2m-2}(\Phi_m) \geq 2$ is easily checked or follows from (v), which is proved independently. Thus, the above complex is exact; hence, Φ_m is indeed the full syzygy matrix.

(iv) Letting M denote the $(m-1) \times m$ top right matrix of Φ_m, we have

$$\Phi_m = \begin{pmatrix} \mathbf{0}_{(m-1)\times(m-1)} & M \\ -M^t & C \end{pmatrix},$$

where C is a certain alternating submatrix.

Given $1 \leq j \leq m-1$, let A denote a $j \times j$ submatrix of M. Then, easy inspection shows that there is an alternating submatrix $C(j)$ of C such that

$$U = \begin{pmatrix} \mathbf{0}_{j\times j} & A \\ -A^t & C(j) \end{pmatrix}$$

is a $2j \times 2j$ alternating submatrix of Φ_m. In particular, $\det A = \operatorname{Pf}(U)$, the Pfaffian of U. Thus,

$$I_j(M) \subset \operatorname{Pf}_{2j}(\Phi_m) \tag{4.4}$$

for $1 \leq j \leq m-1$.

Similarly, let N denote the $m \times (m-1)$ top right matrix of Φ_m. Then, likewise,

$$\Phi_m = \begin{pmatrix} C' & N \\ -N^t & \mathbf{0}_{(m-1)\times(m-1)} \end{pmatrix},$$

where C' is a certain alternating submatrix. By a similar argument,

$$I_j(N) \subset \operatorname{Pf}_{2j}(\Phi_m), \tag{4.5}$$

for $1 \leq j \leq m-1$.

Now, by the nature of $\mathcal{L}$, one clearly has $I_j(\mathcal{L}) = (I_j(M), I_j(N))$, for $1 \leq j \leq m-1$. Therefore, (4.4) and (4.5) yield the inclusion

$$I_j(\mathcal{L}) \subset \operatorname{Pf}_{2j}(\Phi_m). \tag{4.6}$$

Let us now argue conversely, namely, given $1 \leq r_1 < \ldots < r_{2j} \leq 2m-1$, consider the $2j \times 2j$ alternating submatrix U of Φ_m with rows and columns of orders $r_1, \ldots, r_{2j}$.

Let q_1, q_2, q_3 denote the respective numbers of indices r_i in each of the integer intervals $[1, m-1]$, $[m, m]$, $[m+1, 2m-1]$. Clearly, $q_2 = 0$ or 1 and $2j = q_1 + q_2 + q_3$.

We have the following possibilities:

$\underline{q_1 \geq j+1}$. In this case the $2j \times 2j$ alternating submatrix U has the format

$$U = \left(\begin{array}{c|c} \mathbf{0}_{q_1 \times q_1} & A \\ \hline -A^t & * \end{array} \right),$$

where A is a $q_1 \times (2j - q_1)$ matrix. Since $q_1 \geq j + 1$, then $q_1 > 2j - q_1$. Thus, $\mathrm{Pf}(U)^2 = \det U = 0$.

$\underline{q_3 \geq j + 1}$. In this case the $2j \times 2j$ alternating submatrix U has the format

$$U = \left(\begin{array}{c|c} * & A \\ \hline -A^t & \mathbf{0}_{q_3 \times q_3} \end{array} \right).$$

By a similar token, here also $\mathrm{Pf}(U)^2 = \det U = 0$.

$\underline{q_1 = j}$. In this case the $2j \times 2j$ alternating submatrix U has the format

$$U = \left(\begin{array}{c|c} 0_{j \times j} & A \\ \hline -A^t & * \end{array} \right),$$

where A is a $j \times j$ submatrix of M. Thus, $\mathrm{Pf}(U) = \det A \in I_j(M) \subset I_j(\mathcal{L})$.

$\underline{q_3 = j}$. In this case the $2j \times 2j$ alternating submatrix U has the format

$$U = \left(\begin{array}{c|c} * & A \\ \hline -A^t & 0_{j \times j} \end{array} \right),$$

where A is a $j \times j$ submatrix of N. Thus, $\mathrm{Pf}(U) = \det A \in I_j(N) \subset I_j(\mathcal{L})$.

In every case, $\mathrm{Pf}(U) \in I_j(\mathcal{L})$; hence $\mathrm{Pf}_{2j}(\Phi_m) \subset I_j(\mathcal{L})$ as required.

(v) The following relations always hold:

$$I_{2j-1}(\Phi_m) \subset \mathrm{Pf}_{2j}(\Phi_m)$$

and

$$\sqrt{I_{2j}(\Phi_m)} = \sqrt{\mathrm{Pf}_{2j}(\Phi_m)}$$

[4, Corollaries 2.5 and 2.6]. Therefore, the result follows from part (v) and the fact that $I_j(\mathcal{L}_1)$ is a prime ideal. $\square$

Corollary 4.2 $\mathfrak{g}_{m-1}$ *is an ideal of linear type.*

Proof We apply the criterion of [13, Theorem 9.1], by which we need to check that I satisfies the condition G_∞ and the sliding-depth property. The first of these conditions in the present landscape reads as grade $I_t(\Phi_m) \geq 2m - t$, for every $1 \leq t \leq 2m - 2$. This follows from the value obtained in Theorem 4.1 (i).

The second condition follows from the fact that a Gorenstein ideal of codimension 3 over a Gorenstein ground local ring (in particular, over a field) is even strongly Cohen–Macaulay [14, Example 2.2]. $\square$

Remark 4.3 Note that, in general, the linear type property does not carry to an equigenerated subideal: one can easily write down a subset of minimal generators of the equigenerated ideal of submaximal minors that is not of linear type. For example, with $m = 4$, a subideal generated by 7 suitably chosen minors, such that 6 of these minors generate an ideal of linear type.

Conjecture 4.4 $\mathfrak{g}_{m-c}$ is the ideal of $2(m - c)$-Pfaffians of an alternating matrix of odd order if and only if $c = 1$.

One direction is obvious by Theorem 4.1.

The following puzzle looks at close range:

Question Let A be an $n \times n$ alternating matrix whose entries are indeterminates over a field or zeros. What is the maximum number of nonzero entries in A such that $\mathrm{Pf}_4(A)$ is generated by 2×2 minors? $\qquad\qquad\square$

4.1.2   Codimension Four Tight Gorenstein Subideals

A recent revival of the discussion around the classification of codimension four Gorenstein ideals is going on (see, e.g., [6, 17, 18, 20]) inspired by the original work of A. Kustin and M. Miller [15] and dubbed *theory of unprojection* by M. Reid. This short subsection brings up a particular family of such ideals that are moreover equigenerated and sections of ladder Gorenstein ideals of codimension three. Although their free resolutions are fairly easy or obvious, not so much their ideal theoretic details.

With the notation of the previous subsection, fix $m \geq 2$ and $1 \leq c \leq m - 1$. As before, if no confusion arises, we denote the $m \times m$ generic matrix by $\mathbf{X}$, which also stands for the set of its entries and $k[\mathbf{X}]$ for the corresponding ground polynomial ring.

Recall that $I_{m-c}(\mathbf{X}) \subset R := k[\mathbf{X}]$ has height $(m - (m - c) + 1)^2 = (c + 1)^2$, while from Theorem 4.1 (i) the ladder ideal $\mathfrak{g}_{m-c}$ has height $(c + 1)^2 - \binom{c+1}{2}$ in $k[\mathcal{L}_c]$, hence also in $k[\mathbf{X}]$. Therefore, ht $I_{m-c}(\mathbf{X}) =$ ht $\mathfrak{g}_{m-c} + \binom{c+1}{2}$.

This suggests introducing the following:

Definition 4.5 Let $\mathbf{\Delta}_c := \{\Delta_1, \ldots, \Delta_{\binom{c+1}{2}}\}$ denote $(m - c) \times (m - c)$-minors forming a regular sequence on $R/\mathfrak{g}_{m-c}$. The codimension $(c + 1)^2$ Gorenstein ideal $\widetilde{\mathfrak{g}}_{m-c} := (\mathfrak{g}_{m-c}, \mathbf{\Delta}_c)$ will be called a *tight Gorenstein subideal* of $I_{m-c}(\mathbf{X})$ containing $\mathfrak{g}_{m-c}$.

As in the case of $\mathfrak{g}_{m-c}$, the set of generators of $\widetilde{\mathfrak{g}}_{m-c}$ consisting of the canonical generators of $\mathfrak{g}_{m-c}$ and $\mathbf{\Delta}_c$ will be said to be canonical.

We believe that these equigenerated codimension four Gorenstein ideals are ubiquitous in the following sense:

Conjecture 4.6 Let $J \subset I_{m-c}(\mathbf{X}) \subset R := k[\mathbf{X}]$ denote a Gorenstein ideal satisfying the following conditions:

1. J is equigenerated in degree $m - c$.
2. $\operatorname{ht} J = \operatorname{ht} I_{m-c}(\mathbf{X})(= (c+1)^2)$.
3. $\mathfrak{g}_{m-c} \subset J$.

   Then, up to k-linear combinations of generators, J is a tight Gorenstein ideal containing $\mathfrak{g}_{m-c}$.

In other words, its minimal number of generators ought to be

$$\sum_{i=0}^{c} \binom{m-i}{m-c}\binom{m-c+i-1}{m-c-1} + \binom{c+1}{2}.$$

Let us deal with the case where $c = 1$. Then, $\widetilde{\mathfrak{g}}_{m-1} = (\mathfrak{g}_{m-1}, \Delta)$, for some $(m-1)$-minor Δ necessarily involving $x_{m,m}$. To fix ideas, we will throughout assume that Δ is the cofactor of $x_{1,1}$. Note that $\widetilde{\mathfrak{g}}_{m-1}$ has codimension four.

One main piece of this subsection is the following conjectural theorem. We state it in five items, two of which are fully argued and the remaining three more difficult ones are given a suggestive approach.

Theorem 4.7 (Conjectured) *Write $P := I_{m-1}(\mathbf{X})$. Then:*

 (i) *$\widetilde{\mathfrak{g}}_{m-1}$ is a radical ideal. In particular, the P-primary component of $\widetilde{\mathfrak{g}}_{m-1}$ is P.*
 (ii) *The primary component of $\widetilde{\mathfrak{g}}_{m-1}$ complementary to P is the intersection of two non-equigenerated prime ideals (necessarily, of codimension 4) contained in $I_{m-2}(\mathbf{X})$; in particular, $\widetilde{\mathfrak{g}}_{m-1}$ is unmixed.*
 (iii) *The minimal free resolution of $R/\widetilde{\mathfrak{g}}_{m-1}$ has the form*

$$0 \to R(-(3m-2)) \longrightarrow R(-(2m-1))^{2m} \longrightarrow R(-2(m-1))^{2m-1} \oplus R(-m)^{2m-1}$$

$$\longrightarrow R(-(m-1))^{2m} \longrightarrow R$$

 and is a certain mapping cone out of the minimal free resolution of $\mathfrak{g}_{m-1}$ (4.3).
 (iv) *The canonical generators of $\widetilde{\mathfrak{g}}_{m-1}$ are the set of submaximal minors of a square homogeneous matrix if and only if $m = 2$.*
 (v) *$\widetilde{\mathfrak{g}}_{m-1}$ is an ideal of linear type.*

Proof

 (i) (Modicum) Since $\widetilde{\mathfrak{g}}_{m-1}$ is Gorenstein, hence, Cohen–Macaulay, it suffices to prove that it satisfies Serre's condition (R_0). In the present case, this means that the ideal $(I_4(\widetilde{\Theta}), \widetilde{\mathfrak{g}}_{m-1})$ has codimension at least 5, where $\widetilde{\Theta}$ denotes the Jacobian matrix of $\widetilde{\mathfrak{g}}_{m-1}$.

One has

$$\widetilde{\Theta} = \left(\begin{array}{c|c} \Theta & \mathbf{0} \\ \hline * & \delta \end{array}\right),$$

where $\delta = \Delta_{1,m}^{1,m}$. Since $\mathfrak{g}_{m-1}$ is a normal prime, in particular $I_3(\Theta)$ has codimension ≥ 5. Then the above display ought to imply that, more strongly, $I_4(\Theta)$ alone has height ≥ 5.

(ii) (Modicum) Consider the following submatrices of $\mathbf{X}$:

- B_1(respectively, B_2) is the submatrix of $\mathbf{X}$ omitting the first and last columns (respectively, the first and last rows).
- C_1(respectively, C_2) is the submatrix of $\mathbf{X}$ omitting the last row (respectively, the last column).

Let $\tilde{I}_{m-1}(C_1)$ (respectively, $\tilde{I}_{m-1}(C_2)$) denote the subideal of $I_{m-1}(C_1)$ (respectively, $I_{m-1}(C_2)$) fixing the first and last columns (respectively, the first and last rows).

Finally, set $P_i = (I_{m-2}(B_i), \tilde{I}_{m-1}(C_i), i = 1, 2$.

CLAIM 1 P_i is a codimension four prime ideal, for $i = 1, 2$.

Clearly, counting separate codimensions gives that P_i has codimension at most $3 + 2 = 5$.

CLAIM 2 $\widetilde{\mathfrak{g}}_{m-1}: P = P_1 \cap P_2$.

Note that $P_1 \cap P_2$ has one single minimal generator in degree $m - 2$, namely, the central minor with indices $2, \ldots, m - 1$.

(iii) This is a particular event in [15, 1.2 Examples (b)], but we give the details for completeness. Write $\widetilde{\mathfrak{g}}_{m-1} = (\mathfrak{g}_{m-1}, \Delta)$, where $\Delta \in I_{m-1}(\mathbf{X})$ is a minor involving $x_{m,m}$. Consider the map of complexes induced by multiplication by Δ in the complex (4.3):

$$
\begin{array}{ccccccc}
0 \rightarrow R(-(2m-1)) & \xrightarrow{\Delta^t} & R(-(m))^{2m-1} & \xrightarrow{\Phi} & R(-(m-1))^{2m-1} & \xrightarrow{\Delta} & R \\
\uparrow & & \uparrow & & \uparrow & & \uparrow \cdot \Delta \\
0 \rightarrow R(-(3m-2)) & \xrightarrow{\Delta^t} & R(-(2m-1))^{2m-1} & \xrightarrow{\Phi} & R(-2(m-1))^{2m-1} & \xrightarrow{\Delta} & R(m-1)
\end{array}
$$

Then the resulting mapping cone resolves $R/\widetilde{\mathfrak{g}}_{m-1}$ and has the form

$$0 \rightarrow R(-(3m-2)) \longrightarrow R(-(2m-1))^{2m} \longrightarrow R(-2(m-1))^{2m-1}$$

$$\oplus \, R(-m)^{2m-1} \xrightarrow{\mathfrak{s}} R(-(m-1))^{2m} \longrightarrow R$$

as stated. Note that the restriction of the map $\mathfrak{s}$ to the leftmost summand is defined by the Koszul relations of Δ and the generators of $\mathfrak{g}_{m-1}$.

(iv) This is most probably known as well, but we give a special argument in this situation. Supposing otherwise, let N be an $r \times r$ homogeneous matrix whose set of submaximal minors coincides with the canonical generators of $\widetilde{\mathfrak{g}}_{m-1}$. Since the latter have all the same degree $m - 1$ and N is homogeneous, then it is easy to see that the columns of N must then have the same (standard) degree, say, d. Then $m - 1 = (r - 1)d$. On the other hand, we must have $2m = r^2$. Therefore, multiplication by 2 yields $r^2 - 2dr + 2(d - 1) = 0$. The discriminant of this equation in r is a square if and only if $d = 1$, which implies that $r = m$. Therefore, $m = 2$. Conversely, if $m = 2$, then $\widetilde{\mathfrak{g}}_{m-1} = I_{m-1}(\mathbf{X}) = (x_{1,1}, x_{1,2}, x_{2,1}, x_{2,2})$.

 (v) (Modicum) The proof of this item may well fall within a general environment of cases of ideals of the form $I := (J, f)$, where $J \in k[x_1, \ldots, x_{n-1}]$ is a d-equigenerated ideal of linear type and $f \in k[x_1, \ldots, x_{n-1}][x_n]$ is a strict x_n-monoid of degree d (see [19, Lemma 7.3.10] for an easy case of monomial ideals).

A direct approach through saturating the defining ideal of the symmetric algebra of I should take into account that it is generated by the forms either coming from the syzygies of J (in our situation, $J = \mathfrak{g}_{m-1}$) or from the Koszul relations of f (in our situation, $f = \Delta$) with the generators of J. Since one can saturate by any generator g of $\mathfrak{g}_{m-1}$, one could use that the saturation of $I_1(\mathbf{Y}K)$ by g returns $I_1(\mathbf{Y}\Phi)$, where K denotes the syzygies coming from Koszul relations of the generators of $\mathfrak{g}_{m-1}$. $\square$

Remark 4.8 One could rightfully ask when a subset of minimal generators of the ideal of submaximal minors generates an ideal of linear type. In general, the answer is negative, even for $m = 4$. It would be interesting to find such examples that, moreover, miss some entry. A quick search for negative cases containing a Gorenstein ideal of codimension 3 "missing" one variable was unsuccessful.

4.2 Hollow Type Linear Sections

We look at a certain linear section $\mathbf{L}$ of the $m \times m$ generic matrix $\mathbf{X}$ such that $\mathrm{codim}_{\mathbf{X}}(\mathbf{L})$ is small, while the spanning entries are strategically arranged. For this, we draw upon the technique of decomposition endowment, as discussed in Section 2.1.3, by picking the component $\mathbf{X}_1$ of a peculiar shape. The present material seems to be new in the current literature as far as we could verify. As such, it will employ a few techniques from birational theory. Some of its features will have impact in the chapter on dual varieties.

We assume that $m \geq 2$ throughout.

4.2.1 Notation

We deal with a decomposition of $\mathbf{X}$ by stressing instead the corresponding set of indices $\mathbf{I} := \{(i, j)|1 \leq i, j \leq m\}$. Then, if $\mathfrak{J}$ is any subset of $\mathbf{I}$, one denotes by $\mathbf{X}_{\mathfrak{J}}$ the corresponding set of entries of $\mathbf{X}$. Similarly, for any linear section $\mathbf{L}$ of $\mathbf{X}$.

Consider the decomposition $\mathbf{I} = \mathcal{I} \cup \mathcal{J}$, with $\mathcal{I} = \mathcal{I}_1 \cup \mathcal{I}_2 \cup \mathcal{I}_3$, where

$$\mathcal{I}_1 = \{(1, j) \mid 1 \leq j \leq m\}, \ \mathcal{I}_2 = \{(i, 1) \mid 1 \leq i \leq m\}, \ \mathcal{I}_3 = \{(i, j) \mid i+j = m+2\}.$$

Associated to such a decomposition, there are a few distinguished linear sections that will be given names for reference convenience.

Definition 4.9 Let $\mathbf{L} = (\ell_{i,j})$ be a linear section of $\mathbf{X}$. Then:

1. $\mathbf{L}$ is *hollow-filled* if it is endowed with the above decomposition and $\mathbf{L}_{\mathcal{I}} = \mathbf{X}_{\mathcal{I}}$.
2. $\mathbf{L}$ is *hollow* if it is hollow-filled and strictly sparse, i.e., $\ell_{i,j} = 0$ for every $(i, j) \in \mathcal{J}$.
3. $\mathbf{L}$ is *semi-hollow of order r*, for a given integer $0 \leq r \leq m - 2$, if it is hollow-filled and strictly sparse along $\mathcal{J} \cap \{(i, j)|i + j \geq 2m - r + 1\}$.

Obviously, the terminology is borrowed from the second in the above list, but it is easier to define hollow-filled first. We stress that these notions are relative to the generic matrix. In this case, a hollow section and a semi-hollow section are distinguished cases of a sparse matrix in the sense of [9] and [1].

These notions admit analogues in the case where $\mathbf{L}$ is a symmetric linear section of the $m \times m$ generic symmetric matrix, along the lines of discussion in Section 2.1.3. In this case $\mathbf{L}$ will be referred to as a hollow (respectively, a hollow-filled, a semi-hollow) symmetric linear section of the generic symmetric matrix. See Section 5.3 for the details.

As we will see here and in the subsequent chapters, hollow matrices tend to interact with homaloidal properties, while hollow-filled and semi-hollow intertwine with suitable ladder ideals.

We denote the hollow linear section of $\mathbf{X}$ by $\mathfrak{H}_m$. Thus,

$$\mathfrak{H}_m = \begin{pmatrix} x_{1,1} & x_{1,2} & \cdots & & x_{1,m-1} & x_{1,m} \\ x_{2,1} & 0 & \cdots & & 0 & x_{2,m} \\ x_{3,1} & 0 & \cdots & & x_{3,m-1} & 0 \\ \vdots & \vdots & & \cdot^{\cdot^{\cdot}} & \vdots & \vdots \\ x_{m-1,1} & 0 & x_{m-1,3} & & 0 & 0 \\ x_{m,1} & x_{m,2} & 0 & & 0 & 0 \end{pmatrix}. \tag{4.7}$$

Similarly, the typical semi-hollow linear section of order $0 \leq r \leq m - 2$ of $\mathbf{X}$ has the form

$$
\begin{pmatrix}
x_{1,1} & x_{1,2} & x_{1,3} & \cdots & x_{1,r+1} & x_{1,r+2} & \cdots & x_{1,m-1} & x_{1,m} \\
x_{2,1} & \ell_{2,2} & \ell_{2,3} & \cdots & \ell_{2,r+1} & \ell_{2,r+2} & \cdots & \ell_{2,m-1} & x_{2,m} \\
x_{3,1} & \ell_{3,2} & \ell_{3,3} & \cdots & \ell_{3,r+1} & \ell_{3,r+2} & \cdots & x_{3,m-1} & \ell_{3,m} \\
\vdots & \vdots & \vdots & \ddots & \vdots & \vdots & \ddots & \vdots & \vdots \\
x_{m-r,1} & \ell_{m-r,2} & \ell_{m-r,3} & \cdots & \ell_{m-r,r+1} & x_{m-r,r+2} & \cdots & \ell_{m-r,m-1} & \ell_{m-r,m} \\
x_{m-r+1,1} & \ell_{m-r+1,2} & \ell_{m-r+1,3} & \cdots & x_{m-r+1,r+1} & \ell_{m-r+1,r+2} & \cdots & \ell_{m-r+1,m-1} & 0 \\
x_{m-r+2,1} & \ell_{m-r+2,2} & \ell_{m-r+2,3} & \cdots & \ell_{m-r+2,m-r} & \ell_{m-r+2,r+2} & \cdots & 0 & 0 \\
\vdots & \vdots & \vdots & \ddots & \vdots & \vdots & \ddots & \vdots & \vdots \\
x_{m-1,1} & \ell_{m-1,2} & x_{m-1,3} & \cdots & \ell_{m-1,m-r} & \ell_{m-1,m-r+1} & \cdots & 0 & 0 \\
x_{m,1} & x_{m,2} & \ell_{m,3} & \cdots & \ell_{m,m-r} & 0 & \cdots & 0 & 0
\end{pmatrix} , \qquad (4.8)
$$

where $\ell_{i,j}$ is a linear form in $k[x_{i,j}]$ not involving any entry indexed by $\mathcal{I}$, while the hollow-filled version is obtained by replacing the right corner zeros by similar linear forms.

4.2.2 Generic Hollow Sections

Let $\mathfrak{H}_m$ as in the previous subsection denote the hollow linear section of the $m \times m$ generic matrix. Set $\mathfrak{f}_m := \det \mathfrak{H}_m$.

Consider the set $(\mathfrak{H}_m)_{\mathcal{I}_1 \cup \mathcal{I}_2 \cup \{(2,m)\}}$ in the notation of Section 4.2.1, and set:

- $\partial(\mathfrak{f}_m) :=$ the set of partial derivatives of $\mathfrak{f}$ with respect to $(\mathfrak{H}_m)_{\mathcal{I}_1 \cup \mathcal{I}_2 \cup \{(2,m)\}}$.
- $\Theta(\partial(\mathfrak{f}_m)) :=$ the $2m \times 2m$ Jacobian matrix of $\partial(\mathfrak{f}_m)$ with respect to the variables in $(\mathfrak{H}_m)_{\mathcal{I}_1 \cup \mathcal{I}_2 \cup \{(2,m)\}}$.

Proposition 4.10 *Let $\mathfrak{f}_m$ be as above. Then*:

(a) $\mathfrak{f}_m$ *is an irreducible $x_{1,1}$-monoid.*
(b) $\mathfrak{f}_m$ *is homaloidal.*
(c) $\mathfrak{f}_m$ *is not a factor of* $\det \Theta(\partial(\mathfrak{f}_m))$.

Proof

(a) Denote $\zeta_m := \prod_{i+j=m+2} x_{i,j} = x_{2,m} x_{3,m-1} \cdots x_{m,2}$. We give an explicit form of $\mathfrak{f}_m$:

CLAIM $\mathfrak{f}_m = x_{1,1} \zeta_m - \sum_{i+j=m+2} x_{1,j} x_{i,1} \frac{\zeta_m}{x_{i,j}}$ up to a sign.

We induct on m. The expression is obvious for $m = 2$ as $\mathfrak{H}_2$ is the generic matrix. Let then $m \geq 3$ and apply Laplace expansion along the last row to get up to a sign

$$f_m = x_{m,1}x_{1,2}\frac{\zeta_m}{x_{m,2}} - x_{m,2}\tilde{f}_{m-1},$$

with $\tilde{f}_{m-1} = \det\tilde{\mathfrak{H}}_{m-1}$, where the latter stands for the hollow linear section of the $(m-1)\times(m-1)$ generic matrix in the ordered subset of the variables $\mathbf{X}\setminus\{x_{1,2}, x_{m,1}, x_{m,2}\}$. Substituting for g_{m-1} by the inductive assumption and collecting terms yield the sought expression.

So much for the claim.

Therefore, f_m is a strict $x_{1,1}$-monoid with no proper common factor on the two sides, hence is irreducible.

(b) Since m is fixed in the discussion, to simplify the notation, write $\mathfrak{H} = \mathfrak{H}_m$, $f = f_m$, and $f_{i,j} := \partial f/\partial x_{i,j}$, for $(i, j) \in \mathcal{I}_1\cup\mathcal{I}_2\cup\mathcal{I}_3$. By Proposition 2.10, the adjugate matrix $\mathrm{adj}(\mathfrak{H})$ of $\mathfrak{H}$ has the following shape:

$$\mathrm{adj}(\mathfrak{H}) = \begin{pmatrix} f_{1,1} & f_{2,1} & \cdots & f_{m-1,1} & f_{m,1} \\ f_{1,2} & * & \cdots & * & f_{m,2} \\ f_{1,3} & * & \cdots f_{m-1,3} & * \\ \vdots & \vdots & \ddots & \vdots & \vdots \\ f_{1,m} & f_{2,m} & \cdots & * & * \end{pmatrix}, \tag{4.9}$$

where each $*$ denotes some unspecified entry. Thus, the adjugate formula (2.1) for $\mathfrak{H}$ gives the following linear relations among the partial derivatives of f

$$x_{1,1}f_{1,1}+\cdots+x_{1,m}f_{1,m}-x_{i,1}f_{i,1}-x_{i,m+2-i}f_{i,m+2-i} = 0, \quad i = 2,\ldots, m \tag{4.10}$$

as the expression of equating two different terms that repeat the same entry along the diagonal of $f\,\mathbb{I}_{m\times m}$, and

$$x_{1,j}f_{1,1} + x_{m-j+2,j}f_{m-j+2,1} = 0, \quad j = 2,\ldots, m \tag{4.11}$$

$$x_{i,1}f_{1,1} + x_{i,m-i+2}f_{1,m-i+2} = 0 \quad i = 2,\ldots, m, \tag{4.12}$$

as a result of the zero entries in $f\,\mathbb{I}_{m\times m}$.

These relations give rise to the following $(3m - 2) \times 3(m - 1)$ submatrix of the syzygy matrix of the partial derivatives $f_{i,j}$

$$
\mathcal{Z} =
\left(
\begin{array}{cccc|cccc|cccc}
x_{1,1} & x_{1,1} & \cdots & x_{1,1} & x_{1,2} & x_{1,3} & \cdots & x_{1,m} & x_{2,1} & x_{3,1} & \cdots & x_{m,1} \\
x_{1,2} & x_{1,2} & \cdots & x_{1,2} & 0 & 0 & \cdots & 0 & 0 & 0 & \cdots & x_{m,2} \\
\vdots & \vdots & \ddots & \vdots & \vdots & \vdots & \ddots & \vdots & \vdots & \vdots & \ddots & \vdots \\
x_{1,m-1} & x_{1,m-1} & \cdots & x_{1,m-1} & 0 & 0 & \cdots & 0 & 0 & x_{3,m-1} & \cdots & 0 \\
x_{1,m} & x_{1,m} & \cdots & x_{1,m} & 0 & 0 & \cdots & 0 & x_{2,m} & 0 & \cdots & 0 \\
\hline
-x_{2,1} & 0 & \cdots & 0 & 0 & 0 & \cdots & x_{2,m} & 0 & 0 & \cdots & 0 \\
\vdots & \vdots & \ddots & \vdots & \vdots & \vdots & \ddots & \vdots & \vdots & \vdots & \ddots & \vdots \\
0 & 0 & \cdots & 0 & 0 & x_{m-1,3} & \cdots & 0 & 0 & 0 & \cdots & 0 \\
0 & 0 & \cdots & -x_{m,1} & x_{m,2} & 0 & \cdots & 0 & 0 & 0 & \cdots & 0 \\
\hline
-x_{2,m} & 0 & \cdots & 0 & 0 & 0 & \cdots & 0 & 0 & 0 & \cdots & 0 \\
\vdots & \vdots & \ddots & \vdots & \vdots & \vdots & \ddots & \vdots & \vdots & \vdots & \ddots & \vdots \\
0 & 0 & \cdots & -x_{m,2} & 0 & 0 & \cdots & 0 & 0 & 0 & \cdots & 0
\end{array}
\right).
$$

It can be shown that $\mathcal{Z}$ has full rank. If we knew at this stage that the partial derivatives are algebraically independent over k (e.g., if the ideal they generate is of linear type), we would conclude by Theorem 2.27. Instead, by a quirk, we argue directly upon the submatrix of the Jacobian dual matrix [19, Definition 3.2.18] induced by $\mathcal{Z}$. Namely, introducing "dual" variables

$$
\mathbf{Y} = y_{1,1}, \ldots, y_{1,m};\ y_{2,1}, \ldots, y_{m,1};\ y_{2,m}, y_{3,m-1}, \ldots, y_{m,2},
$$

the transpose of the following matrix is a submatrix of the Jacobian dual matrix of the partial derivatives

$$
B =
\left(
\begin{array}{cccc|cccc|cccc}
y_{1,1} & y_{1,1} & \cdots & y_{1,1} & 0 & 0 & \cdots & 0 & 0 & 0 & \cdots & 0 \\
y_{1,2} & y_{1,2} & \cdots & y_{1,2} & y_{1,1} & 0 & \cdots & 0 & 0 & 0 & \cdots & 0 \\
y_{1,3} & y_{1,3} & \cdots & y_{1,3} & 0 & y_{1,1} & \cdots & 0 & 0 & 0 & \cdots & 0 \\
\vdots & \vdots & \ddots & \vdots & \vdots & \vdots & \ddots & \vdots & \vdots & \vdots & \ddots & \vdots \\
y_{1,m} & y_{1,m} & \cdots & y_{1,m} & 0 & 0 & \cdots & y_{1,1} & 0 & 0 & \cdots & 0 \\
\hline
-y_{2,1} & 0 & \cdots & 0 & 0 & 0 & \cdots & 0 & y_{1,1} & 0 & \cdots & 0 \\
0 & -y_{3,1} & \cdots & 0 & 0 & 0 & \cdots & 0 & 0 & y_{1,1} & \cdots & 0 \\
\vdots & \vdots & \ddots & \vdots & \vdots & \vdots & \ddots & \vdots & \vdots & \vdots & \ddots & \vdots \\
0 & 0 & \cdots & -y_{m,1} & 0 & 0 & \cdots & 0 & 0 & 0 & \cdots & y_{1,1} \\
\hline
-y_{2,m} & 0 & \cdots & 0 & y_{2,1} & 0 & \cdots & 0 & y_{1,m} & 0 & \cdots & 0 \\
0 & -y_{3,m-1} & \cdots & 0 & 0 & y_{3,1} & \cdots & 0 & 0 & y_{1,m-1} & \cdots & 0 \\
\vdots & \vdots & \ddots & \vdots & \vdots & \vdots & \ddots & \vdots & \vdots & \vdots & \ddots & \vdots \\
0 & 0 & \cdots & -y_{m,2} & 0 & 0 & \cdots & y_{m,1} & 0 & 0 & \cdots & y_{1,2}
\end{array}
\right).
$$

Let B_1 be the $(3m-3) \times (3m-3)$ submatrix of B obtained by omitting the first row. By [19, Theorem 2.18], it suffices to show that $\det B_1$ does not vanish when the entries of B_1 are evaluated via $y_{i,j} \mapsto \mathfrak{f}_{i,j}$. And indeed, expanding $\det B_1$ along the last $m-1$ rows yields

$$\det B_1 = y_{1,1}^{2m-2} y_{2,m} y_{3,m-1} \cdots y_{m,2} + T,$$

where T is a sum of monomials of degree at most $m-2$ in the variables $y_{2,m}, y_{3,m-1}, \ldots, y_{m,2}$. Evaluating as indicated gives the expression

$$\mathfrak{f}_{1,1}^{2m-2} \mathfrak{f}_{2,m} \mathfrak{f}_{3,m-1} \cdots \mathfrak{f}_{m,2} + T(\mathfrak{f}_{1,1}, \ldots, \mathfrak{f}_{1,m}, \mathfrak{f}_{2,1}, \ldots, \mathfrak{f}_{m,1}, \mathfrak{f}_{2,m}, \mathfrak{f}_{3,m-1}, \ldots, \mathfrak{f}_{m,2}),$$

which is non-vanishing because, e.g., the degree of $x_{1,1}$ in the first summand is $m-1$, whereas in the second summand its degree is at most $m-2$.

(c) We will give the explicit form of $\Theta(\partial(\mathfrak{f}))$.

Drawing upon the expression of $\mathfrak{f}$ obtained in the proof of item (a), we derive the following equalities:

$$\frac{\partial^2 \mathfrak{f}}{\partial x_{i,1} \partial x_{1,j}} = \frac{\partial^2 \mathfrak{f}}{\partial x_{1,j} \partial x_{i,1}} = \zeta / x_{i,j}, \quad \text{if } i + j = m + 2$$

$$\frac{\partial^2 \mathfrak{f}}{\partial x_{i,1} \partial x_{1,j}} = 0, \quad \text{if } 1 \le i, j \le m \text{ and } i + j \ne m + 2$$

$$\frac{\partial^2 \mathfrak{f}}{\partial x_{1,1} \partial x_{2,m}} = \frac{\partial^2 \mathfrak{f}}{\partial x_{2,m} \partial x_{1,1}} = \frac{\zeta}{x_{2,m}}$$

$$\frac{\partial^2 \mathfrak{f}}{\partial x_{1,1} \partial x_{1,1}} = \frac{\partial^2 \mathfrak{f}}{\partial x_{2,m} \partial x_{2,m}} = 0.$$

Ordering the list of variables and of the partial derivatives as follows

$$\{x_{1,2}, \ldots, x_{1,m}, x_{2,1}, \ldots, \ldots, x_{m,1}, x_{1,1}, x_{2,m}\}$$

$$\left\{ \frac{\partial \mathfrak{f}}{\partial x_{m,1}}, \ldots, \frac{\partial \mathfrak{f}}{\partial x_{2,1}}, \frac{\partial \mathfrak{f}}{\partial x_{1,m}}, \ldots, \frac{\partial \mathfrak{f}}{\partial x_{1,2}}, \frac{\partial \mathfrak{f}}{\partial x_{1,1}}, \frac{\partial \mathfrak{f}}{\partial x_{2,m}} \right\}$$

implies the following format:

$$\Theta(\partial(\mathfrak{f})) =$$

$$
\left(
\begin{array}{cccc|cccc|cc}
-\dfrac{\zeta}{x_{m,2}} & 0 & \cdots & 0 & 0 & 0 & \cdots & 0 & 0 & * \\[2mm]
0 & -\dfrac{\zeta}{x_{m-1,3}} & \cdots & 0 & 0 & 0 & \cdots & 0 & 0 & * \\[2mm]
\vdots & \vdots & \ddots & \vdots & \vdots & \vdots & \ddots & \vdots & \vdots & \vdots \\[2mm]
0 & 0 & \cdots & -\dfrac{\zeta}{x_{2,m}} & 0 & 0 & \cdots & 0 & 0 & * \\[2mm]
\hline
0 & 0 & \cdots & 0 & -\dfrac{\zeta}{x_{m,2}} & 0 & \cdots & 0 & 0 & * \\[2mm]
0 & 0 & \cdots & 0 & 0 & -\dfrac{\zeta}{x_{m-1,3}} & \cdots & 0 & 0 & * \\[2mm]
\vdots & \vdots & \ddots & \vdots & \vdots & \vdots & \ddots & \vdots & \vdots & \vdots \\[2mm]
0 & 0 & \cdots & 0 & 0 & 0 & \cdots & -\dfrac{\zeta}{x_{2,m}} & 0 & * \\[2mm]
\hline
0 & 0 & \cdots & 0 & 0 & 0 & \cdots & 0 & 0 & \dfrac{\zeta}{x_{2,m}} \\[2mm]
* & * & \cdots & * & * & * & \cdots & * & \dfrac{\zeta}{x_{2,m}} & 0
\end{array}
\right)
$$

Thus, up to a sign,

$$
\det \Theta(\partial(\mathfrak{f})) = \frac{\zeta^{2m}}{(x_{2,m}\zeta)^2} = \frac{\zeta^{2m-2}}{x_{2,m}^2} \notin (\mathfrak{f}),
$$

since $\dfrac{\zeta^{2m-2}}{x_{2,m}^2}$ is a nonzero monomial. $\qquad\qquad\square$

Remark 4.11 Since $\mathfrak{f}$ is homaloidal, its partial derivatives are algebraically independent over k and so are the partial derivatives in the subset $\partial(\mathfrak{f})$. Item (c) above shows that in arbitrary characteristic, the rank of the Hessian matrix of $\mathfrak{f}$ is at least $2m$ – one expects this rank to be one less than the maximum $3m - 2$ in any characteristic.

Item (c) of the above proposition also holds for the case of a hollow-filled linear section. Borrowing the previous notation, one has:

Corollary 4.12 *Let* $\mathbf{L}$ *denote a hollow-filled linear section of the* $m \times m$ *generic matrix and* $f := \det \mathbf{L}$. *Set* $\partial(f) := \{\partial f/\partial x_{i,j} \mid x_{i,j} \in (\mathfrak{H}_m)_{I_1 \cup I_{2\cup\{(2,m)\}}}\}$ *and let* $\Theta(\partial(f))$ *stand for the* $2m \times 2m$ *Jacobian matrix of* $\partial(f)$ *with respect to the variables in* $(\mathfrak{H}_m)_{I_1 \cup I_{2\cup\{(2,m)\}}}$. *Then* $\det \Theta(\partial(f)) \neq 0 \bmod f$.

Proof Let $\mathfrak{f}$ be as above the determinant of the hollow matrix $\mathfrak{H}_m$. Note that f is still irreducible since $\mathbf{L}$ is a deformation of $\mathfrak{H}_m$. Write $f = \mathfrak{f} + T$, where T belongs to the ideal $\mathfrak{n} := (\mathbf{L}_{\mathcal{J}})$ in the notation of Section 4.2.1. In particular,

$$
\Theta(\partial(f)) \equiv \Theta(\partial(\mathfrak{f})) \bmod \mathfrak{n}.
$$

Thus, it is enough to show that $\det \Theta(\partial(\mathfrak{f})) \not\equiv 0 \mod \mathfrak{f}$, which is the content of Proposition 4.10 (c). $\qquad\square$

Remark 4.13 The reader may wonder why, in the hollow case, we consider only partial derivatives of $\mathfrak{f}$ with respect to $(\mathfrak{H}_m)_{\mathcal{I}_1 \cup \mathcal{I}_2 \cup \{(2,m)\}}$ instead of the whole gradient ideal of $\mathfrak{f}$ – the same question regarding the hollow-filled environment. One reason is that one seeks for submatrices of the Hessian matrix that do not vanish modulo $\mathfrak{f}$, as in item (c) of Proposition 4.10, while the whole Hessian determinant admits $\mathfrak{f}$ as a factor – this is a consequence, via Proposition 2.7, of the dimension of the dual variety obtained later in Proposition 8.2.

Without further ado, we pose:

Conjecture 4.14 The gradient ideal of $\mathfrak{f}$ is of linear type.

4.3 Codimension One Linear Section

As before, $\mathbf{X} := (x_{i,j})$ stands for an $m \times m$ generic matrix.

Favoring codimension to dimension, the simplest class of linear sections that come up is that of codimension one, that is, the member $\mathbf{L}$ of the class is such that $\mathrm{codim}_{\mathbf{X}}(\mathbf{L}) = 1$ in the notation of Section 1.5.3. Since we are naturally interested in those results that are stable by conjugation, we can assume that the entries of $\mathbf{L}$ are those of $\mathbf{X}$ except for a pivot entry, say, $x_{m,m}$, which is replaced by a linear form $\lambda \in k[\mathbf{X}]$ not involving $x_{m,m}$.

Any other linear section can be seen as an iteration of the codimension one pivotal case (of course, the subsequent relative codimension will deceive our eyes). And yet, even in codimension one, one has to deal with essentially three cases, as to whether λ is a general form, simply non-vanishing or vanishing.

Set $\lambda := \sum_{(i,j)\neq(m,m)} \lambda_{i,j} x_{i,j}$ and consider the $m \times m$ section of $\mathbf{X}$

$$\mathbf{L}(m)_\lambda := \begin{pmatrix} x_{1,1} & x_{1,2} & \cdots & x_{1,m-1} & x_{1,m} \\ x_{2,1} & x_{2,2} & \cdots & x_{2,m-1} & x_{2,m} \\ \vdots & \vdots & \cdots & \vdots & \vdots \\ x_{m-1,1} & x_{m-1,2} & \cdots & x_{m-1,m-1} & x_{m-1,m} \\ x_{m,1} & x_{m,2} & \cdots & x_{m,m-1} & \lambda \end{pmatrix}. \tag{4.13}$$

The symbol m in the notation is a mnemonic for the size of the generic matrix $\mathbf{X}$ fixed once for all, along with the associated polynomial ring $k[\mathbf{X}]$ of its entries.

The following notation will be fixed throughout the entire section, with the additional assumption that $m \geq 3$:

- $R := k[\mathbf{X} \setminus \{x_{m,m}\}]$.
- $\mathbf{Y}$ is the $m \times m$ generic matrix in the dual variables of $\mathbf{X}$ and $k[\mathbf{Y}]$ is the corresponding polynomial ring.

- $f := \det \mathbf{L}(m)_\lambda$.
- $P := I_{m-1}(\mathbf{L}(m)_\lambda)$, the ideal of the submaximal minors of $\mathbf{L}(m)_\lambda$.
- $f_{i,j} := \partial f / \partial x_{i,j}$, for $(i, j) \neq (m, m)$.
- $\mathbf{f} := \{f_{i,j} \mid (i, j) \neq (m, m)\}$.
- $J :=$ the gradient ideal of f.

Quite generally, given a square matrix M, the symbol $\Delta_{i,j}(M)$ denotes the signed (i, j)-cofactor of M. In the generic case, $M = \mathbf{X}$ or $M = \mathbf{Y}$, we write $\Delta_{i,j}(\mathbf{X})$ and $\Delta_{i,j}(\mathbf{Y})$, respectively.

By the relation (2.6) in the proof of Proposition 2.10,

$$f_{i,j} = \Delta_{i,j}(\mathbf{L}(m)_\lambda) + \lambda_{i,j} \Delta_{m,m}(\mathbf{L}(m)_\lambda) \tag{4.14}$$

for every $1 \leq i, j \leq m$ such that $(i, j) \neq (m, m)$. In particular, $J \subset P = (J, \Delta_{m,m}(\mathbf{L}(m)_\lambda))$, and one has a homogeneous inclusion of graded k-algebras

$$k[\mathbf{f}] \subset k[\mathbf{\Delta}] = k[\mathbf{f}, \Delta_{m,m}(\mathbf{L}(m)_\lambda)], \tag{4.15}$$

where $\mathbf{\Delta} = \{\Delta_{i,j}(\mathbf{L}(m)_\lambda), 1 \leq i, j \leq m\}$.

Proposition 4.15 *With the above notation, one has*:

(a) *f is irreducible.*
(b) *$m^2 - 2 \leq \dim k[\mathbf{f}] \leq m^2 - 1$.*
(c) *The matrix $\mathbf{L}(m)_\lambda$ is 1-generic if and only if*

$$\operatorname{rank} \begin{pmatrix} -\lambda_{1,1} & \cdots & -\lambda_{1,m-1} & -\lambda_{1,m} \\ \vdots & \ddots & \vdots & \vdots \\ -\lambda_{m-1,1} & \cdots & -\lambda_{m-1,m-1} & -\lambda_{m-1,m} \\ -\lambda_{m,1} & \cdots & -\lambda_{m,m-1} & 1 \end{pmatrix} \geq 2.$$

In particular, this is the case when the $\lambda_{i,j}$'s are general and also when all of them vanish except one on the main diagonal.

Proof

(a) We can prove this in two ways. The first, more sophisticated, appeals to the argument that, since $\operatorname{codim}_{\mathbf{X}} \mathbf{L}(m)_\lambda = 1 \leq (m-1) - 1$ and $\mathbf{X}$ is 1-generic, then Theorem 1.26(ii) implies that f is irreducible.

A second proof, simpler by theory, draws upon showing that f is a monoid of a special nature. Namely, write $\Delta_{i,j} := \Delta_{i,j}(\mathbf{L}_\lambda)$. By a Laplace expansion along the last row, f has an $x_{m,1}$-monoid shape:

$$f = x_{m,1}\left(\Delta_{m,1} + \lambda_{m,1}\Delta_{m,m}\right) + \sum_{j=2}^{m-1} x_{m,j}\Delta_{m,j} + (\lambda - \lambda_{m,1}x_m,)\Delta_{m,m}$$

$$:= x_{m,1}g + h.$$

Clearly, f is irreducible if and only if $\gcd(g, h) = 1$. To show the latter, note that $\Delta_{m,1}, \ldots, \Delta_{m,m}$ are the maximal minors of the $(m-1) \times m$ generic submatrix $\tilde{\mathbf{X}}$ of the $m-1$ first rows of $\mathbf{L}_\lambda$. As such, they generate a codimension two (Proposition 1.10) prime ideal (Proposition 1.12). In particular, their syzygies are generated by the rows of $\tilde{\mathbf{X}}$.

From this immediately follows that g and h are nonzero. Moreover, since $\deg(g) = m - 1$, it must be irreducible. Thus, saying that $\gcd(g, h) \neq 1$ would mean that $\ell g = h$ for some linear form ℓ. But, this equality yields the relation

$$\ell \Delta_{m,1} - \sum_{j=2}^{m-1} x_{m,j} \Delta_{m,j} + (\ell - \lambda + \lambda_{m,1} x_{m,1}) \Delta_{m,m} = 0,$$

that is, $(\ell, -x_{m,2}, \ldots, -x_{m,m-1}, \ell - \lambda + \lambda_{m,1} x_{m,1})$ is a syzygy of $\Delta_{m,1}, \ldots, \Delta_{m,m}$. Since the coordinate entry $-x_{m,2}$ does not belong to the ideal of entries of $\tilde{\mathbf{X}}$, this is in contradiction to the fact that the syzygies of $\Delta_{m,1}, \ldots, \Delta_{m,m}$ are generated by the rows of $\tilde{\mathbf{X}}$.

(b) Set L and K for the respective fields of fractions of the k-algebras in (4.15). One has

$$\dim k[\mathbf{f}] = \mathrm{trdeg}_k L = \mathrm{trdeg}_k K - \mathrm{trdeg}_L(K) = \ell(P) - \mathrm{trdeg}_L(K)$$
$$= m^2 - 1 - \mathrm{trdeg}_L(K). \tag{4.16}$$

By (4.14), $K = L(\Delta_{m,m}(\mathbf{L}(m)_\lambda))$; hence $0 \leq \mathrm{trdeg}_L(K) \leq 1$. Thus, $m^2 - 2 \leq \dim k[\mathbf{f}] \leq m^2 - 1$ as claimed.

(c) Computing the corresponding trace-orthogonal matrix according to Definition 1.20 yields the following matrix up to a factor $y_{m,m}$:

$$\begin{pmatrix} -\lambda_{1,1} & \cdots & -\lambda_{1,m-1} & -\lambda_{1,m} \\ \vdots & \ddots & \vdots & \vdots \\ -\lambda_{m-1,1} & \cdots & -\lambda_{m-1,m-1} & -\lambda_{m-1,m} \\ -\lambda_{m,1} & \cdots & -\lambda_{m,m-1} & 1 \end{pmatrix}. \tag{4.17}$$

By Proposition 1.21, $\mathbf{L}(m)_\lambda$ is 1-generic if and only if some 2-minor of this numerical matrix does not vanish. $\qquad\square$

4.3.1 The Submaximal Minors

Next are the main features of the ideal P generated by the submaximal minors as in the previous subsection.

Proposition 4.16 *With the preceding notation, one has:*

(a) *R/P is a Gorenstein ring of codimension 4 with minimal graded free resolution*

$$0 \to R(-2m+2) \to R(-m+1)^{m^2} \to R(-m)^{2(m^2-1)}$$
$$\to R(-(m-1))^{m^2} \to R \to R/P \to 0.$$

(b) *If $m \geq 4$, then R/P is a normal domain.*

(c) *The analytic spread $\ell(P)$ of P is $m^2 - 1$.*

(d) *Set $\mathbb{P}^{m^2-2} = \mathrm{Proj}(k[\mathbf{X} \setminus \{x_{m,m}\}])$ and $(\mathbb{P}^{m^2-1})^{\vee} = \mathrm{Proj}(k[\mathbf{Y}])$ (dual variables). The rational map $\Psi : \mathbb{P}^{m^2-2} \dashrightarrow \mathbb{P}^{m^2-1}$ defined by a minimal set of generators of P is a birational map onto the hypersurface with defining equation*

$$\Delta_{m,m}(\mathbf{Y}) - \sum_{(i,j)\neq(m,m)} \lambda_{i,j}\Delta_{i,j}(\mathbf{Y}) \in k[\mathbf{Y}]$$

up to a k-linear transformation of $k[\mathbf{Y}]$.

Proof

(a) It is well-known that $I_{m-1}(\mathbf{X})$ is a Gorenstein ideal of codimension 4 with minimal graded free resolution

$$0 \to S(-2m+2) \to S(-m+1)^{m^2} \to S(-m)^{2(m^2-1)}$$
$$\to S(-(m-1))^{m^2} \to S \to S/I_{m-1}(\mathbf{X}) \to 0,$$

where $S = R[x_{m,m}]$ ([11], also [2, Chapter 2, Section D] and, more recently, [21, Section 6.4.2]). Since $\lambda - x_{m,m} \neq 0$ and $I_{m-1}(\mathbf{X})$ is a prime ideal, it specializes.

(b) For $m \geq 5$ the assertion follows from Theorem 1.26 (iii). Indeed, with the notation there, set $\mathbb{L} := \mathbf{X}$ and $\mathbf{L} := \mathbf{L}(m)_{\lambda}$, and take $s = m - 2$, so $m > s + 1$ and $\mathrm{codim}_{\mathbf{X}}\mathbf{L}(m)_{\lambda} = 1 \leq m - 4 = s - 2$, as $m \geq 5$.

For $m = 4$ one draws upon a subtle refinement as found in the discussion immediately following the statement of [8, Theorem 2.1 (3)] and its proof in [8, Proposition 3.1 and ff.]. Namely, in the notation there set $M' = H := \mathbf{X}$ and $M := \mathbf{L}(m)_{\lambda}$. Then, since we have $\mathrm{codim}_{\mathbf{X}}\mathbf{L}(m)_{\lambda} = 1 \leq (m - 1) - 2 = m - 3$ for $m = 4$, it suffices to prove that $\mathrm{codim}\, I_{m-2}(\mathbf{L}(m)_{\lambda})/P \geq 2$. But, since $\mathbf{X}$ is 3-generic, then, by Theorem 1.26 (i), $I_{m-2}(\mathbf{L}(m)_{\lambda})$ has codimension $(m-(m-2)+1)(m-(m-2)+1) = 9$. Hence, $\mathrm{codim}\, I_{m-2}(\mathbf{L}(m)_{\lambda})/P = 5 \geq 2$.

As a side we note that for $m = 3$ and arbitrary λ the ideal P may even fail to be prime, as is the case when $\lambda = 0$ or, similarly, when $\lambda = x_{i,m}$ or $\lambda = x_{m,j}$, with $i, j \neq m$.

(c) Since $\det \mathbf{L}(m)_\lambda \neq 0$, then, by Proposition 2.3, the inclusion of k-algebras $k[\mathbf{\Delta}] \subset R$ extends to an algebraic extension of the respective fields of fractions. Hence, $\ell(P) = \dim k[\mathbf{\Delta}] = \dim R = m^2 - 1$.

(d) Since P is linearly presented and has maximal analytic spread $m^2 - 1$, then, by Theorem 2.27, the map Ψ is birational onto a hypersurface. To get the defining equation of this hypersurface, we assume that Ψ is defined by the minimal set of generators of P consisting of the entries of the adjugate matrix of $\mathbf{L}(m)_\lambda$.

CLAIM Let $\mathfrak{J} \subset k[\mathbf{Y}]$ denote the homogeneous defining ideal of the image of Ψ. Then

$$\Delta_{m,m}(\mathbf{Y}) - \sum_{(i,j)\neq(m,m)} \lambda_{i,j}\Delta_{i,j}(\mathbf{Y}) \in \mathfrak{J}. \tag{4.18}$$

For this consider the algebraic formulation of Ψ in terms of the k-homomorphism $\varphi : y_{i,j} \mapsto \Delta_{i,j}(\mathbf{L}(m)_\lambda)$. Then $\mathfrak{J} = \ker \varphi$.

The formula (2.2) for $\mathbf{L}(m)_\lambda$ implies the following equalities:

$$\varphi(\Delta_{j,i}(\mathbf{Y})) = \begin{cases} (\det \mathbf{L}(m)_\lambda)^{m-2} x_{ij}, & \text{if } (i, j) \neq (m, m) \\ (\det \mathbf{L}(m)_\lambda)^{m-2} \lambda, & \text{otherwise.} \end{cases}$$

These equalities as applied to the values $(i, j) = (m, m)$ and $(i, j) \neq (m, m)$ give

$$\varphi(\Delta_{m,m}(\mathbf{Y})) = \sum_{(i,j)\neq(m,m)} \lambda_{i,j}\varphi(\Delta_{i,j}(\mathbf{Y})),$$

which proves the claim.

To conclude we observe that $\Delta_{m,m}(\mathbf{Y}) - \sum_{(i,j)\neq(m,m)} \lambda_{i,j}\Delta_{i,j}(\mathbf{Y})$ is nonzero because the cofactors of the generic matrix $\mathbf{Y}$ are algebraically independent over k, and it is an irreducible form because it has degree $m - 1$ and belongs to the prime ideal $I_{m-1}(\mathbf{Y})$. $\qquad\square$

4.3.2 The Gradient Ideal

Recall that J denotes the gradient ideal of the determinant $f = \det \mathbf{L}(m)_\lambda$. As usual, the homological properties of J are way more difficult than those of the ideal $P = I_{m-1}(\mathbf{L}(m)_\lambda)$ of submaximal minors.

Theorem 4.17 *With the standing notation of the section, if $m \geq 3$, then:*

(i) $m^2 - 2 \leq \ell(J) \leq m^2 - 1$.

(ii) *Let*

$$\bigoplus R(-j)^{\beta_{1,j}} \to R(-(m-1))^{m^2-1} \to J \to 0$$

stand for the minimal graded free presentation of J. Then:

(a) $\beta_{1,m} \geq m^2 - 1$.
(b) *If λ is a general linear form, then $\beta_{1,m} = m^2 - 1$.*
(c) *If $\beta_{1,m} \leq 2m^2 - 7$, then the unmixed part of J is $P = I_{m-1}(\mathbf{L}(m)_\lambda)$ and J
     has codimension 4.*

Proof

(i) This is Proposition 4.15.
(ii) (a) The assertion is about the linear syzygies of J. Since by (4.14), J is related
     to P as $P = (J, \Delta_{m,m}(\mathbf{L}(m)_\lambda))$, it is natural to look at the syzygies of the
     latter. Let $\mathcal{Z}$ denote the syzygy matrix of P with respect to its minimal set of
     generators $\{f_{i,j}\,((i,j) \neq (m,m)), \Delta_{m,m}(\mathbf{L}(m)_\lambda)\}$. By Proposition 4.16 (a), $\mathcal{Z}$
     is an $m^2 \times 2(m^2 - 1)$ linear matrix. Let $\mathfrak{d}$ be the dimension of the k-vector space
     V spanned by the last row of $\mathcal{Z}$. Then, up to elementary column operations, $\mathcal{Z}$
     has the following shape:

$$\left(\begin{array}{c|c} \mathfrak{K} & * \\ \hline \mathbf{0} & \mathfrak{L} \end{array}\right),\tag{4.19}$$

where $\mathbf{0}$ is the $1 \times (2(m^2 - 1) - \mathfrak{d})$ null matrix and the entries of $\mathfrak{L}$ are a k-
basis of V. In this way, the latter matrix is the syzygy matrix of a minimal set
of generators of P, given by certain k-linear combinations of the original ones,
while $\mathfrak{K}$ gives the generating linear syzygies of the corresponding elementally
modified $f_{i,j}$'s. Note that $\mathfrak{K}$ is an $(m^2 - 1) \times (2(m^2 - 1) - \mathfrak{d})$ matrix; hence
$\beta_{1,m} = 2(m^2 - 1) - \mathfrak{d}$. Since $\mathfrak{d} \leq \dim_k R_1 = m^2 - 1$, then $\beta_{1,m} \geq m^2 - 1$.
     (b) This is the hardest piece of the item. For it, start with the $1 \times m^2$ matrix

$$\Lambda = \left(\underbrace{[-\lambda_{1,1} \cdots -\lambda_{m,1}]}_{=:\Lambda_1} \cdots \underbrace{[-\lambda_{1,m} \cdots -\lambda_{m-1,m}\ 1]}_{=:\Lambda_m}\right),\tag{4.20}$$

whose entries are the entries of the matrix (4.17) listed along its rows.
     By the equality in (4.14) and the notation of (4.15), one can write

$$\Delta = [\mathbf{f}\ \Delta_{m,m}(\mathbf{L}(m)_\lambda)]\left(\frac{I_{m^2-1}\quad \mathbf{0}}{\Lambda}\right),\tag{4.21}$$

where $\mathbf{0}$ denotes the null $m^2 - 1 \times 1$ matrix. A sufficient scrutiny of the linear
relations afforded by the adjugate formula (2.1) for $\mathbf{L}(m)_\lambda$ gives the following

$m^2 \times (m^2 - 1)$ submatrix $\mathfrak{S}$ of the syzygy matrix $\mathcal{Z}$ of P with respect to the sequence $\mathbf{\Delta}$ of its generators

$$
\left(
\begin{array}{ccc|cc}
\begin{matrix}
x_{2,1} & \cdots & x_{m,1} \\
x_{2,2} & \cdots & x_{m,2} \\
\vdots & \ddots & \vdots \\
x_{2,m} & \cdots & \lambda
\end{matrix} & & & \begin{matrix} x_{1,1} & \cdots & x_{m-1,1} \end{matrix} & \\
& \ddots & & & \\
& \begin{matrix}
x_{1,1} & \cdots & \widehat{x_{i,1}} & \cdots & x_{m,1} \\
x_{1,2} & \cdots & \widehat{x_{i,2}} & \cdots & x_{m,2} \\
\vdots & \ddots & \vdots & \ddots & \vdots \\
x_{1,m} & \cdots & \widehat{x_{i,m}} & \cdots & \lambda
\end{matrix} & & \begin{matrix} x_{1,i} & \cdots & x_{m-1,i} \end{matrix} & \\
& & \ddots & & \\
& & \begin{matrix}
x_{1,1} & \cdots & x_{m-1,1} \\
x_{1,2} & \cdots & x_{m-1,2} \\
\vdots & \ddots & \vdots \\
x_{1,m} & \cdots & x_{m-1,m}
\end{matrix} & \begin{matrix} x_{1,m} & \cdots & x_{m-1,m} \end{matrix} &
\end{array}
\right)
$$

Here, the hat symbol over an entry indicates its deletion.

Therefore, (4.21) implies that $\left(\dfrac{I_{m^2-1} \quad \mathbf{0}}{\mathbf{\Lambda}} \right) \mathfrak{S}$ is an $m^2 \times (m^2 - 1)$ submatrix of $\mathcal{Z}$. In particular, in the notation of the proof of item (a), the entries of $\mathbf{\Lambda}\mathfrak{S}$ span a k-vector subspace $W \subset V$, so $\mathfrak{d} \geq \dim_k W$.

Claim $\mathfrak{d} = m^2 - 1$.

By item (a) and the above inequality $\mathfrak{d} \geq \dim_k W$, it is enough to show that $\dim_k W \geq m^2 - 1$.

We can group the set of entries of the product $\mathbf{\Lambda}\mathfrak{S}$ in the following m subsets, according to the partitions in (4.20):

$$
\mathcal{Q}_i := \left[\Lambda_j \ell_i^t, \ (-\lambda_{m,1} \cdots - \lambda_{m,m-1} \ 1)\ell_i^t \right]_{j \in \{1,\ldots,m\}\setminus\{i\}}
$$

for $1 \leq i \leq m - 1$, and

$$
\mathcal{Q}_m := \left[\Lambda_j \ell_m^t \right]_{j \in \{1,\ldots,m-1\}} .
$$

Here ℓ_i^t, stands for the transpose of the ith row ℓ_i of $\mathbf{L}(m)_\lambda$.

For each $1 \leq i \leq m - 1$, the Jacobian matrix of the linear forms of the block $\mathcal{Q}_i$, with respect to the variables that are entries of ℓ_i, is

$$A_i = \begin{pmatrix} -\lambda_{1,1} & \cdots & -\lambda_{1,i-1} & -\lambda_{1,i+1} & \cdots & -\lambda_{1,m-1} & -\lambda_{1,m} & -\lambda_{m,1} \\ -\lambda_{2,1} & \cdots & -\lambda_{2,i-1} & -\lambda_{2,i+1} & \cdots & -\lambda_{2,m-1} & -\lambda_{2,m} & -\lambda_{m,2} \\ \vdots & \ddots & \vdots & \vdots & \ddots & \vdots & \vdots & \vdots \\ -\lambda_{m-1,1} & \cdots & -\lambda_{m-1,i-1} & -\lambda_{m-1,i+1} & \cdots & -\lambda_{m-1,m-1} & -\lambda_{m-1,m} & -\lambda_{m,m-1} \\ -\lambda_{m,1} & \cdots & -\lambda_{m,i-1} & -\lambda_{m,i+1} & \cdots & -\lambda_{m,m-1} & 1 & 1 \end{pmatrix}$$

and the one of the linear forms of the block Q_m, with respect to all variables, is

$$B = \begin{pmatrix} -\lambda_{1,1} - \lambda_{m,1}\lambda_{m,1} & \cdots & -\lambda_{1,m-1} - \lambda_{m,m-1}\lambda_{m,1} \\ -\lambda_{2,1} - \lambda_{m,1}\lambda_{m,2} & \cdots & \lambda_{2,m-1} - \lambda_{m,m-1}\lambda_{m,2} \\ \vdots & \ddots & \vdots \\ -\lambda_{m-1,1} - \lambda_{m,1}\lambda_{m,m-1} & \cdots & -\lambda_{m-1,m-1} - \lambda_{m,m-1}\lambda_{m,m-1} \end{pmatrix}.$$

Therefore, the total Jacobian matrix of all linear forms that are entries of $\Lambda\mathfrak{S}$ is

$$\Theta = \begin{pmatrix} A_1 & & & & & * \\ & \ddots & & & & * \\ & & A_i & & & * \\ & & & \ddots & & * \\ & & & & A_{m-1} & * \\ & & & & & B \end{pmatrix}.$$

It is high time to use the assumption that λ is a general linear form. Namely, we claim that for a general choice of the coefficients $\lambda_{i,j}$'s, the rank of Θ is $m^2 - 1$. As usual, it is sufficient to display a particular choice of $\lambda_{i,j}$'s such that $\det \Theta \neq 0$. Set $\lambda_{m,1} = \cdots = \lambda_{m,m-1} = 0$. For this choice, one has

$$A_i = \left(\begin{array}{c|c} C_i & \mathbf{0}_{(m-1)\times 1} \\ \hline \mathbf{0}_{1\times(m-2)} \;\; 1 & 1 \end{array} \right),$$

where

$$C_i = \begin{pmatrix} -\lambda_{1,1} & \cdots & -\lambda_{1,i-1} & -\lambda_{1,i+1} & \cdots & -\lambda_{1,m-1} & -\lambda_{1,m} \\ -\lambda_{2,1} & \cdots & -\lambda_{2,i-1} & -\lambda_{2,i+1} & \cdots & -\lambda_{2,m-1} & -\lambda_{2,m} \\ \vdots & \ddots & \vdots & \vdots & \ddots & \vdots & \vdots \\ -\lambda_{m-1,1} & \cdots & -\lambda_{m-1,i-1} & -\lambda_{m-1,i+1} & \cdots & -\lambda_{m-1,m-1} & -\lambda_{m-1,m} \end{pmatrix}$$

and

$$
B = \begin{pmatrix}
-\lambda_{1,1} & \cdots & -\lambda_{1,m-1} \\
-\lambda_{2,1} & \cdots & -\lambda_{2,m-1} \\
\vdots & \ddots & \vdots \\
-\lambda_{m-1,1} & \cdots & -\lambda_{m-1,m-1}
\end{pmatrix}.
$$

Since C_i and B are specializations of corresponding generic matrices over k, for a general choice of their entries, $\det C_i \neq 0$ and $\det B \neq 0$, hence, $\det \Theta \neq 0$ as well, as claimed.

Now, since the entries of $\Lambda \mathfrak{S}$ are linear forms, the dimension of the k-vector subspace W is the dimension of the k-subalgebra of the polynomial ring R generated by these linear forms. Then (3.2) gives $\dim_k W \geq \operatorname{rank} \Theta$; hence, $\dim_k W \geq m^2 - 1$, as required in the claim.

As a consequence, in the notation of the proof of (a), $\beta_{1,m} = 2(m^2 - 1) - \mathfrak{d} = m^2 - 1$ as was to be shown.

(c) Referring to (4.19), clearly $I_1(\mathfrak{L})\Delta_{m,m}(\mathbf{L}(m)_\lambda) \subset J$. In particular, $I_1(\mathfrak{L}) \subset J : P$. Since $\operatorname{ht} I_1(\mathfrak{L}) = \mathfrak{d} = 2(m^2 - 1) - \beta_{1,m} \geq 5$ by assumption, and $\operatorname{codim} P = 4$, then J has no component in codimension ≤ 3 and its P-primary component must be P itself. Thus, we are through. $\square$

Remark 4.18 We emphasize the subtlety of Proposition 4.17, (b). Thus, e.g., having that $\mathbf{L}(m)_\lambda$ is just 1-generic is not enough as we will see in the case where $\lambda = x_{m-1,m-1}$ (Theorem 4.21).

Besides, $\beta_{1,m}$ is the number of independent linear syzygies of J, not the linear rank of J. For the latter we pose:

Conjecture 4.19 If λ is a general linear form, then the linear rank of J is $< m^2 - 2$.

4.4 Case Study: One Term Linear Form

In this section we focus on a particular case of the linear section of the previous section, obtained by taking the linear form λ to be non-vanishing simplest, by assuming that $\lambda = x_{m-1,m-1}$. That is to say, the basic linear section in this part has the form

$$
\mathbf{L}(m) := \begin{pmatrix}
x_{1,1} & x_{1,2} & \cdots & x_{1,m-1} & x_{1,m} \\
x_{2,1} & x_{2,2} & \cdots & x_{2,m-1} & x_{2,m} \\
\vdots & \vdots & \cdots & \vdots & \vdots \\
x_{m-1,1} & x_{m-1,2} & \cdots & x_{m-1,m-1} & x_{m-1,m} \\
x_{m,1} & x_{m,2} & \cdots & x_{m,m-1} & x_{m-1,m-1}
\end{pmatrix}. \tag{4.22}
$$

We have removed the subscript $\lambda = x_{m-1,m-1}$ in the notation, hoping no confusion arises.

4.4.1 The Submaximal Minors and the Gradient Ideal

As previously, in the notation established in Section 4.3, let $P = I_{m-1}(\mathbf{L}(m))$ denote the ideal of $(m-1)$-minors of $\mathbf{L}(m)$, and let J denote the gradient ideal of $f = \det \mathbf{L}(m)$. In this part we will allow for the case $m = 3$.

Parts of the following theorem will be straightforwardly derived from Proposition 4.16, but there are special results of their own.

Theorem 4.20 *Let P and J denote, respectively, the ideal of $(m-1)$-minors of $\mathbf{L}(m)$ and the gradient ideal of $f = \det \mathbf{L}(m)$. If $m \geq 3$, then:*

(i)   *R/P is a Gorenstein normal domain of codimension 4.*
(ii) *J has codimension 4, and P is the minimal primary component of J in R.*
(iii) *The colon ideal $J : P$ is generated by the $4m - 5$ entries of $\mathbf{L}(m)$ complementary to the upper left $(m-2) \times (m-2)$ submatrix. In particular, $J : P$ is a prime ideal.*
(iv) *J defines a double structure on the variety defined by P, admitting one single embedded component, the latter being a linear space of codimension $4m - 5$.*
(v)   *Set $\mathbb{P}^{m^2-2} = \mathrm{Proj}(k[\mathbf{X} \setminus \{x_{m,m}\}])$ and $\mathbb{P}^{m^2-1} = \mathrm{Proj}(k[\mathbf{Y}])$ (dual variables). The rational map $\Psi : \mathbb{P}^{m^2-2} \dashrightarrow \mathbb{P}^{m^2-1}$ defined by a minimal set of generators of P is a birational map onto the hypersurface with defining equation*

$$\Delta_{m,m}(\mathbf{Y}) - \Delta_{m-1,m-1}(\mathbf{Y}) \in k[\mathbf{Y}]$$

up to a k-linear transformation of $k[\mathbf{Y}]$. In addition, the inverse map is defined by the linear system spanned by $\{\Delta_{i,j}(\mathbf{Y}) \mid (i, j) \neq (m, m)\}$ modulo $\Delta_{m,m}(\mathbf{Y}) - \Delta_{m-1,m-1}(\mathbf{Y})$.
(vi) *J is not a reduction of P.*

Proof

(i)   If $m \geq 4$, this is Proposition 4.16, (b) and (c).

The case $m = 3$ seems to require a direct intervention. We prove that R/P is normal – and, hence, a domain as P is a homogeneous ideal. Since R/P is a Gorenstein ring, it suffices to show that R/P is locally regular in codimension one. For this consider the Jacobian matrix of P:

$$\begin{pmatrix}
x_{2,2} & -x_{2,1} & 0 & -x_{1,2} & x_{1,1} & 0 & 0 & 0 \\
x_{2,3} & 0 & -x_{2,1} & -x_{1,3} & 0 & x_{1,1} & 0 & 0 \\
0 & x_{2,3} & -x_{2,2} & 0 & -x_{1,3} & x_{1,2} & 0 & 0 \\
x_{3,2} & -x_{3,1} & 0 & 0 & 0 & 0 & -x_{1,2} & x_{1,1} \\
x_{2,2} & 0 & -x_{3,1} & 0 & x_{1,1} & 0 & -x_{1,3} & 0 \\
0 & x_{2,2} & -x_{3,2} & 0 & x_{1,2} & 0 & 0 & -x_{1,3} \\
0 & 0 & 0 & x_{3,2} & -x_{3,1} & 0 & -x_{2,2} & x_{2,1} \\
0 & 0 & 0 & x_{2,2} & x_{2,1} & -x_{3,1} & -x_{2,3} & 0 \\
0 & 0 & 0 & 0 & 2x_{2,2} & -x_{3,2} & 0 & -x_{2,3}
\end{pmatrix}.$$

Direct inspection, or alternatively, computer calculation, yields that the following pure powers are (up to sign) 4-minors of this matrix: $x_{1,3}^4$, $x_{2,1}^4$, $x_{2,2}^4$, $x_{2,3}^4$, $x_{3,1}^4$, and $x_{3,2}^4$. Therefore, the ideal of 4-minors of the Jacobian matrix has codimension at least $6 = 4 + 2$, thus ensuring that R/P satisfies Serre's property (R_1).

(ii) By item (i), P is a prime ideal of codimension 4. As in the proof of Theorem 4.17 (ii) (b) and its notation, it suffices to prove that $J : P$ has codimension > 4, which is accomplished by showing that ht $I_1(\mathfrak{L}) \geq 5$.

Here we take some leverage of the fact that $\lambda = x_{m-1,m-1}$. First, set $\Delta_{i,j} = \Delta_{i,j}(\mathbf{L}(m))$ for short, and recall that $P = (J, \Delta_{m,m})$. From the adjugate formula (2.1), we read the following relations:

$$\sum_{j=1}^{m} x_{k,j} \Delta_{j,m} = 0, \quad \text{for } k = 1, \ldots, m-1,$$

$$\sum_{i=1}^{m} x_{i,k} \Delta_{m,i} = 0, \quad \text{for } k = 1, \ldots, m-1$$

and

$$x_{m-1,m-1} \Delta_{m,m} = \sum_{j=1}^{m} x_{1,j} \Delta_{j,1} - \sum_{j=1}^{m-1} x_{m,j} \Delta_{m,j}.$$

Therefore, the entries of the mth column and of the mth row all belong to the ideal $I_1(\mathfrak{L}) \subset J : \Delta_{m,m} = J : P$. In particular, the codimension of $J : P$ is at least 5, as needed.

(iii) As seen in item (ii), P is the unmixed part of J and $I \subset J : P$, where I is generated by the $4m - 5$ entries of $\mathbf{L}(m)$ complementary to the upper left $(m-2) \times (m-2)$ submatrix.

We now prove the reverse inclusion thanks to a trick (due to R. Cunha). Write $I = I' + I''$ as sum of two prime ideals, where I' (respectively, I'') is the ideal generated by the variables on the $(m-1)$th row and on the $(m-1)$th column of $\mathbf{L}(m)$ (respectively, by the variables on the mth row and on the mth column of $\mathbf{L}(m)$). Observe that $\Delta_{i,j} \in I''$ for all $(i,j) \neq (m,m)$ and $\Delta_{i,j} \in I'$ for all $(i,j) \neq (m-1, m-1)$. Clearly, then $\Delta_{m,m} \notin I''$ and $\Delta_{m-1,m-1} \notin I'$.

Let $b \in J : P = J : \Delta_{m,m}$, say,

$$b\,\Delta_{m,m} = \sum_{(i,j)\neq(m-1,m-1)} a_{i,j} f_{i,j} + a f_{m-1,m-1}$$

$$= \sum_{(i,j)\neq(m-1,m-1)} a_{i,j}\Delta_{j,i} + a(\Delta_{m-1,m-1} + \Delta_{m,m}) \quad (4.23)$$

for certain $a_{i,j}, a \in R$. Then

$$(b-a)\Delta_{m,m} = \sum_{(i,j)\neq(m-1,m-1)} a_{i,j}\Delta_{j,i} + a\Delta_{m-1,m-1} \in I''.$$

Since I'' is a prime ideal and $\Delta_{m,m} \notin I''$, we have $c := b-a \in I''$. Substituting for $a = b - c$ in (4.23) gives

$$(-b+c)\Delta_{m-1,m-1} = \sum_{(i,j)\neq(m-1,m-1)} a_{i,j}\Delta_{j,i} - c\Delta_{m,m} \in I'.$$

By a similar token, since $\Delta_{m-1,m-1} \notin I'$, then $-b+c \in I'$. Therefore

$$b = c - (-b+c) \in I'' + I' = I,$$

as required.

(iv) In particular, $J : P$ is a prime ideal that is necessarily an associated prime of prime of R/J. As pointed out, $P \subset J : P$; hence, $J : P$ is an embedded prime of R/J. Moreover, this also gives $P^2 \subset J$; hence, J defines a double structure on the irreducible variety defined by P.

Let Q denote the embedded component of J with radical $J : P$ and let Q' denote the intersection of the remaining embedded components of J. From $J = P \cap Q \cap Q'$, we get

$$J : P = (Q : P) \cap (Q' : P),$$

in particular, passing to radicals, $J : P \subset \sqrt{Q'}$. This shows that Q is the unique embedded component of codimension $\leq 4m - 5$, and the corresponding geometric component is supported on a linear subspace.

(v) The assertion about the rational map is a special case of Proposition 4.16 (d).

For the assertion regarding the inverse map, by the same token, from the adjugate formula (2.2) for $\mathbf{L}(m)$, one deduces that the inverse map is defined by the stated linear system.

(vi) It follows from (iv) that the reduction number of a minimal reduction of P is $m - 2$. Thus, to conclude, it suffices to prove that $P^{m-1} \not\subset J P^{m-2}$.

As in the previous items, set $\Delta_{i,j} = \Delta_{i,j}(\mathbf{L}(m))$. We will show that $\Delta_{m,m}^{m-1} \in P^{m-1}$ does not belong to $J P^{m-2}$.

Recall that J is generated by the cofactors $\Delta_{i,j}$, with $(i, j) \neq (m-1, m-1)$, (m, m) and the additional form $\Delta_{m,m} + \Delta_{m-1,m-1}$. Thus, if $\Delta_{m,m}^{m-1} \in J P^{m-2}$, we can write

$$\Delta_{m,m}^{m-1} = \left(\sum_{\substack{(i,j)\neq(m-1,m-1) \\ (i,j)\neq(m,m)}} \Delta_{i,j} Q_{i,j} \right) + (\Delta_{m,m} + \Delta_{m-1,m-1})Q, \qquad (4.24)$$

where $Q_{i,j}$ and Q are homogeneous polynomial expressions of degree $m - 2$ in the set of generators $\{f_{i,j}, ((i, j) \neq (m, m)), \Delta_{m,m}\}$ of P.

Clearly, this gives a polynomial relation of degree $m - 1$ on the generators of P, so the corresponding form of degree $m - 1$ in $k[\mathbf{Y}]$ is a scalar multiple of the defining equation $\mathfrak{e} := \Delta_{m,m}(\mathbf{Y}) - \Delta_{m-1,m-1}(\mathbf{Y})$ obtained in the previous item. We now argue that such a relation is impossible, since the nonzero terms of $\mathfrak{e}$ are squarefree.

For this, we look separately at the two summands on the right side of (4.24). Clearly, the first of these summands does not involve any nonzero terms of the form $\alpha \Delta_{m,m}^{m-1}$ or $\beta \Delta_{m-1,m-1} \Delta_{m,m}^{m-2}$, for certain α, β in k.

Now, if such two terms appear in $(\Delta_{m,m} + \Delta_{m-1,m-1})Q$, they must have the same scalar coefficient, say, $c \in k \setminus 0$. In this case, bring the first of these to the left-hand side of (4.24) to get a polynomial relation of P having a term $(1 - c)y_{m,m}^{m-1}$. If $c \neq 1$, this is a contradiction due to the squarefree nature of $\mathfrak{e}$.

On the other hand, if $c = 1$, then we still have a polynomial relation of P having a nonzero term $y_{m-1,m-1}y_{m,m}^{m-2}$. Now, if $m > 3$, this is again a contradiction vis-à-vis the nature of $\mathfrak{e}$ as the nonzero terms of the latter are squarefree monomials of degree $m - 1 > 3 - 1 = 2$. Finally, if $m = 3$, a direct checking shows that the monomial $y_{2,2}y_{3,3}$ cannot be the support of a nonzero term in $\mathfrak{e}$. This concludes the proof of the statement. $\square$

4.4.2   The Polar Map

We keep the notation of the previous subsection. Let $\boldsymbol{\Delta}$ denote the set of the cofactors of $\mathbf{L}(m)$. In order to avoid changing sets of generators within an argument, we assume once for all that the elements of this set are taken in the order in which they appear as entries of the corresponding adjugate matrix along the rows. With this convention, $\mathbf{f} \subset \boldsymbol{\Delta}$ denotes the subset of the $x_{i,j}$-derivatives of f for $(i, j) \neq (m, m)$ respecting the above ordering, while J is the ideal generated by $\mathbf{f}$.

Recall that the polar map of f refers to the rational map defined by the elements of $\mathbf{f}$, while the analytic spread $\ell(J)$ is the Krull dimension of $k[\mathbf{f}]$.

Theorem 4.21 ($\mathrm{char}(k) \neq 2$) *With the above notation, one has:*

(a) $\ell(J) = m^2 - 1$.

(b) f *is homaloidal.*

(c) *The linear rank of J is $m^2 - 2$ (maximum).*

Proof

(a) Let $L \subset K$ denote the respective fraction fields of the k-algebras $k[\mathbf{f}] \subset k[\boldsymbol{\Delta}]$. By the equality in (4.16), it suffices to show that $K|L$ is an algebraic extension. We actually prove that $K = L$.

By Proposition 4.16(e) and its proof, $\Delta_{m,m}(\mathbf{Y}) - \Delta_{m-1,m-1}(\mathbf{Y})$ is a relation of the set $\boldsymbol{\Delta}$ whose elements are ordered as explained above. Laplace expansion of the minors $\Delta_{m,m}(\mathbf{Y})$ and $\Delta_{m-1,m-1}(\mathbf{Y})$ along the respective last rows gives

$$\Delta_{m,m}(\mathbf{Y}) = y_{m-1,m-1}\delta + h_1 \quad \text{and} \quad \Delta_{m-1,m-1}(\mathbf{Y}) = y_{m,m}\delta + h_2,$$

where $\{\delta, h_1, h_2\} \subset k[\mathbf{Y}]$ are polynomials that do not depend on the variables $y_{m-1,m-1}, y_{m,m}$. Subtracting,

$$\Delta_{m,m}(\mathbf{Y}) - \Delta_{m-1,m-1}(\mathbf{Y}) = 2\delta \cdot y_{m,m} - (y_{m,m} + y_{m-1,m-1})\delta + h_1 - h_2. \qquad (4.25)$$

Now, recalling that $f_{i,j} = \Delta_{i,j}$ for $(i, j) \neq (m - 1, m - 1)$ and $f_{m-1,m-1} = \Delta_{m,m} + \Delta_{m-1,m-1}$, evaluating (4.25) by the structural presentation map $\mathbf{Y} \mapsto \boldsymbol{\Delta}$ yields

$$2\delta(\mathbf{f}) \cdot \Delta_{m,m} - f_{m-1,m-1}\delta(\mathbf{f}) + h_1(\mathbf{f}) - h_2(\mathbf{f}) = 0.$$

Since $\mathrm{char}(k) \neq 2$, one has $\Delta_{m,m} \in L = k(\mathbf{f})$, hence $K = L$ as claimed.

(b) By Proposition 4.16(e) and its proof, the set $\boldsymbol{\Delta} \subset R = k[x_{i,j}]$ defines a birational map onto the image in $\mathbb{P}^{m^2-1}$; hence, one has a field equality $K = k([R]_{m-1})$ ([19, Proposition 3.2.12]). By the proof of (a), one has $K = L$. By the same citation, the resulting equality $L = k([R]_{m-1})$ implies that the set of partial derivatives $\mathbf{f}$ defines a birational map onto the image, i.e., the polar map is a Cremona map.

(c) Once again, the adjugate formula (2.1) for $\mathbf{L}(m)$ yields by expansion a set of linear relations involving the cofactors of $\mathbf{L}(m)$:

$$\sum_{j=1}^{m} x_{i,j} \Delta_{j,k} = 0, \quad \text{for } 1 \leq i \leq m-1 \text{ and } 1 \leq k \leq m-2 \ (k \neq i) \tag{4.26}$$

$$\sum_{j=1}^{m-1} x_{m,j} \Delta_{j,k} + x_{m-1,m-1} \Delta_{m,k} = 0, \quad \text{for } 1 \leq k \leq m-2 \tag{4.27}$$

$$\sum_{j=1}^{m} x_{i,j} \Delta_{j,i} = \sum_{j=1}^{m} x_{i+1,j} \Delta_{j,i+1}, \quad \text{for } 1 \leq i \leq m-3 \tag{4.28}$$

$$\sum_{i=1}^{m} x_{i,k} \Delta_{j,i} = 0, \quad \text{for } 1 \leq j \leq m-3 \text{ and } j < k \leq j+2. \tag{4.29}$$

$$\sum_{i=1}^{m} x_{i,m-1} \Delta_{m-2,i} = 0. \tag{4.30}$$

$$\sum_{i=1}^{m-1} x_{i,m} \Delta_{m-2,i} + x_{m-1,m-1} \Delta_{m-2,m} = 0. \tag{4.31}$$

Counting the number of relations above, we have a total of

$$
\begin{aligned}
m^2 - 5 &= (m-1)(m-2) - (m-2), \quad \text{for } (4.26) \text{ as } k \neq i \\
&+ m - 2, \quad \text{for } (4.27) \\
&+ m - 3, \quad \text{for } (4.28) \\
&+ 2(m-3), \quad \text{for } (4.29) \\
&+ 1, \quad \text{for } (4.30) \\
&+ 1, \quad \text{for } (4.31).
\end{aligned}
$$

Since $f_{j,i} = \Delta_{j,i}$ for every $(j,i) \neq (m-1, m-1)$ and the above relations involve neither $\Delta_{m-1,m-1}$ nor $\Delta_{m,m}$, then they give linear syzygies of the partial derivatives of f.

Further, (2.1) yields the following additional linear relations involving $\Delta_{m-1,m-1}$ and $\Delta_{m,m}$:

$$\sum_{j=1}^{m-1} x_{m-1,j} \Delta_{j,m} + x_{m-1,m} \Delta_{m,m} = 0 \tag{4.32}$$

$$\sum_{i=1}^{m-2} x_{i,m}\Delta_{m-1,i} + x_{m-1,m}\Delta_{m-1,m-1} + x_{m-1,m-1}\Delta_{m-1,m} = 0 \qquad (4.33)$$

$$\sum_{i=1}^{m-1} x_{i,m-1}\Delta_{m,i} + x_{m,m-1}\Delta_{m,m} = 0 \qquad (4.34)$$

$$\sum_{j=1}^{m-2} x_{m,j}\Delta_{j,m-1} + x_{m,m-1}\Delta_{m-1,m-1} + x_{m-1,m-1}\Delta_{m,m-1} = 0 \qquad (4.35)$$

$$\sum_{j=1,j\neq m-1}^{m} x_{m-1,j}\Delta_{j,m-1} + x_{m-1,m-1}\Delta_{m-1,m-1} = \sum_{j=1}^{m} x_{m-2,j}\Delta_{j,m-2} \qquad (4.36)$$

$$\sum_{j=1}^{m-1} x_{m,j}\Delta_{j,m} + x_{m-1,m-1}\Delta_{m,m} = \sum_{j=1}^{m} x_{m-2,j}\Delta_{j,m-2}. \qquad (4.37)$$

The latter are not linear relations of the partial derivatives of f, but since $f_{m-1,m-1} = \Delta_{m-1,m-1} + \Delta_{m,m}$, adding (4.32) to (4.33), (4.34) to (4.35), and (4.36) to (4.37) outputs three new linear syzygies of the partial derivatives of f. Thus one has a total of $m^2 - 5 + 3 = m^2 - 2$ linear syzygies of J.

It remains to show that the coefficients line up as the columns of a matrix of rank $m^2 - 2$.

For this we order the set of partial derivatives $f_{i,j}$ with respect to the following ordered list of the entries $x_{i,j}$:

$$x_{1,1}, x_{1,2}, \ldots, x_{1,m} \rightsquigarrow x_{2,1}, x_{2,2}, \ldots, x_{2,m} \rightsquigarrow \cdots \rightsquigarrow x_{m-2,1}, x_{m-2,2} \ldots, x_{m-2,m},$$

$$x_{m-1,1}, x_{m,1} \rightsquigarrow x_{m-1,2}, x_{m,2} \rightsquigarrow \cdots \rightsquigarrow x_{m-1,m-1}, x_{m,m-1} \rightsquigarrow x_{m-1,m}.$$

Here we traverse the entries along the matrix rows, left to right, starting with the first row and stopping prior to the row having $x_{m-1,m-1}$ as an entry; then start traversing the last two rows along its columns top to bottom, until exhausting all variables.

CLAIM With the above ordering, the previous $m^2 - 2$ linear relations can be grouped into the following block matrix of linear syzygies of the set of partial derivatives $f_{i,j}$:

$$
\left(
\begin{array}{cccc|cccc|ccc}
\varphi_1 \\
\mathbf{0} & \varphi_2 & \cdots \\
\vdots & \vdots & \ddots & \vdots \\
\mathbf{0} & \mathbf{0} & \cdots & \varphi_{m-2} \\
\hline
\mathbf{0}_2^{m-1} & \mathbf{0}_2^m & \cdots & \mathbf{0}_2^m & \varphi_2^3 \\
\mathbf{0}_2^{m-1} & \mathbf{0}_2^m & \cdots & \mathbf{0}_2^m & \mathbf{0}_2^2 & \varphi_3^4 \\
\vdots & \vdots & \cdots & \vdots & \vdots & \vdots & \ddots \\
\mathbf{0}_2^{m-1} & \mathbf{0}_2^m & \cdots & \mathbf{0}_2^m & \mathbf{0}_2^2 & \mathbf{0}_2^2 & \cdots & \varphi_{m-2}^{m-1} \\
\mathbf{0}_2^{m-1} & \mathbf{0}_2^m & \cdots & \mathbf{0}_2^m & \mathbf{0}_2^2 & \mathbf{0}_2^2 & \cdots & \mathbf{0}_2^2 & \varphi_{m-1}^m \\
\hline
\mathbf{0}_1^{m-1} & \mathbf{0}_1^m & \cdots & \mathbf{0}_1^m & \mathbf{0}_1^2 & \mathbf{0}_1^2 & \cdots & \mathbf{0}_1^2 & \mathbf{0}_1^2 & x_{m-1,m} & x_{m,m-1} & x_{m-1,m-1} \\
\mathbf{0}_1^{m-1} & \mathbf{0}_1^m & \cdots & \mathbf{0}_1^m & \mathbf{0}_1^2 & \mathbf{0}_1^2 & \cdots & \mathbf{0}_1^2 & \mathbf{0}_1^2 & 2x_{m-1,m-1} & \mathbf{0} & x_{m,m-1} \\
\mathbf{0}_1^{m-1} & \mathbf{0}_1^m & \cdots & \mathbf{0}_1^m & \mathbf{0}_1^2 & \mathbf{0}_1^2 & \cdots & \mathbf{0}_1^2 & \mathbf{0}_1^2 & \mathbf{0} & 2x_{m-1,m-1} & x_{m-1,m}
\end{array}
\right)
$$

Here the blocks are as follows:

- φ_1 is the matrix obtained from the transpose $\mathbf{L}(m)^t$ of $\mathbf{L}(m)$ by omitting the first column.
- $\varphi_2, \ldots, \varphi_{m-2}$ are each a copy of $\mathbf{L}(m)^t$ (up to column permutation).
- $\varphi_r^{r+1} = \begin{pmatrix} x_{m-1,r} & x_{m-1,r+1} \\ x_{m,r} & x_{m,r+1} \end{pmatrix}, \ r = 2, \ldots, m-2; \ \varphi_{m-1}^m = \begin{pmatrix} x_{m-1,m-1} & x_{m-1,m} \\ x_{m,m-1} & x_{m-1,m-1} \end{pmatrix}.$
- Each $\mathbf{0}$ under φ_1 is an $m \times (m-1)$ block of zeros, and each $\mathbf{0}$ under φ_i is an $m \times m$ block of zeros, for $i = 2, \ldots, m-3$.
- $\mathbf{0}_r^c$ denotes an $r \times c$ block of zeros, for $r = 1, 2$ and $c = 2, m-1, m$.
- The empty slots are filled with unspecified entries.

To see the claim, first, as already observed, the relations (4.26) through (4.37) yield linear syzygies of the partial derivatives of f. Setting $k = 1$ in the relations (4.26) and (4.27), the resulting expressions can be written, respectively, as $\sum_{j=1}^m x_{i,j} f_{1,j} = 0$ for all $i = 2, \ldots, m-1$ and $\sum_{j=1}^{m-1} x_{m,j} f_{1,j} + x_{m-1,m-1} f_{1,m} = 0$. Ordering the set of partial derivatives $f_{i,j}$ as explained before, the coefficients of these relations form the first matrix above

$$
\varphi_1 := \begin{pmatrix}
x_{2,1} & x_{3,1} & \cdots & x_{m-1,1} & x_{m,1} \\
\vdots & \vdots & \cdots & \vdots & \vdots \\
x_{2,m-1} & x_{3,m-1} & \cdots & x_{m-1,m-1} & x_{m,m-1} \\
x_{2,m} & x_{3,m} & \cdots & x_{m-1,m} & x_{m-1,m-1}
\end{pmatrix}.
$$

Note that φ_1 coincides indeed with the submatrix of $\mathbf{L}(m)^t$ obtained by omitting its first column.

Getting φ_k, for $k = 2, \ldots, m-2$, is similar, namely, use again relations (4.26) and (4.27) retrieving a submatrix of $\mathbf{L}(m)^t$ excluding the kth column and replacing it with an extra column that comes from relation (4.28) taking $i = k-1$.

Continuing, for each $r = 2, \ldots, m - 2$, the block φ_r^{r+1} comes from the relation (4.29) (setting $j = r - 1$) and φ_{m-1}^m comes from the relations (4.30) and (4.31). Finally, the lower right corner 3×3 block of the matrix of linear syzygies comes from the three last relations obtained by adding (4.32) to (4.33), (4.34) to (4.35), and (4.36) to (4.37).

This proves the claim about the large matrix above. Counting through the sizes of the various blocks, one sees that this matrix is $(m^2 - 1) \times (m^2 - 2)$. Omitting its first row gives a block diagonal submatrix of size $(m^2 - 2) \times (m^2 - 2)$, where each block has nonzero determinant. Thus, the linear rank of J attains the maximum. $\square$

Remark 4.22

(1) Not only the proof of item (a) of Theorem 4.21 requires $\mathrm{char}(k) \neq 2$. The result itself crumbs down in characteristic two. For example, for $m = 3$, the partial derivatives of f are algebraically dependent over k.

(2) A more conceptual proof of item (c) above is certainly desirable. If $\mathrm{char}(k) \neq 2$, by Theorem 4.21, (a), the partial derivatives of f are algebraically independent over k. A short argument for item (c) would follow provided we show, more strongly, that the ideal J is of linear type. In fact, since f is homaloidal by (b), then one must have maximal linear rank ([19, Corollary 3.2.27.]). We bet on the linear type property for J, but at the moment evidence is purely computational. For a couple of reasons, we also believe that the ideal $P = (\boldsymbol{\Delta})$ is of fiber type and actually ask whether this would trigger the linear type property for J.

Yet, we warn against blind generalizations of this particular environment, as the following iterated procedure dashes both of the above expectations:

$$\begin{pmatrix} x_{1,1} & x_{1,2} & x_{1,3} \\ x_{2,1} & x_{1,1} & x_{2,3} \\ x_{3,1} & x_{3,2} & x_{1,1} \end{pmatrix}.$$

Here J still has codimension 4. It admits a nontrivial relation of bidegree $(1, s)$, with $s \geq 2$, and hence is not of linear type. Moreover, this relation along with the relations coming from the syzygies of J generates the presentation ideal of the Rees algebra of J. On the other hand, its linear rank is 2. Then, although J has maximal analytic spread, f is not homaloidal either. We note that, in this example, $\boldsymbol{\Delta}$ defines a birational map onto a codimension two complete intersection and is still of fiber type. Thus, the above expected implication, if true, will be quite tight.

Corollary 4.23 $(\mathrm{char}(k) = 0)$ *If* $\lambda \in k[x_{i,j}; 1 \leq i, j \leq m, (i, j) \neq (m, m)]$ *is a general 1-form and* $J = J_f$, *with* $f = \det \mathbf{L}(m)_\lambda$, *then* $\ell(J) = m^2 - 1$.

Proof Write $\lambda := \sum_{(i,j) \neq (m,m)} \lambda_{i,j} x_{i,j}$ and represent the set of coefficients by a point $(\lambda_{i,j}) = (\lambda_{1,1} : \cdots : \lambda_{m-1,m-1})$ in parameter space $\mathbb{P}^{m^2-2}$. Since the Hessian of the corresponding $f = \det \mathbf{L}(m)_\lambda$ has the form $\sum_{|\alpha|=\deg h} p_\alpha(\lambda) \mathbf{X}^\alpha$ for suitable homogeneous polynomials in the $\lambda_{i,j}$'s, it follows that its vanishing corresponds to a

Zariski closed set in parameter space $\mathbb{P}^{m^2-2}$. Therefore, it suffices to find one single point $(\lambda_{i,j}) \in \mathbb{P}^{m^2-2}$ not belonging to this closed set. Theorem 4.21 provides us with such a point, namely, $(0 : \cdots : 0 : 1)$. $\qquad\square$

Remark 4.24

(a) It may be worth pointing out that the R-submodule generated by the linear syzygies of J is not R-free, i.e., its minimal number of generators is larger than its rank $m^2 - 2$.
(b) The above corollary fails for special values of λ, e.g., $\lambda = 0$, and, by the same token, $\lambda = x_{m,j}$ or $\lambda = x_{1,m}$, with $i, j \neq m$.

4.4.3 Diagonal One Term Iteration

In this short part we explain in which direction the previous results could be extended, for which we prove a few preliminary facts. Unfortunately, some of the expected deeper results are in a conjectural form.

For every $1 \leq t \leq m$, consider the following linear section of the $m \times m$ generic matrix:

$$\mathbf{L}(m, t) = \begin{pmatrix}
x_{1,1} & \cdots & x_{1,m-2} & x_{1,m-1} & x_{1,m} \\
\vdots & \ddots & \vdots & \vdots & \vdots \\
x_{t-2,1} & \cdots & x_{t-2,m-2} & x_{t-2,m-1} & x_{t-2,m} \\
x_{t-1,1} & \cdots & x_{t-1,m-2} & x_{t-1,m-1} & x_{t-1,m} \\
\hline
x_{t,1} & \cdots & x_{t,m-2} & x_{t,m-1} & x_{t,m} \\
x_{t+1,1} & \cdots & x_{t+1,m-2} & x_{t+1,m-1} & x_{t,m-1} \\
\vdots & \ddots & \vdots & \vdots & \vdots \\
x_{m-1,1} & \cdots & x_{m-1,m-2} & x_{m-1,m-1} & x_{m-2,m-1} \\
x_{m-1,1} & \cdots & x_{m-1,m-2} & x_{m,m-1} & x_{m-1,m-1}
\end{pmatrix}. \tag{4.38}$$

Note that for $t = m$ we just have the fully generic matrix, while for $t = m - 1$ we recover the one term section $\mathbf{L}(m)_{x_{m,m}}$.

We assume that $m \geq 3$ throughout and fix the following notation:

- R denotes the polynomial ring over k in the distinct entries of $\mathbf{L}(m, t)$ (in particular, $\dim R = m(m - 1) + t$).
- $f := \det \mathbf{L}(m, t)$.
- J denotes the gradient ideal of f.
- $f_{i,j}$ denotes the $x_{i,j}$-derivative of f.
- $\mathbf{f} = \{f_{i,j} \mid (i, j)\}$.
- $\Delta_{i,j}$ stands for the (i, j)th entry of the adjugate matrix $\operatorname{adj} \mathbf{L}(m, t)$.
- $\boldsymbol{\Delta} = \{\Delta_{i,j} \mid 1 \leq i, j \leq m\}$.
- $P := I_{m-1}(\mathbf{L}(m, t)) = (\boldsymbol{\Delta}) \subset R$.

Proposition 4.25 *With the above notation, one has*:

(i) *R/P is a Gorenstein ring of codimension* 4 *with minimal graded free resolution*

$$0 \to R(-2m) \to R(-(m+1))^{m^2} \to R(-m)^{2(m^2-1)}$$
$$\to R(-(m-1))^{m^2} \to R \to R/P \to 0.$$

(ii) *P has maximal analytic spread.*

(iii) *The rational map $\mathbb{P}^{m^2-t} \dashrightarrow \mathbb{P}^{m^2-1}$ defined by the entries $\Delta_{i,j}$ of the adjugate matrix* $\operatorname{adj} \mathbf{L}(m,t)$ *is a birational map onto the projective variety whose homogeneous defining ideal contains the forms*

$$\Delta_{j,m-1}(\mathbf{Y}) - \Delta_{j+1,m}(\mathbf{Y}) \in k[y_{i,j};\ 1 \le i, j \le m] = k[\mathbf{Y}]$$

for every $t \le j \le m-1$, where $\Delta_{i,j}(\mathbf{Y})$ is the (i,j)th cofactor of the $m \times m$ generic matrix $\mathbf{Y} = (y_{i,j})_{1\le i,j\le m}$.

Proof

(i) Iterating, it suffices to consider the extreme case where $t = 1$. Up to column and variable permutation, $\mathbf{L}(m,1) = C(m, m-1)$ in the notation of Section 7.3, in which case P is a Gorenstein ideal of codimension 4 by Theorem 7.6 (ii) in that section (which is proved independently). The shape of the resolution is clear since it is the same as that of the fully generic case.

(ii) This follows from Proposition 2.3 since $f \ne 0$ (by, e.g., Corollary 7.4, because again it suffices to consider the case where $t = 1$).

(iii) The assertion proceeds as in the claim of the proof of Proposition 4.16 (d), one for each repeated term, and by observing the particular form of the terms in this situation.   □

Proposition 4.26 (Conjectured) *With the above notation, one has*:

(i) *If $m \ge 4$, then R/P is a normal domain.*
(ii) *f is homaloidal.*

Note that, in this range, the domain part of the above conjecture is actually conjectured as well in Theorem 7.6 (iv).

4.5   Semi-Hollow Linear Sections

In this section we deal with a special class of semi-hollow linear sections of the square generic matrix. By their shape, they could be called *sub-generic semi-hollow*, but to ease up on terminology we keep calling them simply as semi-hollow. As such, they will also be special cases of the hollow-filled matrices introduced in Section 4.2.

To wit, fixing integers m, r with $1 \le r \le m - 2$, consider the following hollow-filled linear section of the $m \times m$ generic matrix $\mathbf{X}$:

$$
\begin{pmatrix}
x_{1,1} & \cdots & x_{1,m-r} & x_{1,m-r+1} & x_{1,m-r+2} & \cdots & x_{1,m-1} & x_{1,m} \\
\vdots & \ddots & \vdots & \vdots & \vdots & \ddots & \vdots & \vdots \\
x_{m-r,1} & \cdots & x_{m-r,m-r} & x_{m-r,m-r+1} & x_{m-r,m-r+2} & \cdots & x_{m-r,m-1} & x_{m-r,m} \\
x_{m-r+1,1} & \cdots & x_{m-r+1,m-r} & x_{m-r+1,m-r+1} & x_{m-r+1,m-r+2} & \cdots & x_{m-r+1,m-1} & 0 \\
x_{m-r+2,1} & \cdots & x_{m-r+2,m-r} & x_{m-r+2,m-r+1} & x_{m-r+2,m-r+2} & \cdots & 0 & 0 \\
\vdots & \ddots & \vdots & \vdots & \vdots & \ddots & \vdots & \vdots \\
x_{m-1,1} & \cdots & x_{m-1,m-r} & x_{m-1,m-r+1} & 0 & \cdots & 0 & 0 \\
x_{m,1} & \cdots & x_{m,m-r} & 0 & 0 & \cdots & 0 & 0
\end{pmatrix}. \tag{4.39}
$$

This matrix will be denoted $\mathbf{L}(m)[r]$, the generic matrix $\mathbf{X}$ being subsumed in the discussion. Note that, due to the special positioning of the zeros, the transpose matrix is of the same kind.

Remark 4.27 We observe that $\mathbf{L}(m)[r]$ is a distinguished shape among the so-called sparse matrices. The latter have been the subject of previous work. The priority in investigating this sort of problem is attributed to [9] (see also [12, Section 6], and [1, 5] for more recent developments with a focus on the maximal minors).

In principle, the distribution of zeros (sparsing) can be random like. However, for the sake of further insight into subtler algebraic and geometric features, such as Fitting ideals and polar map, the distribution may better be strategically located.

4.5.1 Preliminaries

Throughout, $R = k[\mathbf{X}] := k[x_{1,1}, \ldots, x_{m,m-r}]$, with $1 \le r \le m - 2$.

Proposition 4.28 *Let* $f := \det \mathbf{L}(m)[r]$ *and* $P := I_{m-1}(\mathbf{L}(m)[r])$. *Then:*

(i) *f is irreducible.*

(ii) *P is a Gorenstein ideal of codimension 4 and maximal analytic spread.*

(iii) *The $(m - 1)$-minors of $\mathbf{L}(m)[r]$ define a birational map $\mathbb{P}^{m^2 - \binom{r+1}{2} - 1} \dashrightarrow (\mathbb{P}^{m^2 - 1})^{\vee}$ onto its image.*

(iv) *Let $\mathcal{L}_r(m)$ denote the following ladder of $\mathbf{L}(m)[r]$:*

$$
\begin{pmatrix}
x_{1,1} & \cdots & x_{1,m-r} & x_{1,m-r+1} & \cdots & x_{1,m-1} & x_{1,m} \\
\vdots & \vdots & \vdots & \vdots & \vdots & \vdots & \vdots \\
x_{m-r,1} & \cdots & x_{m-r,m-r} & x_{m-r,m-r+1} & \cdots & x_{m-r,m-1} & x_{m-r,m} \\
x_{m-r+1,1} & \cdots & x_{m-r+1,m-r} & x_{m-r+1,m-r+1} & \cdots & x_{m-r+1,m-1} & 0 \\
\vdots & \vdots & \vdots & \vdots & \ddots & \iddots & \vdots \\
x_{m-1,1} & \cdots & x_{m-1,m-r} & x_{m-1,m-r+1} & \cdots & 0 & 0 \\
x_{m,1} & \cdots & x_{m,m-r} & 0 & \cdots & 0 & 0
\end{pmatrix}
$$

Then the ideal $I_{m-r}(\mathcal{L}_r(m))$ of $(m-r)$ minors of $\mathbf{L}(m)[r]$ involving but the entries of $\mathcal{L}_r(m)$ is a Gorenstein prime ideal of height $\binom{r+1}{2}$.

Proof

(i) Note that for an arbitrary $1 \le r \le m-2$, $f = \det \mathbf{L}(m)[r]$ maps homomorphically to $\det \mathbf{L}(m)[m-2]$ by an obvious map of the defining ground ring. Thus, it suffices to show that $f = \det \mathbf{L}(m)[m-2]$ is irreducible and nonzero. Laplace expansion of the determinant $f = \det \mathbf{L}(m)[m-2]$ along the first row yields the following $x_{1,1}$-monoid:

$$f = \left(\prod_{i+j=m+2} x_{i,j} \right) x_{1,1} + H.$$

Thus, we need to show that no variable $x_{i,j}$ with $i+j = m+2$ divides H. The latter is however clear by the following modular expression:

$$f \equiv H \equiv \prod_{i+j=m+1} x_{i,j} \not\equiv 0 \mod (x_{i,j} \mid i+j = m+2),$$

from which also obviously follows that $f \ne 0$.

(ii) That P is Gorenstein of codimension 4 follows from Proposition 2.4, which shows that P is a specialization of the ideal of generic submaximal minors, provided we argue that the set of entries of the generic $m \times m$ matrix that degenerated in zeros in $\mathbf{L}(m)[r]$ is a subset of the set X as in that proposition. But this is immediate because of the assumption $r \le m-2$.

As for the analytic spread, the assertion follows immediately from Proposition 2.3.

(iii) As in the proof of (ii), P is a specialization of the ideal of submaximal minors in the generic case; in particular, it is linearly presented. By (ii), its analytic spread is maximal. Therefore, by Theorem 2.27, the minors define a birational map onto the image.

(iv) First, check the height. Let $\mathcal{L}_{r-1}(m-1)$ denote the same ladder considered as a substructure of the $(m-1) \times (m-1)$ semi-hollow matrix $\mathbf{L}(m-1)[r-1]$ with entries in the polynomial subring $k[x_{1,1}, \ldots, x_{m-1,m-r+1}] \subset k[\mathbf{X}]$. Note that this ladder has $r-1$ inside corners; hence in the notation of Section 4.1.1, one has

$$I_{m-1-(r-1)}(\mathcal{L}_{r-1}(m-1)) = \mathfrak{g}_{m-1-(r-1)}.$$

But the latter has height $\binom{(r-1)+2}{2} = \binom{r+1}{2}$ by Theorem 4.1 (i). To conclude, one notes that the extended ideal of $\mathcal{L}_{r-1}(m-1)$ in the ring $k[\mathbf{X}]$ is $I_{m-r}(\mathcal{L}_r(m))$; hence, the height does not change.

Primality and the Gorenstein property have also been given in
Theorem 4.1. $\square$

Remark 4.29 The statement of item (ii) in Proposition 4.28 depends not only on
the number of the entries forming a regular sequence but also on their mutual
position. Thus, for example, if more than r of the entries belong to one same column
or row, it may be the case that I has codimension strictly less than 4.

4.5.2 The Structure of the Polar Image

Recall from Section 3.1 that $\mathfrak{P}(f)$ denotes the polar image of a homogeneous
polynomial f in a standard graded polynomial ring over a field. In the following
theorem we take $f := \det \mathbf{L}(m)[r]$, so $\mathfrak{P}(f) \subset (\mathbb{P}^{m^2 - \binom{r+1}{2} - 1})^{\vee}$, the projective space
in the dual variables $y_{i,j}$.

Theorem 4.30 *Let* $f := \det \mathbf{L}(m)[r]$ *and let* $J = J_f \subset R$ *denote the gradient
ideal of* f. *Then*:

(a) *The homogeneous defining ideal of* $\mathfrak{P}(f) \subset (\mathbb{P}^{m^2 - \binom{r+1}{2} - 1})^{\vee}$ *is* $I_{m-r}(\mathcal{L}_r(m))$
read in the dual variables.

(b) *J has maximal linear rank.*

Proof (a) Let $\mathfrak{J} \subset S := k[y_{1,1}, \ldots, y_{m,m-r}]$ denote the homogeneous defining
ideal of the polar image $\mathfrak{P}(f)$ in its embedding in $\mathbb{P}^{m^2 - \binom{r+1}{2} - 1}$.
  We argue as follows.
  CLAIM 1 $I_{m-r}(\mathcal{L}_r(m)) \subset \mathfrak{J}$.
  To see this, let $x_{i,j}$ denote a nonzero entry of $\mathbf{L}(m)[r]$. Since the nonzero entries
of the matrix are independent variables, it follows from Proposition 2.10 that the
$x_{i,j}$-derivative $f_{i,j}$ of f coincides with the cofactor of $x_{i,j}$, heretofore denoted $\Delta_{i,j}$.
  Given integers $1 \leq i_1 < i_2 < \ldots < i_{m-r} \leq m - 1$, consider the following
submatrix of the transpose matrix of cofactors:

$$
F = \begin{pmatrix}
\Delta_{i_1,1} & \Delta_{i_1,2} & \Delta_{i_1,3} & \cdots & \Delta_{i_1, m-i_{m-r}+(m-r-1)} \\
\Delta_{i_2,1} & \Delta_{i_2,2} & \Delta_{i_2,3} & \cdots & \Delta_{i_2, m-i_{m-r}+(m-r-1)} \\
\vdots & \vdots & \vdots & \cdots & \vdots \\
\Delta_{i_{m-r},1} & \Delta_{i_{m-r},2} & \Delta_{i_{m-r},3} & \cdots & \Delta_{i_{m-r}, m-i_{m-r}+(m-r-1)}
\end{pmatrix}.
$$

Letting

$$C = \begin{pmatrix} x_{1,i_{m-r}+1} & x_{1,i_{m-r}+2} & \cdots & x_{1,m-1} & x_{1,m} \\ \vdots & \vdots & \cdots & \vdots & \vdots \\ x_{m-r,i_{m-r}+1} & x_{m-r,i_{m-r}+2} & \cdots & x_{m-r,m-1} & x_{m-r,m} \\ x_{m-r+1,i_{m-r}+1} & x_{m-r+1,i_{m-r}+2} & \cdots x_{m-r+1,m-1} & 0 \\ \vdots & \vdots & \cdots & \vdots & \vdots \\ x_{m-i_{m-r}+(m-r-2),i_{m-r}+1} & x_{m-i_{m-r}+(m-r-2),i_{m-r}+2} & \cdots & 0 & 0 \\ x_{m-i_{m-r}+(m-r-1),i_{m-r}+1} & 0 & \cdots & 0 & 0 \end{pmatrix},$$

the adjugate relation $\mathrm{adj}\, \mathbf{L}(m)[r]\, \mathbf{L}(m)[r] = \det \mathbf{L}(m)[r]\, \mathbb{I}_m$ yields the relation

$$FC = 0.$$

Since the columns of C are linearly independent, it follows that the rank of F is at most $m - i_{m-r} + (m - r - 1) - (m - i_{m-r}) = (m - r) - 1$. In other words, the maximal minors of the following matrix

$$\begin{pmatrix} y_{i_1,1} & y_{i_1,2} & y_{i_1,3} & \cdots & y_{i_1,m-i_{m-r}+(m-r-1)} \\ y_{i_2,1} & y_{i_2,2} & y_{i_2,3} & \cdots & y_{i_2,m-i_{m-r}+(m-r-1)} \\ \vdots & \vdots & \vdots & \cdots & \vdots \\ y_{i_{m-r},1} & y_{i_{m-r},2} & y_{i_{m-r},3} & \cdots & y_{i_{m-r},m-i_{m-r}+(m-r-1)} \end{pmatrix}.$$

all vanish on the partial derivatives of f via the map $y_{i,j} \mapsto \partial f/\partial x_{i,j}$, thus proving the claim.

CLAIM 2 $\mathrm{ht}\, \mathfrak{J} \leq \binom{r+1}{2}$.

To see this, let $\mathfrak{D} \subset T$ stand for the homogeneous defining ideal of the image of the rational map defined by the cofactors $\Delta_{i,j}$ of $\mathbf{L}(m)[r]$. By Proposition 2.10, we have $\partial f/\partial x_{i,j} = \Delta_{i,j}$, so $\mathfrak{J}T \subset \mathfrak{D}$ as ideals of T, where T is a polynomial ring over S. Hence,

$$\mathrm{ht}\, \mathfrak{J} = \mathrm{ht}\, \mathfrak{J}T \leq \mathrm{ht}\, \mathfrak{D}.$$

But $\mathrm{ht}\, \mathfrak{D} = m^2 - (m^2 - \binom{r+1}{2}) = \binom{r+1}{2}$, by Proposition 2.3. This proves Claim 2.

The statement of the item is now a consequence of Claims 1 and 2 via Proposition 4.28.

(c) The proof is similar to the one of Theorem 4.21 (c), but there is a numerical diversion and, besides, the cases where $r > m - r - 1$ and $r \leq m - r - 1$ have slight differences.

Let $f_{i,j}$ denote the $x_{i,j}$-derivative of f and let $\Delta_{i,j}$ stand for the cofactor of $x_{i,j}$ in $\mathbf{L}(m)[r]$. We first assume that $r > m - r - 1$. Once more the adjugate formula (2.1) for $\mathbf{L}(m)[r]$ yields by expansion the following three blocks of linear relations involving the cofactors of $\mathbf{L}(m)[r]$:

$$\begin{cases} \sum_{j=1}^{m} x_{i,j}\Delta_{j,k} = 0, & \text{for } 1 \le i \le m-r,\ 1 \le k \le m-r\ (k \ne i) \\ \sum_{j=1}^{m-l} x_{m-r+l,j}\Delta_{j,k} = 0, & \text{for } 1 \le l \le r,\ 1 \le k \le m-r \\ \sum_{j=1}^{m} x_{i,j}\Delta_{j,i} - \sum_{j=1}^{m} x_{i+1,j}\Delta_{j,i+1} = 0, & \text{for } 1 \le i \le m-r-1 \end{cases} \tag{4.40}$$

with $m^2 - rm - 1$ such relations;

$$\begin{cases} \sum_{i=1}^{m} x_{i,j}\Delta_{k,i} = 0, & \text{for } 1 \le j \le m-r,\ 1 \le k \le m-r\ (k \ne j) \\ \sum_{i=1}^{m-l} x_{i,m-r+l}\Delta_{k,i} = 0, & \text{for } 1 \le l \le 2r-m+1,\ 1 \le k \le m-r \end{cases} \tag{4.41}$$

with $(m-r)(m-r-1) + (m-r)(2r-m+1) = r(m-r)$ such relations; and

$$\begin{cases} \sum_{i=1}^{m-l} x_{i,m-r+l}\Delta_{m-r+l,i} - \sum_{j=1}^{m} x_{i,1}\Delta_{1,i} = 0, & \text{for } 1 \le l \le r-1 \\ \sum_{i=1}^{m-k} x_{i,m-r+k}\Delta_{m-r+l,i} = 0, & \text{for } 1 \le l \le r-2,\ l+1 \le k \le r-1 \end{cases} \tag{4.42}$$

with $r(r-1)/2$ such relations.

When $r \le m - r - 1$, the adjugate formula similarly outputs three blocks of linear relations involving the cofactors of $\mathcal{D}$. Here, the first and third blocks are, respectively, exactly as the above ones, while the second one requires some change due to the inequality reversal; namely, we get

$$\begin{cases} \sum_{i=1}^{m} x_{i,j}\Delta_{k,i} = 0 \text{ for } 1 \le j \le r+1,\ 1 \le k \le r\ (k \ne j) \\ \sum_{i=1}^{m} x_{i,j}\Delta_{k,i} = 0 \text{ for } 1 \le j \le r,\ r+1 \le k \le m-r, \end{cases} \tag{4.43}$$

with $r(m-r)$ such relations (as before).

Since $f_{i,j}$ coincides with the cofactor $\Delta_{i,j}$, any of the above relations gives a linear syzygy of the partial derivatives of f. Thus one has a total of $m^2 - rm - 1 + r(m-r) + r(r-1)/2 = m^2 - \binom{r+1}{2} - 1$ linear syzygies of J.

It remains to show that these are independent.

For this, we adopt the same strategy as in the proof of Theorem 4.21 (c), whereby we list the partial derivatives according to the following ordering of the nonzero entries: traverse the first row from left to right, then the second row in the same way, and so on until reaching the last row with no zero entry; thereafter start from the first row having a zero and travel along the columns, from left to right, on each column from top to bottom, till all nonzero entries are counted in.

Thus, the desired ordering is depicted in the following scheme, where we once more used arrows for easy reading:

$$x_{1,1}, x_{1,2}, \ldots, x_{1,m} \rightsquigarrow x_{2,1}, x_{2,2}, \ldots,$$

$$x_{2,m} \rightsquigarrow \cdots \rightsquigarrow x_{m-r,1}, x_{m-r,2} \ldots, x_{m-r,m} \rightsquigarrow$$

$$x_{m-r+1,1}, \ldots, x_{m,1} \rightsquigarrow x_{m-r+1,2}, \ldots, x_{m,2}$$

$$\rightsquigarrow \cdots \rightsquigarrow x_{m-r+1,m-r}, x_{m-r+2,m-r}, \ldots, x_{m,m-r}$$

$$\rightsquigarrow x_{m-r+1,m-r+1}, \ldots, x_{m-1,m-r+1} \rightsquigarrow x_{m-r+1,m-r+2}, \ldots, x_{m-2,m-r+2}$$

$$\rightsquigarrow \cdots \rightsquigarrow x_{m-r+1,m-2}, x_{m-r+2,m-2} \rightsquigarrow x_{m-r+1,m-1}.$$

With this ordering the above linear relations translate into linear syzygies collected in the following block matrix:

$$
\mathcal{M} =
\left(
\begin{array}{cccc|ccc|ccc}
\varphi_1 & \cdots & & & & & & & & \\
\mathbf{0} & \varphi_2 & \cdots & & & & & & & \\
\vdots & \vdots & \ddots & \vdots & & & & & & \\
\mathbf{0} & \mathbf{0} & \cdots & \varphi_{m-r} & & & & & & \\
\hline
\mathbf{0}_r^{m-1} & \mathbf{0}_r^m & \cdots & \mathbf{0}_r^m & \varphi_r^1 & & & & & \\
\mathbf{0}_r^{m-1} & \mathbf{0}_r^m & \cdots & \mathbf{0}_r^m & \mathbf{0}_r^r & \varphi_r^2 & & & & \\
\vdots & \vdots & \cdots & \vdots & \vdots & \vdots & \ddots & & & \\
\mathbf{0}_r^{m-1} & \mathbf{0}_r^m & \cdots & \mathbf{0}_r^m & \mathbf{0}_r^r & \mathbf{0}_r^r & \cdots & \varphi_r^{(m-r)} & & \\
\hline
\mathbf{0}_{r-1}^{m-1} & \mathbf{0}_{r-1}^m & \cdots & \mathbf{0}_{r-1}^m & \mathbf{0}_{r-1}^r & \mathbf{0}_{r-1}^r & \cdots & \mathbf{0}_{r-1}^r & \Phi_1 & \\
\mathbf{0}_{r-2}^{m-1} & \mathbf{0}_{r-2}^m & \cdots & \mathbf{0}_{r-2}^m & \mathbf{0}_{r-2}^r & \mathbf{0}_{r-2}^r & \cdots & \mathbf{0}_{r-2}^r & \mathbf{0}_{r-2}^{r-1} & \Phi_2 \\
\vdots & \vdots & \vdots & \vdots & \vdots & \vdots & \vdots & \vdots & \vdots & \vdots & \ddots \\
\mathbf{0}_1^{m-1} & \mathbf{0}_1^m & \cdots & \mathbf{0}_1^m & \mathbf{0}_1^r & \mathbf{0}_1^r & \cdots & \mathbf{0}_1^r & \mathbf{0}_1^{r-1} & \mathbf{0}_1^{r-2} & \cdots & \Phi_{r-1}
\end{array}
\right),
$$

where:

- φ_1 is the matrix obtained from the transpose $\mathbf{L}(m)[r]^t$ of $\mathbf{L}(m)[r]$ by omitting the first column.
- $\varphi_2, \ldots, \varphi_{m-r}$ are each a copy of $\mathbf{L}(m)[r]^t$ (up to column permutation).
- When $r > m - r - 1$, setting $\eth := 2r - m + 1$, for $i = 1, \ldots, m - r$, one has that φ_r^i is the $r \times r$ submatrix omitting the ith column of the following submatrix of $\mathbf{L}(m)[r]$:

$$
\begin{pmatrix}
x_{m-r+1,1} & \cdots & x_{m-r+1,m-r} & x_{m-r+1,m-r+1} & x_{m-r+1,m-r+2} & \cdots & x_{m-r+1,r+1} \\
\vdots & \vdots & \vdots & \vdots & \vdots & \vdots & \vdots \\
x_{m-\eth,1} & \cdots & x_{m-\eth,m-r} & x_{m-\eth,m-r+1} & x_{m-\eth,m-r+2} & \cdots & x_{m-\eth,r+1} \\
x_{m-\eth+1,1} & \cdots & x_{m-\eth+1,m-r} & x_{m-\eth+1,m-r+1} & x_{m-\eth+1,m-r+2} & \cdots & 0 \\
\vdots & \vdots & \vdots & \vdots & \vdots & \vdots & \vdots \\
x_{m-1,1} & \cdots & x_{m-1,m-r} & x_{m-1,m-r+1} & 0 & \cdots & 0 \\
x_{m,1} & \cdots & x_{m,m-r} & 0 & 0 & \cdots & 0
\end{pmatrix}.
$$

When $r \le m - r - 1$, consider the following submatrix of $\mathbf{L}(m)[r]$:

$$
\begin{pmatrix}
x_{m-r+1,1} & \cdots & x_{m-r+1,r} & x_{m-r+1,r+1} \\
\vdots & \vdots & \vdots & \vdots \\
x_{m,1} & \cdots & x_{m,r} & x_{m,r+1}
\end{pmatrix}.
$$

Then, for $i = 1, \ldots, r$ (respectively, for $i = r + 1, \ldots, m - r$), φ_r^i denotes the $r \times r$ submatrix obtained by omitting the ith column (respectively, the last column).

- Each $\mathbf{0}$ under φ_1 is an $m \times (m - 1)$ block of zeros, and each $\mathbf{0}$ under φ_i is an $m \times m$ block of zeros for $i = 2, \ldots, m - r - 1$.
- $\mathbf{0}_l^c$ denotes an $l \times c$ block of zeros.
- Φ_i is the following $(r - i) \times (r - i)$ submatrix of $\mathbf{L}(m)[r]$:

$$\Phi_i = \begin{pmatrix} x_{m-r+1,m-r+i} & x_{m-r+1,m-r+i+1} & \cdots & x_{m-r+1,m-1} \\ x_{m-r+2,m-r+i} & x_{m-r+2,m-r+i+1} & \cdots & 0 \\ \vdots & \vdots & \vdots & \vdots \\ x_{m-i,m-r+i} & 0 & \cdots & 0 \end{pmatrix}.$$

Next we justify why these blocks make up (linear) syzygies. As already explained, the relations in (4.40), (4.41), and (4.42) yield linear syzygies of the partial derivatives of f. Setting $k = 1$ in the first two relations of (4.40), the latter can be written as $\sum_{j=1}^{m} x_{i,j} f_{1,j} = 0$, for $i = 2, \ldots, m - r$, and $\sum_{j=1}^{m-l} x_{m-r+l,j} f_{1,j} = 0$, for all $l = 1, \ldots, r$. By ordering the set of partial derivatives $f_{i,j}$ as explained before, the coefficients of these relations become the entries of the submatrix of $\mathbf{L}(m)[r]^t$ obtained by omitting its first column, as mentioned above, namely:

$$\varphi_1 = \begin{pmatrix} x_{2,1} & \cdots & x_{m-r,1} & x_{m-r+1,1} & x_{m-r+2,1} & \cdots & x_{m-1,1} & x_{m,1} \\ \vdots & \ddots & \vdots & \vdots & \vdots & \ddots & \vdots & \vdots \\ x_{2,m-r} & \cdots & x_{m-r,m-r} & x_{m-r+1,m-r} & x_{m-r+2,m-r} & \cdots & x_{m-1,m-r} & x_{m,m-r} \\ x_{2,m-r+1} & \cdots & x_{m-r,m-r+1} & x_{m-r+1,m-r+1} & x_{m-r+2,m-r+1} & \cdots & x_{m-1,m-r+1} & 0 \\ x_{2,m-r+2} & \cdots & x_{m-r,m-r+2} & x_{m-r+1,m-r+2} & x_{m-r+2,m-r+2} & \cdots & & 0 & 0 \\ \vdots & \ddots & \vdots & \vdots & \vdots & \ddots & \vdots & \vdots \\ x_{2,m-1} & \cdots & x_{m-r,m-1} & x_{m-r+1,m-1} & 0 & \cdots & 0 & 0 \\ x_{2,m} & \cdots & x_{m-r,m} & 0 & 0 & \cdots & 0 & 0 \end{pmatrix}.$$

Getting φ_k, for $k = 2, \ldots, m - r$, is similar, namely, we use again the first two relations in the block (4.40) retrieving the submatrix of $\mathbf{L}(m)[r]^t$ excluding the kth column and replacing it with an extra column that comes from the last relation in (4.40) by taking $i = k - 1$.

Continuing, for each $i = 1, \ldots, m - r$ the block φ_r^i comes from the relations in the blocks (4.41), if $r > m - r - 1$, or (4.43), if $r \leq m - r - 1$, by setting $k = i$. Finally, for each $i = 1, \ldots, r - 1$, the block Φ_i comes from the relations in (4.42) by setting $l = i$.

This proves the claim about the large matrix above. Adding the sizes along the various blocks, one sees that this matrix is $(m^2 - \binom{r+1}{2}) \times (m^2 - \binom{r+1}{2} - 1)$. Omitting its first row gives a block diagonal square submatrix where each block has nonzero determinant. Thus, the linear rank of J attains the maximum value. $\square$

Remark 4.31 Once again, a more conceptual proof of item (c) above is welcome. Besides, we expect a similar result as in the case of codimension one section, namely, if $r = 1$, then J is of fiber type (best possible since it cannot be of linear type). Dashing expectations of a more general outcome, we note that, if $r = 2$, then J is not anymore of fiber type, admitting a genuine relation of bidegree $(1, 2)$.

4.5.3   Narrowing the Set of Associated Primes

In this part we aim at some understanding of the associated primes of the gradient ideal J of $\det \mathbf{L}(m)[r]$ in its close relationship to the ideal $I_{m-1}(\mathbf{L}(m)[r])$ of submaximal minors. Recall that, for any nonzero entry $x_{i,j}$ of $\mathbf{L}(m)[r]$, the $x_{i,j}$-derivative $f_{i,j}$ of f coincides with the cofactor of $x_{i,j}$ (Proposition 2.10). Understanding the colon ideal $J : I_{m-1}(\mathbf{L}(m)[r])$ plays a role in understanding the associated primes of J away from the associated primes of $I_{m-1}(\mathbf{L}(m)[r])$.

In addition, the following submatrices of $\mathbf{L}(m)[r]$ will also play a substantial role: for every $0 \le j \le m$, consider the submatrix $\mathcal{M}_j$ (respectively, $\mathcal{N}_j$) consisting of its last j columns (respectively, last j rows). The next observation will be crucial in subsequent arguments.

Remark 4.32 For every $1 \le j \le r$, the subset of the entries of $\mathbf{L}(m)[r]$ common to both $\mathcal{N}_j$ and $\mathcal{M}_{r-j+1}$ is the set of entries of the $j \times (r - j + 1)$ submatrix of $\mathbf{L}(m)[r]$, the latter being a null matrix. In particular, the respective sets of nonzero entries of $\mathcal{N}_j$ and $\mathcal{M}_{r-j+1}$ are disjoint.

We will initially focus on a couple of lemmas concerning certain strategic ideals of minors of these submatrices.

Lemma 4.33 *With the above notation, one has*

$$\operatorname{codim} I_j(\mathcal{M}_j) = \operatorname{codim} I_j(\mathcal{N}_j) = m - r,$$

for every $1 \le j \le r$.

Proof We give the argument for $I_j(\mathcal{N}_j)$, the one for $I_j(\mathcal{M}_j)$ being entirely similar. An easy counting gives that $\mathcal{N}_j$ has $r - j + 1$ null columns, so its ideal of j-minors coincides with the ideal of j-minors of its $j \times (m - (r - j + 1))$ submatrix $\mathcal{N}'_j$ with no null columns. Clearly, this ideal of (maximal) minors has codimension at most $(m - (r - j + 1)) - j + 1 = m - r$. Note the shape of $\mathcal{N}'_j$:

$$
\begin{pmatrix}
x_{m-j+1,1} & \cdots & x_{m-j+1,m-r} & & \cdots & x_{m-j+1,m-r+j-3} & x_{m-j+1,m-r+j-2} & x_{m-j+1,m-r+j-1} \\
x_{m-j+2,1} & \cdots & x_{m-j+2,m-r} & & \cdots & x_{m-j+2,m-r+j-3} & x_{m-j+2,m-r+j-2} & 0 \\
x_{m-j+3,1} & \cdots & x_{m-j+3,m-r} & & \cdots & x_{m-j+3,m-r+j-3} & 0 & 0 \\
\vdots & & \vdots & & & \vdots & \vdots & \vdots \\
x_{m-1,1} & \cdots & x_{m-1,m-r} & x_{m-1,m-r+1} & 0 & & \cdots & 0 \\
x_{m,1} & \cdots & x_{m,m-r} & 0 & 0 & & \cdots & 0
\end{pmatrix}.
$$

This matrix specializes all the way down to the $j \times (m - r + j + 1)$ matrix

$$
D := \begin{pmatrix}
0 & \cdots & 0 & x_{m,1} & & \cdots & & \cdots & x_{m,m-r-2} & x_{m,m-r-1} & x_{m,m-r} \\
\vdots & \ddots & & & & & & & x_{m,m-r-1} & x_{m,m-r} & 0 \\
0 & & \ddots & & & & & & & & \vdots \\
x_{m,1} & \cdots & & x_{m,m-r-2} & x_{m,m-r-1} & x_{m,m-r} & 0 & & \cdots & & 0
\end{pmatrix},
$$

with entries in $k[x_{m,1}, \ldots, x_{m,m-1}]$, where it is well-known (see, e.g., [2, p. 15, after Remark 2.13]) that $I_j(D) = (x_{m,1}, \ldots, x_{m,m-1})^j$. Therefore, for even more reason, $I_j(\mathcal{N}_j')$ has codimension at least $m - r$. $\square$

Lemma 4.34 *Let $\mathbf{L}(m)[r]$ be as in (4.39), with $1 \le r \le m - 2$. Then, with the above notation,*

$$
I_j(\mathcal{N}_j)I_{r-j}(\mathcal{M}_{r-j}) \subset J : I_{m-1}(\mathbf{L}(m)[r]),
$$

for every $0 \le j \le r$.

Proof For a fixed $1 \le j \le r$, we write the matrices $\mathbf{L}(m)[r]$ and its adjugate $\mathrm{adj}(\mathbf{L}(m)[r])$ in the following block form:

$$
\mathbf{L}(m)[r] = \begin{pmatrix} \widetilde{\mathcal{N}}_j \\ \mathcal{N}_j \end{pmatrix}, \quad \mathrm{adj}(\mathbf{L}(m)[r]) = \left(\begin{array}{c|c} \Theta_{1,j} & \Theta_{2,j} \\ \hline \Theta_{3,j} & \Theta_{4,j} \end{array} \right), \tag{4.44}
$$

where $\Theta_{1,j}, \Theta_{2,j}, \Theta_{3,j}, \Theta_{4,j}$ stand for submatrices of sizes $(j + m - r) \times (m - j)$, $(j + m - r) \times j$, $(r - j) \times (m - j)$ and $(r - j) \times j$, respectively. Thus, we have

$$
\mathrm{adj}(\mathbf{L}(m)[r])\mathbf{L}(m)[r] = \begin{pmatrix} \Theta_{1,j}\widetilde{\mathcal{N}}_j + \Theta_{2,j}\mathcal{N}_j \\ \Theta_{3,j}\widetilde{\mathcal{N}}_j + \Theta_{4,j}\mathcal{N}_j \end{pmatrix} = f\mathbb{I}_m. \tag{4.45}
$$

with $\mathbb{I}_m$ denoting the identity matrix of order m. Since f belongs to J, then $I_1(\Theta_{1,j}\widetilde{\mathcal{N}}_j + \Theta_{2,j}\mathcal{N}_j) \subset J$. On the other hand, the entries of $\Theta_{1,j}$ are cofactors of the entries on the upper left corner of $\mathbf{L}(m)[r]$ and hence belong to J as well. Therefore $I_1(\Theta_{2,j}\mathcal{N}_j) \subset J$ as well. From this by an easy argument, it follows that

$$I_1(\Theta_{2,j})I_j(\mathcal{N}_j) \subset J \tag{4.46}$$

and, for even more reason,

$$\boxed{I_1(\Theta_{2,j})I_j(\mathcal{N}_j)\,I_{r-j}(\mathcal{M}_{r-j}) \subset J} \tag{4.47}$$

Similarly, writing

$$\mathbf{L}(m)[r] = \left(\widetilde{\mathcal{M}}_{r-j}\big|\mathcal{M}_{r-j}\right),$$

we have

$$\mathbf{L}(m)[r]\mathrm{adj}(\mathbf{L}(m)[r]) = \left(\widetilde{\mathcal{M}}_{r-j}\Theta_{1,j} + \mathcal{M}_{r-j}\Theta_{3,j}\big|\widetilde{\mathcal{M}}_{r-j}\Theta_{2,j} + \mathcal{M}_{r-j}\Theta_{4,j}\right) = f \cdot \mathrm{Id}_m.$$

An entirely analogous reasoning leads to the inclusion $I_1(\Theta_{3,j})I_{r-j}(\mathcal{M}_{r-j}) \subset J$, and for even more reason

$$\boxed{I_1(\Theta_{3,j})I_j(\mathcal{N}_j)\,I_{r-j}(\mathcal{M}_{r-j}) \subset J} \tag{4.48}$$

Arguing now with the second block $\widetilde{\mathcal{M}}_{r-j}\Theta_{2,j} + \mathcal{M}_{r-j}\Theta_{4,j}$, again $I_1(\widetilde{\mathcal{M}}_{r-j}\Theta_{2,j} + \mathcal{M}_{r-j}\Theta_{4,j}) \subset J$, and hence for each $\delta \in I_j(\mathcal{N}_j)$, also $I_1(\delta\widetilde{\mathcal{M}}_{r-j}\Theta_{2,j} + \delta\mathcal{M}_{r-j}\Theta_{4,j}) \subset J$. But, by (4.46), the entries of $\delta\widetilde{\mathcal{M}}_{r-j}\Theta_{2,j}$ belong to J. Thus, the entries of $\delta\mathcal{M}_{r-j}\Theta_{4,j}$ belong to J and consequently

$$\boxed{I_1(\Theta_{4,j})I_j(\mathcal{N}_j)\,I_{r-j}(\mathcal{M}_{r-j}) \subset J}. \tag{4.49}$$

It follows from (4.47), (4.48), and (4.49) that

$$\left(I_1(\Theta_{2,j}),\, I_1(\Theta_{3,j}),\, I_1(\Theta_{4,j})\right)\,I_j(\mathcal{N}_j)I_{r-j}(\mathcal{M}_{r-j})) \subset J. \tag{4.50}$$

Since also $I_1(\Theta_{1,j}) \subset J$, we have

$$(I_1(\Theta_{1,j}),\, I_1(\Theta_{2,j}),\, I_1(\Theta_{3,j}),\, I_1(\Theta_{4,j}))\,I_j(\mathcal{N}_j)I_{r-j}(\mathcal{M}_{r-j})) \subset J. \tag{4.51}$$

From this equality it obtains

$$I_j(\mathcal{N}_j)I_{r-j}(\mathcal{M}_{r-j})I \subset J \tag{4.52}$$

because

$$I_{m-1}(\mathbf{L}(m)[r]) = I_1(\mathrm{adj}(\mathbf{L}(m)[r])) = (I_1(\Theta_{1,j}), I_1(\Theta_{2,j}), I_1(\Theta_{3,j}), I_1(\Theta_{4,j})).$$

This establishes the assertion above – we note that it encompasses as a special case (with $j = 0$ and $j = r$) the separate inclusions $I_r(\mathcal{M}_r)$, $I_r(\mathcal{N}_r) \subset J :$ $I_{m-1}(\mathbf{L}(m)[r])$. $\square$

Corollary 4.35 *With the above notation, one has*

$$\mathrm{codim}\,(I_j(\mathcal{N}_j), I_{r-j+1}(\mathcal{M}_{r-j+1})) = 2(m - r),$$

for every $1 \le j \le r$.

Proof This follows from Lemma 4.33 and Remark 4.32. $\square$

Next is the main theorem of this part. In contrast to previous notation, we will use the letter I to denote the ideal $I_{m-1}(\mathbf{L}(m)[r])$ of submaximal minors because P will be used for a non-specified prime ideal throughout the proof.

Theorem 4.36 *With the notation of* Lemma 4.34, *set* $I := I_{m-1}(\mathbf{L}(m)[r])$. *Then:*

(i) *The image of the birational map* $\mathbb{P}^{m^2-\binom{r+1}{2}-1} \dashrightarrow (\mathbb{P}^{m^2-1})^{\vee}$ *defined by the* $(m-1)$-*minors of* $\mathbf{L}(m)[r]$ *is a cone over the polar image* $\mathfrak{P}(f)$ *with vertex cut by* $\binom{r+1}{2}$ *coordinate hyperplanes.*

(ii) *The colon ideal* $J : I$ *has codimension* $2(m - r) \ge 4$; *in particular*, J *has codimension 4.*

(iii) I *is a prime ideal if and only if* $r \le m - 3$. *If this is the case, then* $I = J^{\mathrm{un}}$, *the unmixed part of* J; *in particular, if* R/J *is Cohen–Macaulay, then* $r = m - 2$.

(iv) *Let* $r = m - 2$. *Then the set of minimal primes of* J *is exactly the set of associated primes of* I *and of the ideals* $(I_j(\mathcal{N}_j), I_{m-1-j}(\mathcal{M}_{m-1-j}))$, *for* $1 \le j \le m - 2$.

Proof

(i) This follows from Proposition 4.28 (ii) and its proof. More explicitly, note the homogeneous inclusion $A := k[J_{m-1}] \subset A' := k[I_{m-1}]$ of k-algebras that are domains, where I is minimally generated by the generators of J and by $\binom{r+1}{2}$ additional generators, say, $g_1, \ldots, g_s$, where $s = \binom{r+1}{2}$, that is, $A' = A[g_1, \ldots, g_s]$. On the other hand, one has $\dim A = m^2 - r(r + 1)$ and $\dim A' = m^2 - \binom{r+1}{2}$. Therefore, $\mathrm{trdeg}_{k(A)}k(A)(g_1, \ldots, g_s) = \dim A' - \dim A = \binom{r+1}{2} = s$, where $k(A)$ denotes the field of fractions of A. This means that $g_1, \ldots, g_s$ are algebraically independent over $k(A)$ and, a fortiori, over A. This shows that A' is a polynomial ring over A in $\binom{r+1}{2}$ indeterminates. Geometrically, the image of the map defined by the $(m - 1)$-minors is a cone over $\mathfrak{P}(f)$ with vertex cut by $\binom{r+1}{2}$ independent hyperplanes.

(ii) The fact that the stated value $2(m - r)$ is an upper bound follows from the following fact: first, since the only cofactors that are not partial derivatives (up to sign) are those corresponding to the zero entries, then J is contained in the ideal Q generated by the variables of the last row and the last column of the matrix. Since Q is prime and $I \not\subset Q$ then clearly $J : I \subset Q$.

To see that $2(m - r)$ is a lower bound as well, we will use the previous lemmata. For this purpose, we focus on the following chains of inclusions:

$$I_1(\mathcal{N}_1) \supset I_2(\mathcal{N}_2) \supset \ldots \supset I_{r-1}(\mathcal{N}_{r-1}) \supset I_r(\mathcal{N}_r) \tag{4.53}$$

and

$$I_1(\mathcal{M}_1) \supset I_2(\mathcal{M}_2) \supset \ldots \supset I_{r-1}(\mathcal{M}_{r-1}) \supset I_r(\mathcal{M}_r). \tag{4.54}$$

Let P denote a prime ideal containing the colon ideal $J : I$. By Lemma 4.34 one has the inclusion $I_1(\mathcal{N})I_{r-1}(\mathcal{M}_{r-1}) \subset P$. Thus:

(A_1) Either $I_1(\mathcal{N}_1) \subset P$, or else
(B_1) $I_1(\mathcal{N}_1) \not\subset J : I$ but $I_{r-1}(\mathcal{M}_{r-1}) \subset P$

If (A_1) is the case, then $(I_1(\mathcal{N}_1),\ I_r(\mathcal{M}_r)) \subset P$, because $I_r(\mathcal{M}_r) \subset J : I$ again by Lemma 4.34 (with $j = 0$). By Corollary 4.35, we then see that the codimension of $J : I$ is at least $2(m - r)$.

If (B_1) takes place, we proceed to the next inclusion moving along the above chains. Precisely, then we consider the inclusion $I_2(\mathcal{N}_2)I_{r-2}(\mathcal{M}_{r-2}) \subset J : I \subset P$ by Lemma 4.34. The latter in turn gives rise to two possibilities according to which:

(A_2) Either $I_2(\mathcal{N}_2) \subset P$, or else
(B_2) $I_2(\mathcal{N}_2) \not\subset P$ but $I_{r-2}(\mathcal{M}_{r-2}) \subset P$.

Again, if (A_2) is the case, then $(I_2(\mathcal{N}_2),\ I_{r-1}(\mathcal{M}_{r-1})) \subset P$ since $I_{r-1}(\mathcal{M}_{r-1}) \subset P$ by hypothesis. Once more, by Corollary 4.35, the codimension of P is at least $2(m - r)$.

If instead (B_2) occurs, then we move up to the inclusion $I_3(\mathcal{N}_3)I_{r-3}(\mathcal{M}_{r-3}) \subset P$ and so on so forth. This way, we may eventually either find an index $1 \leq j \leq r-1$ such that the first alternative (A_j) holds, in which case we are through always by Corollary 4.35, or else, we are in the situation where $I_j(\mathcal{N}_j) \not\subset P$ for every $1 \leq j \leq r - 1$. In particular, $I_{r-1}(\mathcal{N}_{r-1}) \not\subset P$ and $I_1(\mathcal{M}_1) \subset P$. Thus, $(I_r(\mathcal{N}_r),\ I_1(\mathcal{M}_1)) \subset P$, and once and again by Corollary 4.35, P has codimension at least $2(m - r)$. This shows that $J : I$ has codimension at least $2(m - r)$ as was stated.

As to the codimension of J, we know that $\operatorname{codim} J \geq 4$ because if J admits a primary component of codimension ≤ 3, then $J : I$ would also have a primary component of codimension ≤ 3, contradicting the above $\operatorname{codim} J : I = 2(m - r) \geq 4$. Therefore, $\operatorname{codim} J = 4$.

(iii) We first prove the primality assertion.

The proof is by induction on m. For $m = 3$ the matrix $\mathbf{L}(m)[r]$ is the generic matrix; hence, we know that $I_{m-1}(\mathbf{L}(m)[r])$ is prime.

Suppose that $m \geq 4$. Since $r \leq m - 3$, $\mathfrak{X} := \{x_{1,m}, x_{2,m}, x_{3,m}, x_{m,1}, x_{m,2}, x_{m,3}\}$ are entries in the last row and last column of $\mathbf{L}(m)[r]$.

Now, the ideal $I_{m-1}(\mathbf{L}(m)[r])$ extended to $R[x_{1,m}^{-1}]$ is $I_{m-2}(\widetilde{\mathbf{L}}(m-1)[r-1])$, where $\widetilde{\mathbf{L}}(m - 1)[r - 1]$ is the following avatar of $\mathbf{L}(m - 1)[r - 1]$

$$
\begin{pmatrix}
\tilde{x}_{2,1} & \cdots & \tilde{x}_{2,m-r+1} & \tilde{x}_{2,m-r+2} & \tilde{x}_{2,m-r+3} & \cdots & \tilde{x}_{2,m-2} & \tilde{x}_{2,m-1} \\
\vdots & \ddots & \vdots & \vdots & \vdots & \ddots & \vdots & \vdots \\
\tilde{x}_{m-r,1} & \cdots & \tilde{x}_{m-r,m-r+1} & \tilde{x}_{m-r,m-r+2} & \tilde{x}_{m-r,m-r+3} & \cdots & \tilde{x}_{m-r,m-2} & \tilde{x}_{m-r+1,m-1} \\
\tilde{x}_{m-r+1,1} & \cdots & \tilde{x}_{m-r+1,m-r+1} & \tilde{x}_{m-r+1,m-r+2} & \tilde{x}_{m-r+1,m-r+3} & \cdots & \tilde{x}_{m-r+2,m-2} & 0 \\
x_{m-r+2,1} & \cdots & x_{m-r+2,m-r+1} & x_{m-r+2,m-r+2} & x_{m-r+2,m-r+3} & \cdots & 0 & 0 \\
\vdots & \ddots & \vdots & \vdots & \vdots & \ddots & \vdots & \vdots \\
\tilde{x}_{m-2,1} & \cdots & \tilde{x}_{m-2,m-r+1} & \tilde{x}_{m-2,m-r+2} & 0 & \cdots & 0 & 0 \\
\tilde{x}_{m-1,1} & \cdots & \tilde{x}_{m-1,m-r+1} & 0 & 0 & \cdots & 0 & 0
\end{pmatrix}
$$

with

$$
\tilde{x}_{i,j} = \begin{cases} x_{i,j} - x_{1,m}^{-1} x_{i,m} x_{1,j}, & 2 \leq i \leq m - r \\ x_{i,j}, & \text{otherwise.} \end{cases}
$$

Now, the above tilde variables are algebraically independent over k – the argument is the same as the one at the end of the proof of Proposition 1.10. Therefore, by the inductive assumption,

$$
R[x_{1,m}^{-1}]/I_{m-1}(\mathbf{L}(m)[r])R[x_{1,m}^{-1}]
$$

is a domain.

By the same token, this assertion holds for each of the remaining entries belonging to $\mathfrak{X}$, for a suitable avatar of the above matrix. Thus,

$$
R\left[x_{i,j}^{-1}\right]/I_{m-1}(\mathbf{L}(m)[r])R\left[x_{i,j}^{-1}\right]
$$

is a domain for every $x_{i,j} \in \mathfrak{X}$.

Since $R[x_{1,m}^{-1}]/I_{m-1}(\mathbf{L}(m)[r])R[x_{1,m}^{-1}]$ is a domain, there is exactly one associated prime ideal P of $R/I_{m-1}(\mathbf{L}(m)[r])$ such that $x_{1,m} \notin P$. If P is the unique associated prime of $R/I_{m-1}(\mathbf{L}(m)[r])$, then $x_{1,m}$ is $R/I_{m-1}(\mathbf{L}(m)[r])$-regular. In this case, $R/I_{m-1}(\mathbf{L}(m)[r])$ is a domain too.

Otherwise, suppose that there is another associated prime Q of $R/I_{m-1}(\mathbf{L}(m)[r])$ such that $Q \neq P$. In particular, $x_{1,m} \in Q$. Since $\mathrm{ht}(\mathfrak{X}) = 6$ and $I_{m-1}(\mathbf{L}(m)[r])$ is Cohen–Macaulay of height 4, there is a certain $x_{i,j} \in \mathfrak{X}$ such that $x_{i,j} \notin Q$. Necessarily, $x_{i,j} \neq x_{1,m}$. Since $R[x_{i,j}^{-1}]/I_{m-1}(\mathbf{L}(m)[r])R[x_{i,j}^{-1}]$ is also a domain, Q is the unique associated prime of $R/I_{m-1}(\mathbf{L}(m)[r])$ such that $x_{i,j} \notin Q$. Therefore, $x_{i,j} \in P$; hence $PR[x_{1,m}^{-1}]/I_{m-1}(\mathbf{L}(m)[r])R[x_{1,m}^{-1}] = 0$. Thus, we have $x_{i,j}x_{1,m}^{t} \in I_{m-1}(\mathbf{L}(m)[r])$ for some integer $t \geq 0$. Now, consider the R-map $\varphi : R \to k$ such that

$$\varphi(x_{u,v}) = \begin{cases} 1, \text{ if } (u, v) = (1, m) \text{ or } (u, v) = (i, j) \\ 0, \text{ otherwise.} \end{cases}$$

Since φ maps the entire $I_{m-1}(\mathbf{L}(m)[r])$ to zero, this contradicts the last inclusion, hence also the assumed inequality $Q \neq P$.

This concludes the primality assertion.

To see the supplementary assertion, by (ii), if $r \leq m - 3$, then $J : I$ has codimension at least $2(m - r) \geq 6$. This implies that $I \subset J^{\mathrm{un}}$; hence, they coincide since I is prime.

The assertion on the Cohen–Macaulayness of R/J is clear since then J is already unmixed; hence $J = I$, which is impossible since $r \geq 1$.

(iv) Let M denote the cofactor of a nonzero entry in $\mathbf{L}(m)[r]$ and let $1 \leq j \leq r$. Computing $\det M$ by a generalized Laplace expansion along the j-minors of a $j \times (m - 1)$ submatrix of M gives a form expressed in terms of products of generators of $I_j(\mathcal{N}_j)$ and of $I_{m-1-j}(\mathcal{M}_{m-1-j})$. Since every partial derivative of $\det \mathbf{L}(m)[r]$ is a cofactor of $\mathbf{L}(m)[r]$, it follows that $J \subset (I_j(\mathcal{N}_j), I_{m-1-j}(\mathcal{M}_{m-1-j}))$.

Next, since one is assuming that $r = m - 2$, the ideal $(I_j(\mathcal{N}_j), I_{m-1-j}(\mathcal{M}_{m-1-j}))$ has codimension 4 by Corollary 4.35 and Remark 4.32. Also recall from the proof of Lemma 4.33, with the notation there, that $I_j(\mathcal{N}_j) = I_j(\mathcal{N}_j')$ and $I_{m-j-1}(\mathcal{M}_{m-1-j}) = I_{m-1-j}(\mathcal{M}_{m-1-j}')$ where $\mathcal{N}_j'$ and $\mathcal{M}_{m-1-j}'$ are $j \times (j + 1)$ and $(m - j) \times (m - j - 1)$ matrices. In particular, by the Hilbert–Burch theorem, $I_j(\mathcal{N}_j)$ and $I_{m-j-1}(\mathcal{M}_{m-1-j})$ are perfect ideals of codimension 2. In addition, by Remark 4.32, $R/(I_j(\mathcal{N}_j), I_{m-1-j}(\mathcal{M}_{m-1-j}))$ is the tensor product over k of two Cohen–Macaulay k-algebras of finite type. Thus, $R/(I_j(\mathcal{N}_j), I_{m-1-j}(\mathcal{M}_{m-1-j}))$ is Cohen–Macaulay as well.

At the other end, by (i) the ideal I is certainly perfect. Therefore, any associated prime of either $(I_j(\mathcal{N}_j), I_{m-1-j}(\mathcal{M}_{m-1-j}))$ or I is a minimal prime of J.

Conversely, let P denote a minimal prime of J not containing I. Then $J : I \subset P$; hence, the same argument in the proof of (iii) says that P contains some ideal of the form $(I_j(\mathcal{N}_j), I_{m-1-j}(\mathcal{M}_{m-1-j}))$. $\qquad\square$

Slightly on the technical side, we pose:

Conjecture 4.37 Let $r = m - 2$. Then:

(i) $I_j(\mathcal{M}_j)$ and $I_j(\mathcal{N}_j)$ are radical ideals.

More precisely, for any $1 \leq j \leq m - 2$, one has the following primary decomposition:

$$I_j(\mathcal{M}_j)=(x_{j+1,m-j+1},\delta_j)\cap(x_{j,m-j+2},\delta_{i-1})\cap\ldots\cap(x_{3,m-1},\delta_2)\cap(x_{2,m},x_{1,m}),$$

where δ_t denotes the determinant of the $t \times t$ upper submatrix of $\mathcal{M}_t$. (A similar result holds for $I_j(\mathcal{N}_j)$ upon reverting the indices of the entries and replacing δ_t by the determinant γ_t of the $t \times t$ leftmost submatrix of $\mathcal{N}_t$.)

(ii) R/J and $R/J : I$ are reduced Cohen–Macaulay rings and, moreover, one has $J = I \cap (J : I)$.

Exercises

4.38 Consider the following two alternating matrices:

$$A = \begin{pmatrix} 0 & 0 & x_1 & x_2 & x_3 & x_4 \\ 0 & 0 & x_5 & x_6 & x_7 & x_8 \\ -x_1 & -x_5 & 0 & x_9 & x_{10} & x_{11} \\ -x_2 & -x_6 & -x_9 & 0 & -x_{12} & -x_{13} \\ -x_3 & -x_7 & -x_{10} & x_{12} & 0 & 0 \\ -x_4 & -x_8 & -x_{11} & x_{13} & 0 & 0 \end{pmatrix}, \quad B = \begin{pmatrix} 0 & 0 & x_1 & x_2 & x_3 & x_4 \\ 0 & 0 & 0 & x_5 & x_6 & 0 \\ -x_1 & 0 & 0 & x_7 & x_8 & x_9 \\ -x_2 & -x_5 & -x_7 & 0 & x_{10} & x_{11} \\ -x_3 & -x_6 & -x_8 & -x_{10} & 0 & 0 \\ -x_4 & 0 & -x_9 & -x_{11} & 0 & 0 \end{pmatrix}.$$

1. Prove that the ideal $\mathrm{Pf}_4(B)$ is generated by (quadric) binomials, while $\mathrm{Pf}_4(A)$ requires some minimal generators that are (quadric) trinomials.
2. Single out the 4-Pfaffians of A that are trinomials and explain the phenomenon in (1) in the light of Question 4.1.1.
3. Justify why $\mathrm{Pf}_4(B) \neq \mathfrak{g}_{m-2}$, for $m = 4$.
   (HINT: Count the number of ground variables.)

4.39 Give full proofs, for $m = 3$, of items (i), (ii), and (v) of the conjectured Theorem 4.7.

4.40 Prove Conjecture 4.26 for $m = 4$.
   (HINT: Computer assistance is assumed.)

4.41 Consider the linear section $\mathbf{L}(4)[2]$ of Section 4.5. Prove that the gradient ideal J of $\det \mathbf{L}(4)[2]$ is not of fiber type with the following steps:

1. Verify by theory that $d := x[2, 4]x[3, 3]x[4, 2]$ is a minimal generator of J.
2. Let $\mathcal{L}(\mathcal{I})$ denote the set of minimal generators of the defining ideal of the symmetric algebra of J coming from the linear syzygies of J. Show that $\mathcal{L}(\mathcal{I}): d$ contains a generator of bidegree $(1, 2)$.
   (HINT: This may require computer assistance.)
3. Conclude that J is not of fiber type.

4.42 Corroborate Conjecture 4.37 for $m = 3$.

(HINT: Though computer assistance is expected, a good mileage can be covered without it).

References

1. A. Boocher, Free resolutions and sparse determinantal ideals. Math. Res. Lett. **18**, 10001–10017 (2011) 87, 113
2. W. Bruns, U. Vetter, *Determinantal Rings*. Lecture Notes in Mathematics, vol. 1327 (Springer, Berlin, 1988) 96, 121
3. D. Buchsbaum, D. Eisenbud, What makes a complex exact? J. Algebra **25**, 259–268 (1973) 80
4. D. Buchsbaum, D. Eisenbud, Algebraic structures for finite free resolutions, and some structure theorems for ideals of codimension 3. Amer. J. Math. **99**, 447–485 (1977) 82
5. E. Celikbas, E. Dufresne, L. Fouli, E. Gorla, K.-N. Lin, C. Polini, I. Swanson, Rees algebras of sparse determinantal ideals, Trans. Amer. Math. Soc. **377**, 2317–2333 (2024) 113
6. E. Celikbas, J. Laxmi, J. Weyman, Spinor structures on free resolutions of codimension four Gorenstein ideals. Osaka J. Math. **60**, 903–931 (2023) 83
7. A. Conca, Ladder determinantal rings. J. Pure Appl. Algebra **98**, 119–134 (1995) 78, 80
8. D. Eisenbud, Linear sections of determinantal varieties. Amer. J. Math. **110**, 541–575 (1988) 96
9. M. Giusti, M. Merle, Sections des variétés déterminantielles par les plans de coordonnées, in Proceedings of the International Conference on Algebraic Geometry (La Rabida 1981, Spain). Lecture Notes in Mathematics, vol. 961 (Springer, Berlin, 1982), pp. 103–118 87, 113
10. D.J. Glassbrenner, K.E. Smith, Singularities of ladder determinantal varieties. J. Pure Appl. Alg. **101**, 59–75 (1995) 79
11. T.H. Gulliksen, O. Negard, Un complexe résolvant pour certains idéaux déterminantiels. C. R. Acad. Sci. Paris **274**, 16–18 (1972) 96
12. J. Herzog, N.N. Trung, Gröbner bases and multiplicity of determinantal and Pfaffian ideals. Adv. Math. **96**, 1–37 (1992) 78, 79, 113
13. J. Herzog, A. Simis, W. Vasconcelos, Koszul homology and blowing-up rings, in *Proceedings of the Commutative Algebra*, Trento, ed. by S. Greco, G. Valla. Lecture Notes in Pure and Applied Mathematics, vol. 84 (Marcel-Dekker, New York City, 1983), pp. 79–169 82
14. C. Huneke, Linkage and the Koszul homology of ideals. Amer. J. Math. **104**, 1043–1062 (1982) 82
15. A. Kustin, M. Miller, Algebra structures on minimal resolutions of gorenstein rings of embedding codimension four. Math. Z. **173**, 171–184 (1980) 83, 85
16. H. Narasimhan, The irreducibility of ladder determinantal varieties. J. Algebra **102**, 162–185 (1986) 78
17. J. Neves, S. Papadakis, Parallel Kustin–Miller unprojection with an application to Calabi–Yau geometry. Proc. London Math. Soc. **106**, 203–223 (2013) 83
18. S. Papadakis, Towards a general theory of unprojection. J. Math. Kyoto Univ. **47**, 579–598 (2007) 83
19. Z. Ramos, A. Simis, *Graded Algebras in Algebraic Geometry*. Expositions in Mathematics, vol. 70 (De Gruyter, Berlin, 2022) 45, 86, 90, 91, 106, 110
20. M. Reid, Gorenstein in codimension 4: the general structure theory. Adv. Stud. Pure Math. **65**, 201–227 (2015) 83
21. A. Simis, *Commutative Algebra*, 2nd edn. (De Gruyter Graduate, Berlin, 2023) 96
22. R. Stanley, Some combinatorial aspects of the Schubert calculus, in *Combinatoire et Représentation du Groupe Symétrique*. Lecture Notes in Mathematics, vol. 579 (Springer, New York, 1977), pp. 217–251 79

Chapter 5
Symmetry Preserving Linear Sections of the Generic Symmetric Matrix

Abstract This chapter is about linear sections of the generic symmetric matrix that are symmetric themselves. While for linear sections of the $m \times m$ generic matrix, we could use the resources of the m-generic property with all its spin-offs discussed in a previous chapter, here no such help is available. To replace some of those arguments, we resort to the use of initial ideals as based on a couple of basic related lemmas in Chapter 2. One discusses similar families of linear sections as in the previous chapter, stressing the symmetry. Most proofs are redone ab initio due to the peculiarities of symmetry in those linear sections. The discussion of the main algebraic invariants becomes an interesting challenge vis-à-vis the behavior of its analogs in the case of linear sections of the generic matrix. By and large, a good acquaintance with the previous chapter may help advancing through the present chapter, helping to get a grasp of the main similarities and differences in the theory. As a natural fallout, this chapter is shorter than the previous one.

5.1 Preliminaries

Throughout k denotes an algebraically closed field of characteristic $\neq 2$.

With $m \geq 3$, consider the $m \times m$ generic symmetric matrix:

$$
S = S_m := \begin{pmatrix}
x_{1,1} & x_{1,2} & \cdots & x_{1,m-1} & x_{1,m} \\
x_{1,2} & x_{2,2} & \cdots & x_{2,m-1} & x_{2,m} \\
\vdots & \vdots & \vdots & \vdots & \vdots \\
x_{1,m-1} & x_{2,m-1} & \cdots & x_{m-1,m-1} & x_{m-1,m} \\
x_{1,m} & x_{2,m} & \cdots & x_{m-1,m} & x_{m,m}
\end{pmatrix},
\tag{5.1}
$$

where the entries are indeterminates over k. Let $R = k[x_{i,j} \mid 1 \leq i \leq j \leq m]$ denote the polynomial ring over k generated by the entries of S. We will only consider symmetry preserving linear sections of S. For example, the following section

© The Author(s), under exclusive license to Springer Nature Switzerland AG 2024

Z. Ramos, A. Simis, *Determinantal Ideals of Square Linear Matrices*,

https://doi.org/10.1007/978-3-031-55284-7_5

$$\begin{pmatrix} a_{1,1} & a_{1,2} \\ 0 & a_{2,2} \end{pmatrix} \quad \text{of} \quad \begin{pmatrix} a_{1,1} & a_{1,2} \\ a_{1,2} & a_{2,2} \end{pmatrix}$$

with $a_{1,2} \neq 0$ is not a symmetry preserving linear section of the 2×2 generic symmetric matrix no matter what values the entries have – although it is a perfectly acceptable linear section of the 2×2 generic matrix. Although slightly annoying, this proviso will play a role.

The discussion of the various special cases as considered in Chapter 4 also applies in the present landscape. Yet, as it will become clear along it, there are some genuine differences that require appropriate care in the symmetric environment, not to mention the discomfort of no longer being free to identify entries with variables in a bijective way. Besides, the numerical invariants in the two situations will often diverge as one naturally expects. Last, but not least, the generic symmetric matrix is 1-generic, but not t-generic for $t \geq 2$. This is a big handicap as compared to the square generic matrix. For this reason we will refrain from any blind reference to the arguments in Chapter 4 and rather reinstate proofs ab initio whenever appropriate.

As in Chapter 4, we will consider two basic models of symmetry preserving linear sections, one where its determinant is homaloidal and the other where its Hessian vanishes. The second of these models is pretty much similar to the one in Section 4.5. The diversity will come along as we look at the structure of the corresponding dual variety. Although the two cases are as crudely apart as they could be, the difference between the structure of the corresponding dual varieties is quite subtler as will be seen in Chapter 8. In contrast to the fully generic case, there will be quite a bit of diversity on some of the basic algebraic structures (see Exercise 5.20).

As yet another symmetry preserving model, one could pick an entry $x_{i,j}$ off the main diagonal of S, such that $i + j$ is even, and replicate it onto the $(\frac{i+j}{2}, \frac{i+j}{2})$-slot pivot on the main diagonal, while fixing the remaining variables. Repeating a similar procedure along each anti-diagonal will eventually land us on a Hankel matrix. This procedure entangles a totally different situation, to be focused on in Section 10.3.

5.2   Codimension One Diagonal Section

In this part the linear section only affects $x_{m,m}$ and replaces it by $x_{m-1,m-1}$ – more generally, up to row/column elementary operations, the two entries could equally be any two distinct entries along the main diagonal – this is as far as it goes mimicking the fully generic environment and explains the present terminology. This is clearly a symmetry preserving linear section, to be denoted $\mathbf{L}(S)$ throughout.

Let R denote the polynomial ring generated over the field k by the entries of $\mathbf{L}(S)$. Note that the ground polynomial ring of the symmetric generic matrix S is then $R[x_{m,m}]$. Throughout, char$(k) \neq 2$. When dealing in Chapter 4 with the square generic matrix, we assumed that $m \geq 3$. Here instead we will assume throughout that $m \geq 4$ because the case where $m = 3$ is a bit unstable – e.g., the ideal of 2-minors is not even prime.

5.2.1 The Submaximal Minors

The moral of this part is that whereas for the analogous linear section of the $m \times m$ generic matrix we could use the resources of the m-generic property, here no such help is available. We will argue instead via initial ideals. The next lemmas form the technical core to handle the situation.

In the sequel, given a set of indices $\mathbf{i} = \{1 \leq i_1 < \cdots < i_u \leq m\}$, we denote by $\delta_{\mathbf{i},\mathbf{i}}$ the corresponding central minor based on $\mathbf{i}$.

Lemma 5.1 *Let R as above be the ground ring of $\mathbf{L}(S)$. Consider the lexicographic order of the monomials in R induced from the ordering of the variables respecting rows*

$$x_{1,m} > x_{1,m-1} > \cdots > x_{1,1} > x_{2,m} > x_{2,m-1} > \cdots > x_{2,2} > \cdots$$

$$\cdots > x_{m-1,m} > x_{m-1,m-1}.$$

For any set of indices $\mathbf{i} = \{1 \leq i_1 < \cdots < i_u \leq m\}$, one has $\mathrm{in}(\delta_{\mathbf{i},\mathbf{i}}) = \prod_{s+t=u+1} x_{i_s,j_t}.$

Proof The proof is the same as in Proposition 2.5, by noting that the (m, m)-entry does not disrupt the argument. $\qquad\square$

Lemma 5.2 *Given indices i, j, let $\Delta_{i,j}$ stand for the (i, j)th cofactor of $\mathbf{L}(S)$, and let $\mathfrak{c}(i, j)$ denote the set of complementary indices to i, j. Then:*

(a) *The Jacobian matrix of $\Delta_{1,1}$, $\Delta_{1,2}$, $\Delta_{2,2}$ with respect to the variables $x_{1,1}, x_{1,2}, x_{2,2}$ is*

$$\begin{pmatrix} 0 & 0 & \delta_{\mathfrak{c}(\{1,2\}),\mathfrak{c}(\{1,2\})} \\ 0 & -\delta_{\mathfrak{c}(\{1,2\}),\mathfrak{c}(\{1,2\})} & 0 \\ \delta_{\mathfrak{c}(\{1,2\}),\mathfrak{c}(\{1,2\})} & 0 & 0 \end{pmatrix}.$$

(b) *For every $1 \leq i \leq m - 2$ the Jacobian matrix of $\Delta_{m-1,m-1}$, $\Delta_{m-1,m}$, $\Delta_{m,m}$ with respect to the variables $x_{i,m-1}, x_{m-1,m-1}, x_{m-1,m}$ has the following shape*

$$\begin{pmatrix} 0 & * & 2\,\widetilde{\Delta}_{i,m-1} \\ \delta_{\mathfrak{c}(\{m-1,m\}),\mathfrak{c}(\{m-1,m\})} & 0 & \delta_{\mathfrak{c}(\{m-1,m\}),\mathfrak{c}(\{m-1,m\})} \\ 0 & \delta_{\mathfrak{c}(\{m-1,m\}),\mathfrak{c}(\{m-1,m\})} & 0 \end{pmatrix},$$

for a suitable entry $*$ and where $\widetilde{\Delta}_{i,j}$ stands for the (i, j)th cofactor of the $(m - 1) \times (m - 1)$ generic symmetric matrix $(x_{i,j})_{1 \leq i \leq j \leq m-1}$.

Proof

(a) Laplace block expansion along the first two rows yields

$$f = \det \begin{pmatrix} x_{1,1} & x_{1,2} \\ x_{1,2} & x_{2,2} \end{pmatrix} \cdot \delta_{c(\{1,2\}),\,c(\{1,2\})} + T,$$

where T has degree at most one in the variables $x_{1,1}, x_{1,2}, x_{2,2}$. With this and the fact that $\delta_{c(\{1,2\}),\,c(\{1,2\})}$ does not depend on the variables $x_{1,1}, x_{1,2}, x_{2,2}$, one has

$$\frac{\partial f}{\partial x_{1,1} \partial x_{1,1}} = \frac{\partial f}{\partial x_{2,2} \partial x_{2,2}} = \frac{\partial f}{\partial x_{1,1} \partial x_{1,2}} = \frac{\partial f}{\partial x_{1,2} \partial x_{2,2}} = 0 \tag{5.2}$$

$$\frac{\partial f}{\partial x_{1,1} \partial x_{2,2}} = \delta_{c(\{1,2\}),\,c(\{1,2\})} \quad \text{and} \quad \frac{\partial f}{\partial x_{1,2} \partial x_{1,2}} = -2\delta_{c(\{1,2\}),\,c(\{1,2\})}. \tag{5.3}$$

Since $\Delta_{1,1} = \frac{\partial f}{\partial x_{1,1}}$, $\Delta_{1,2} = \frac{1}{2}\frac{\partial f}{\partial x_{1,2}}$ and $\Delta_{2,2} = \frac{\partial f}{\partial x_{2,2}}$, (5.2) and (5.3) imply the assertion.

(b) Let $\mathcal{S}_1$ (respectively, $\mathcal{S}_2$) denote the square submatrix of $\mathbf{L}(\mathcal{S})$ with row and column indices $\{1, \ldots, m - 1\}$ (respectively, $\{1, \ldots, m - 2, m\}$).

Note that these are symmetric *generic* matrices; in particular,

$$\frac{\partial \Delta_{m,m}}{\partial x_{m-1,m-1}} = \frac{\partial \Delta_{m-1,m-1}}{\partial x_{m-1,m-1}} = \delta_{c(\{m-1,m\}),\,c(\{m-1,m\})}, \quad \frac{\partial \Delta_{m,m}}{\partial x_{i,m-1}} = 2\widetilde{\Delta}_{i,m}. \tag{5.4}$$

Also, since $x_{m-1,m}$ is not an entry of $\mathcal{S}_1$ and neither $x_{i,m-1}, x_{m-1,m}$ are entries of $\mathcal{S}_2$, then

$$\frac{\partial \Delta_{m-1,m-1}}{\partial x_{i,m-1}} = \frac{\partial \Delta_{m-1,m-1}}{\partial x_{m-1,m}} = \frac{\partial \Delta_{m,m}}{\partial x_{m-1,m}} = 0. \tag{5.5}$$

Finally, introduce the following auxiliary $(m - 1) \times (m - 1)$ matrix

$$\mathcal{A} = \begin{pmatrix} \mathbf{x} & \mathbf{u}^t \\ \mathbf{v} & x_{m-1,m} \end{pmatrix},$$

where $\mathbf{u} = (x_{1,m-1} \cdots x_{m-2,m-1})$ and $\mathbf{v} = (x_{1,m} \cdots x_{m-2,m})$. Then $\Delta_{m-1,m} = \det \mathcal{A}$. Since $x_{m-1,m}$ occurs in a single slot of $\mathcal{A}$ and $x_{m-1,m-1}$ is not an entry of $\mathcal{A}$, one has

$$\frac{\partial \Delta_{m-1,m}}{\partial x_{m-1,m-1}} = \delta_{c(\{m-1,m\}),\,c(\{m-1,m\})} \quad \text{and} \quad \frac{\partial \Delta_{m-1,m}}{\partial x_{m-1,m-1}} = 0. \tag{5.6}$$

Now, (5.4), (5.5), and (5.6) give the desired matrix. $\qquad\square$

In contrast to a previous notation having the ideal of submaximal minors denoted by the letter P, in the following we make an exception since we wish to keep P to denote an unspecified prime ideal.

Proposition 5.3 (char$(k) \neq 2$, $m \geq 4$) *Let* $I := I_{m-1}(\mathbf{L}(S))$:

(i) *R/I is a Cohen–Macaulay normal domain of codimension* 3.
(ii) *Set* $\mathbb{P}^{\binom{m+1}{2}-2} = \mathrm{Proj}(R)$ *and* $(\mathbb{P}^{\binom{m+1}{2}-1})^{\vee} = \mathrm{Proj}(k[\mathbf{Y}])$ *(dual variables). The rational map* $\Psi : \mathbb{P}^{\binom{m+1}{2}-2} \dashrightarrow (\mathbb{P}^{\binom{m+1}{2}-1})^{\vee}$ *defined by a minimal set of generators of I is a birational map onto the hypersurface with defining equation*

$$\Delta_{m,m}(\mathbf{Y}) - \Delta_{m-1,m-1}(\mathbf{Y}) \in k[\mathbf{Y}]$$

up to a k-linear transformation of $k[\mathbf{Y}]$, where $\Delta_{m,m}(\mathbf{Y})$ and $\Delta_{m-1,m-1}(\mathbf{Y})$ denote the cofactors of $y_{m,m}$ and $y_{m-1,m-1}$, respectively, in the $m \times m$ generic symmetric matrix $(y_{i,j})_{1 \leq i \leq j \leq m}$ in the dual variables.

Proof

(i) It is well-known that the ideal of submaximal minors of the generic symmetric matrix is a codimension 3 Cohen–Macaulay ring ([4, 5] also [7, Section 6.4.3]). Since I is a specialization of the latter, it follows that R/I is a Cohen–Macaulay ring of codimension 3.

As I is homogeneous, normality of R/I implies that I is prime. Therefore, it suffices to prove the first. To show normality, Serre's property (S_2) is automatic since R/I is Cohen–Macaulay. Therefore, it remains to prove that it satisfies condition (R_1).

For this, let Θ denote the Jacobian matrix of the generators of I with respect to the variables of R. We proceed to show that codim $(I_3(\Theta), I) \geq 5$.

For $m = 4$, computer-assisted calculation gives the monomials $x_{3,3}^6$, $x_{2,4}^6$, $x_{2,3}^6$, $x_{1,4}^6$, and $x_{1,3}^6$ in $\mathrm{in}(I_3(\Theta))$.

Now, assume that $m \geq 5$. By Lemma 5.2 and the notation there, the elements

$$\delta^3_{\mathfrak{c}(\{1,2\}),\mathfrak{c}(\{1,2\})}, \quad \delta^2_{\mathfrak{c}(\{m-1,m\}),\mathfrak{c}(\{m-1,m\})} \, \widetilde{\Delta}_{i,m} \quad (1 \leq i \leq m-2) \qquad (5.7)$$

belong to the ideal $(I_3(\Theta), I)$.

Now, let P be a prime containing $(I_3(\Theta), I)$. If $\delta_{\mathfrak{c}(\{m-1,m\}),\mathfrak{c}(\{m-1,m\})}$ is contained in P, then the latter contains the subideal

$$(f, \Delta_{1,1}, \Delta_{m,m}, \delta_{\mathfrak{c}(\{1,2\}),\mathfrak{c}(\{1,2\})}, \delta_{\mathfrak{c}(\{m-1,m\}),\mathfrak{c}(\{m-1,m\})}).$$

We contend that the generators of this ideal form a regular sequence. To see this, we appeal to the respective initial ideals as in Lemma 5.1:

$$\mathrm{in}(f) = \prod_{i+j=m+1} x_{i,j}, \ \mathrm{in}(\Delta_{1,1}) = \prod_{i+j=m+2} x_{i,j}, \ \mathrm{in}(\Delta_{m,m}) = \prod_{i+j=m} x_{i,j}$$

$$\mathrm{in}(\delta_{\mathfrak{c}(\{1,2\}),\mathfrak{c}(\{1,2\})}) = \prod_{i+j=m+3} x_{i,j}, \ \mathrm{in}(\delta_{\mathfrak{c}(\{m-1,m\}),\mathfrak{c}(\{m-1,m\})}) = \prod_{i+j=m-1} x_{i,j}.$$

Clearly, these monomials form a regular sequence. Therefore, $\mathrm{ht}\,P =$ $\mathrm{ht}\,\mathrm{in}(P) \geq 5$.

If, on the other hand, $\delta_{\mathfrak{c}(\{m-1,m\}),\mathfrak{c}(\{m-1,m\})} \notin P$, then $\widetilde{\Delta}_{i,m-1} \in P$ for every $1 \leq i \leq m$ as implied by (5.7). Hence, since

$$\Delta_{m,m-1} = x_{m-1,m}\delta_{\mathfrak{c}(\{m-1,m\}),\mathfrak{c}(\{m-1,m\})} + \sum_{i=1}^{m-2} x_{i,m}\widetilde{\Delta}_{i,m} \in P$$

and $\Delta_{m,m-1} \in I \subset P$, then $x_{m-1,m} \in P$. Thus, P contains the ideal

$$(f, \Delta_{1,1}, \Delta_{m,m}, \delta_{\mathfrak{c}(\{1,2\}),\mathfrak{c}(\{1,2\})}, x_{m-1,m}).$$

But, since $m \geq 5$, the generators of this ideal form, as before, a regular sequence. Hence, $\mathrm{ht}\,P \geq 5$.

Having obtained that, in all cases P has height at least 5, we conclude that $\mathrm{ht}\,(I_3(\Theta), I) \geq 5$, as was to be shown.

(ii) As I specializes from the ideal of submaximal minors in the symmetric generic case and the latter is linearly presented (see, e.g., [4]), then it is linearly presented as well. In addition, the extension of the k-algebra inclusion $k[\Delta_{i,j}, 1 \leq i \leq j \leq m] \subset R$ to the respective fields of fractions is algebraic (Proposition 2.3), hence, $k[\Delta_{i,j}, 1 \leq i \leq j \leq m]$ has maximal dimension. This gives $\ell(I) = \dim k[\Delta_{i,j}, 1 \leq i \leq j \leq m] = \dim R = \binom{m+1}{2} - 1$.

Therefore, by Theorem 2.27, the submaximal minors define a birational map onto the image, where the latter is a hypersurface. To get the defining equation of this hypersurface, we may assume that Ψ is defined by the minimal set Δ of generators of I consisting of the entries of the adjugate matrix of $\mathbf{L}(S)$.

Consider the cofactors $\Delta_{m,m}(\mathbf{Y})$ and $\Delta_{m-1,m-1}(\mathbf{Y})$ as in the statement, and apply the matrix identity from (2.2) to get

$$\mathrm{adj}(\mathrm{adj}(\mathbf{L}(S))) = f^{m-2}\mathbf{L}(S). \tag{5.8}$$

Looking at the right-hand side matrix of (5.8) in more detail, one sees that the entries in slots $(m-1, m-1)$ and (m, m) are one and the same element, namely, $f^{m-2}x_{m-1,m-1}$. Since the corresponding entries on the left-hand side matrix are $\Delta_{m-1,m-1}(\mathbf{\Delta})$ and $\Delta_{m,m}(\mathbf{\Delta})$ evaluated via $y_{i,j} \mapsto \Delta_{i,j}$, respectively, one gets $(\Delta_{m,m}(\mathbf{Y}) - \Delta_{m-1,m-1}(\mathbf{Y}))(\mathbf{\Delta}) = 0$, as required.

But $\Delta_{m,m}(\mathbf{Y}) - \Delta_{m-1,m-1}(\mathbf{Y})$ is clearly an irreducible polynomial in the target coordinate ring $k[y_{i,j} | 1 \leq i \leq j \leq m]$. Therefore, we are through.   □

5.2.2   *The Gradient Ideal Versus the Submaximal Minors*

We now collect in the following theorem the main properties of the gradient ideal J as related to the ideal P of submaximal minors. Some of the assertions follow the same pattern of argument as the generic case.

Theorem 5.4 (char$(k) \neq 2$, $m \geq 4$) *Let J denote the gradient ideal of $f = \det \mathbf{L}(S)$ and let $P = I_{m-1}(\mathbf{L}(S))$:*

   (i) *The minimal component of the primary decomposition of J is irreducible and coincides with P. In particular, J has codimension three.*

 (ii) *$J : P$ is the ideal generated by the variables of the two rightmost columns of $\mathbf{L}(S)$. In particular, $J : P$ is a prime ideal.*

(iii) *J defines a double structure on the variety defined by P, with a unique embedded component of codimension $2(m - 1)$ supported on a linear space and no other embedded component of codimension $\leq 2(m - 1)$.*

(iv) *J is not a reduction of P.*

Proof

   (i) By and large, the argument tracks the model of the proof of Theorem 4.20 (ii). By Proposition 5.3 (i), P is a prime ideal of codimension 3. We first show that codim $(J : P) > 3$, which ensures that the radical of the unmixed part of J has no primes of codimension < 3 and coincides with P.

     Since $f_{m-1,m-1} = \Delta_{m-1,m-1} + \Delta_{m,m}$ and any other partial derivative $f_{i,j}$ coincides with $\Delta_{i,j}$, up to multiplication by ± 2, we can write $P = (J, \Delta_{m,m})$ and $P = (J, \Delta_{m-1,m-1})$. In particular $J : P = J : \Delta_{m,m}$ and $J : P = J : \Delta_{m-1,m-1}$.

The adjugate formula (2.1) for $\mathbf{L}(S)$ yields the following relations:

$$\sum_{j=1}^{m} x_{k,j} \Delta_{j,m} = 0, \ \text{ for } k = 1, \ldots, m - 1, \text{ and}$$

$$\sum_{i=1}^{m-1} x_{i,m} \Delta_{i,m} + x_{m-1,m-1} \Delta_{m,m} = \sum_{j=1}^{m} x_{1,j} \Delta_{1,j}.$$

By the preceding observation, the above relations imply that the entries of the mth column of $\mathbf{L}(S)$ belong to the ideal $J : \Delta_{m,m} = J : P$.

In addition, we read the following relations of the adjugate formula:

$$\sum_{j=1, j\neq m-1}^{m} x_{k,j}\Delta_{j,m-1} + x_{k,m-1}\Delta_{m-1,m-1} = 0,$$

for $k = 1, \ldots, m, \ (k \neq m - 1)$, and

$$\sum_{j=1, j\neq m-1}^{m} x_{m-1,j}\Delta_{j,m-1} + x_{m-1,m-1}\Delta_{m-1,m-1} = \sum_{j=1}^{m} x_{1,j}\Delta_{j,1}.$$

Then by a similar argument as above, the entries of the $(m - 1)$th column of $\mathbf{L}(S)$ belong to the ideal $J : \Delta_{m-1,m-1} = J : P$.

From this, the variables of the two rightmost columns of $\mathbf{L}(S)$ conduct P into J. In particular, the codimension of $J : P$ is at least 4, as needed.

Now, since P has codimension 3, then $J : P \not\subset P$. Picking an element $a \in J : P \setminus P$ shows that $P_P \subset J_P$. Therefore P is the unmixed part of J.

To conclude that P is the minimal primary component of J, we observe that, by symmetry, the entries of the last two columns are the same as those of the last two rows. As is clear that P is contained in the ideal generated by these variables, it follows that $P^2 \subset J$. Therefore, the radical of J – i.e., the radical of the minimal primary part of J – is P.

(ii) Let I denote the ideal generated by the $m + m - 2 = 2(m - 1)$ variables of the two rightmost columns of $\mathbf{L}(S)$. By (i) and its proof, P is the minimal component of a primary decomposition of J and $I \subset J : P$.

To prove the reverse inclusion, we proceed as in the argument for item (iii) of Theorem 4.20, in the present case dealing with the subideals $I' \subset I$ (respectively, $I'' \subset I$) generated by the variables on the $(m - 1)$th column (respectively, by the variables on the mth column). We go over the details once more to provide for a full reading.

Clearly, $I = I' + I''$. Now, $\Delta_{m,m} \notin I''$, while $\Delta_{i,j} \in I''$ for all $(i, j) \neq (m, m)$, since $\Delta_{i,j}$ includes a row or a column with entries in I''. Similarly, $\Delta_{m-1,m-1} \notin I'$, while $\Delta_{i,j} \in I'$ for all $(i, j) \neq (m - 1, m - 1)$.

Recall that J is generated by the cofactors

$$\Delta_{l,h}, \ \text{ with } (l, h) \neq (m - 1, m - 1), (l, h) \neq (m, m)$$

and the additional form $\Delta_{m,m} + \Delta_{m-1,m-1}$. Therefore, given, say, $b \in J : P = J : \Delta_{m,m}$, one has

$$b\,\Delta_{m,m} = \sum_{(i,j)\neq(m-1,m-1)} a_{i,j}\Delta_{j,i} + a(\Delta_{m-1,m-1} + \Delta_{m,m}) \tag{5.9}$$

for certain $a_{i,j}, a \in R$. Then

$$(b - a)\Delta_{m,m} = \sum_{(i,j) \neq (m-1,m-1)} a_{i,j}\Delta_{j,i} + a\Delta_{m-1,m-1} \in I''.$$

Since I'' is a prime ideal and $\Delta_{m,m} \notin I''$, we have $c := b-a \in I''$. Substituting for $a = b - c$ in (5.9) gives

$$(-b + c)\Delta_{m-1,m-1} = \sum_{(i,j) \neq (m-1,m-1)} a_{i,j}\Delta_{j,i} - c\Delta_{m,m} \in I'.$$

By a similar argument, since $\Delta_{m-1,m-1} \notin I'$, then $-b + c \in I'$. Therefore

$$b = c - (-b + c) \in I'' + I' = I,$$

thus proving the claim.

(iii) Since $J : P$ is a prime ideal, it is necessarily an associated prime of R/J. As pointed out at the end of the proof of (i), $P \subset J : P$, hence $J : P$ is an embedded prime of R/J. Moreover, this also gives $P^2 \subset J$, hence J defines a double structure on the irreducible variety defined by P.

Let Q denote the embedded component of J with radical $J : P$ and let Q' denote the intersection of the remaining embedded components of J. From $J = P \cap Q \cap Q'$, we get

$$J : P = (Q : P) \cap (Q' : P),$$

in particular, passing to radicals, $J : P \subset \sqrt{Q'}$. This shows that Q is the unique embedded component of codimension $\leq 2(m - 1)$, while the corresponding geometric component is supported on a linear subspace.

(iv) The argument is pretty much that of item (vi) of Theorem 4.20, but we go over the details again for the sake of completeness. As already observed in the proof of item (ii), J is generated by the cofactors $\{\Delta_{l,h} | (l, h) \neq (m - 1, m - 1), (l, h) \neq (m, m)\}$ and the additional form $\Delta_{m,m} + \Delta_{m-1,m-1}$.

Now, by Proposition 5.3 (ii), the reduction number of a minimal reduction of P is $m - 2$. Thus, to conclude, it suffices to prove that $P^{m-1} \notin JP^{m-2}$. We will show that the element $\Delta_{m,m}^{m-1}$ of P^{m-1} does not belong to JP^{m-2}. Supposing otherwise, we can write a polynomial relation of degree $m - 1$ on the generators of P, namely,

$$\Delta_{m,m}^{m-1} = \left(\sum_{\substack{1 \leq l \leq h \leq m \\ (l,h) \neq (m-1,m-1),(m,m)}} \Delta_{l,h}Q_{l,h}(\mathbf{\Delta}) \right) + (\Delta_{m-1,m-1} + \Delta_{m,m})Q(\underline{\Delta}),$$

$$(5.10)$$

where $Q_{l,h}(\mathbf{\Delta})$ and $Q(\underline{\Delta})$ are homogeneous polynomial expressions of degree $m - 2$ in the set $\mathbf{\Delta} = \{\Delta_{i,j} \mid 1 \leq i \leq j \leq m\}$ of the cofactors (generators of P).

So the corresponding form of degree $m - 1$ in $k[y_{i,j} \mid 1 \leq i \leq j \leq m]$ is a scalar multiple of the polynomial $\mathbf{H} := \Delta_{m,m}(\mathbf{Y}) - \Delta_{m-1,m-1}(\mathbf{Y})$ obtained in the previous item. We argue that this is impossible.

Observe that the sum

$$\sum_{\substack{(l,h)\neq(m-1,m-1)\\(l,h)\neq(m,m)}} \Delta_{l,h}\, Q_{l,h}(\mathbf{\Delta})$$

does not contain any nonzero terms of the form $\alpha\Delta_{m,m}^{m-1}$ or $\beta\Delta_{m-1,m-1}\Delta_{m,m}^{m-2}$. Now, if these two terms appear in the second summand $(\Delta_{m-1,m-1} + \Delta_{m,m})Q(\mathbf{\Delta})$, they must have the same scalar coefficient, say, $c \in k$. Bring the first of these to the left-hand side of (5.10) to get a polynomial relation of P having a term $(1-c)y_{m,m}^{m-1}$. If $c \neq 1$, this is a contradiction because any term of $\mathbf{H}$ has degree at most 1 in the variable $y_{m,m}$.

On the other hand, if $c = 1$, then we still have a polynomial relation of P having a term $y_{m-1,m-1}y_{m,m}^{m-2}$. Now, if $m > 3$, this is again a contradiction due to the nature of $\mathbf{H}$ as the nonzero term of the latter has degree ≤ 1 in the variable $y_{m,m}$. Finally, if $m = 3$, a direct checking shows that the monomial $y_{2,2}y_{3,3}$ cannot be the support of a nonzero term in $\mathbf{H}$. This concludes the proof of the statement. $\qquad\square$

Remark 5.5 For $m = 3$ the ideal $P = I_2(\mathbf{L}(S))$ is no longer prime, but J still has codimension three as an easy verification shows that the monomials $x_{1,2}^2$, $x_{2,2}^2$ and $x_{1,3}^3$ belong to $\mathrm{in}(J)$ in the reverse lexicographic order of the variables (actually, $\sqrt{\mathrm{in}(J)} = \sqrt{\mathrm{in}(P)} = (x_{1,2}, x_{1,3}, x_{2,2})$).

5.2.3   Complements on the Gradient Ideal

Let $\mathbf{L}(S)$ be as above, with $m \geq 3$.

In this subsection we look at additional properties of the gradient ideal, as confronted with the generic case of the previous chapter. The following theorem is the symmetric analog of parts of Theorem 4.21.

Theorem 5.6 (char$(k) \neq 2$) *Let $J \subset R$ denote the gradient ideal of* $\det \mathbf{L}(S)$. *Then*:

(i) *J has maximal analytic spread and defines a homaloidal map.*
(ii) *J has maximal linear rank.*

Proof

(i) This item proceeds along the same lines of argument of Theorem 4.21 (a) and (b), with tiny adjustments.

(ii) The argument follows pretty much the scheme of the proof of Theorem 4.21 (c), but here there are a few minor subtleties due to the symmetric environment.

The adjugate formula (2.1) for $\mathbf{L}(S)$ gives the following linear relations involving cofactors of $\mathbf{L}(S)$:

$$\sum_{j=1}^{m} x_{i,j}\Delta_{j,1} = 0, \ \text{for } 2 \le i \le m-1; \tag{5.11}$$

$$\sum_{j=1}^{m} x_{i,j}\Delta_{j,k} = 0, \ \text{for } 2 \le k \le m-2 \text{ and } k-1 \le i \le m-1 \ (k \ne i); \tag{5.12}$$

$$\sum_{j=1}^{m-1} x_{m,j}\Delta_{j,k} + x_{m-1,m-1}\Delta_{m,k} = 0, \ \text{for } 1 \le k \le m-2; \tag{5.13}$$

$$\sum_{i=1}^{m-1} x_{i,m-1}\Delta_{i,m} + x_{m-1,m}\Delta_{m,m} = 0; \tag{5.14}$$

$$\sum_{i=1}^{m-2} x_{i,m}\Delta_{i,m-1} + x_{m-1,m}\Delta_{m-1,m-1} + x_{m-1,m-1}\Delta_{m,m-1} = 0; \tag{5.15}$$

$$\sum_{i=1}^{m-2} x_{i,m-1}\Delta_{i,m-1} + x_{m-1,m-1}\Delta_{m-1,m-1} + x_{m-1,m}\Delta_{m,m-1} = \sum_{j=1}^{m} x_{1,j}\Delta_{j,1}; \tag{5.16}$$

$$\sum_{j=1}^{m-1} x_{j,m}\Delta_{j,m} + x_{m-1,m-1}\Delta_{m,m} = \sum_{j=1}^{m} x_{1,j}\Delta_{j,1}. \tag{5.17}$$

Since, as already remarked, one has $f_{i,j} = 2\Delta_{i,j}$ for $1 \le i < j \le m$ and $f_{i,i} = \Delta_{i,i}$ for $1 \le i \le m-2$, then (5.11), (5.12), and (5.13) give linear syzygies of the partial derivatives. Moreover, since $f_{m-1,m-1} = \Delta_{m-1,m-1} + \Delta_{m,m}$, adding (5.14) to (5.15) and (5.16) to (5.17) outputs two additional linear syzygies of the partial derivatives of f. Thus one has counted a total of $(m-1)+(m-1)+(m-2)+\ldots+3+2 = \binom{m+1}{2} - 2$ linear syzygies of J. In order to see that they are moreover independent, we order the set of partial derivatives $f_{i,j}$ in accordance with the following ordered list of the entries $x_{i,j}$:

$$x_{1,1}, x_{1,2}, \ldots, x_{1,m} \rightsquigarrow x_{2,2}, \ldots, x_{2,m} \rightsquigarrow \ldots \rightsquigarrow x_{m-2,m-2}, x_{m-2,m-1}x_{m-2,m}$$

$$\rightsquigarrow x_{m-1,m-1}, x_{m-1,m}.$$

We now claim that, by ordering in this way, the above sets of linear relations can be grouped into the following block matrix of linear syzygies of the set of partial derivatives $f_{i,j}$:

$$
\left(
\begin{array}{cccccc|cc}
\Phi_1 & \cdots & & & & & & \\
\mathbf{0}^{m-1}_{m-1} & \Phi_2 & \cdots & & & & & \\
\mathbf{0}^{m-1}_{m-2} & \mathbf{0}^{m-1}_{m-2} & \Phi_3 & & & & & \\
\vdots & \vdots & \vdots & \ddots & & & & \\
\mathbf{0}^{m-1}_{4} & \mathbf{0}^{m-1}_{4} & \mathbf{0}^{m-2}_{4} & \cdots & \Phi_{m-3} & & & \\
\mathbf{0}^{m-1}_{3} & \mathbf{0}^{m-1}_{3} & \mathbf{0}^{m-2}_{3} & \cdots & \mathbf{0}^{4}_{3} & \Phi_{m-2} & & \\
\hline
\mathbf{0}^{m-1}_{1} & \mathbf{0}^{m-1}_{1} & \mathbf{0}^{m-2}_{1} & \cdots & \mathbf{0}^{4}_{1} & \mathbf{0}^{3}_{1} & x_{m-1,m} & x_{m-1,m-1} \\
\mathbf{0}^{m-1}_{1} & \mathbf{0}^{m-1}_{1} & \mathbf{0}^{m-2}_{1} & \cdots & \mathbf{0}^{4}_{1} & \mathbf{0}^{3}_{1} & x_{m-1,m-1} & x_{m-1,m}
\end{array}
\right).
$$

Here:

- Φ_1 is the matrix obtained from $(\mathbf{L}(\mathcal{S}))^t$ by multiplying the first row by 2 and omitting the first column.
- Φ_2 is the matrix obtained from $(\mathbf{L}(\mathcal{S}))^t$ by multiplying the second row by 2 and omitting the second column and the first row.
- And so on, for $l = 3, \ldots, m - 2$, Φ_l is the matrix obtained from $(\mathbf{L}(\mathcal{S}))^t$ by multiplying the lth row by 2 and omitting the columns $1, \ldots, l-2, l$ and the rows $1, \ldots, l - 1$.
- Finally, $\mathbf{0}^c_r$ denotes a block of zeros of size $r \times c$.

Justification is as follows.

First, as already observed, the relations (5.11) through (5.17) yield linear syzygies of the partial derivatives of f.

Using the relation between partial derivatives and cofactors of Proposition 2.10, (5.11) can be written as

$$
2\, x_{i,1}\, f_{1,1} + \sum_{j=2}^{m} x_{i,j}\, f_{1,j} = 0,
$$

for $i = 2, \ldots, m - 1$. Moreover, setting $k = 1$ in (5.13) yields

$$
2\, x_{m,1}\, f_{1,1} + \sum_{j=2}^{m-1} x_{m,j}\, f_{1,j} + x_{m-1,m-1}\, f_{1,m} = 0.
$$

Reading off the coefficients of these relations, one gets

$$
\Phi_1 := \begin{pmatrix}
2x_{1,2} & 2x_{1,3} & \cdots & 2x_{1,m-1} & 2x_{1,m} \\
x_{2,2} & x_{2,3} & \cdots & x_{2,m-1} & x_{2,m} \\
\vdots & \vdots & \cdots & \vdots & \vdots \\
x_{2,m-1} & x_{3,m-1} & \cdots & x_{m-1,m-1} & x_{m-1,m} \\
x_{2,m} & x_{3,m} & \cdots & x_{m-1,m} & x_{m-1,m-1}
\end{pmatrix}.
$$

Note that Φ_1 coincides indeed with the submatrix of $\mathbf{L}(\mathcal{S})^t$ obtained by multiplying the first row by 2 and omitting the first column.

Continuing, for each $l = 2, \ldots, m - 2$ the block Φ_l comes from the relation (5.12) and (5.13) (setting $k = l$). Finally, the lower right corner 2×2 block of the matrix of linear syzygies comes from the last two relations obtained by adding (5.14) to (5.15) and (5.16) to (5.17).

So much for justification.

Now, counting through the sizes of the various blocks, one sees that this matrix is $\left(\binom{m+1}{2} - 1\right) \times \left(\binom{m+1}{2} - 2\right)$. Omitting its first row obtains a block diagonal submatrix of size $\left(\binom{m+1}{2} - 2\right) \times \left(\binom{m+1}{2} - 2\right)$, where each block has nonzero determinant. Thus, the linear rank of J attains the maximum.

(v) By (iii) the polar map of f is dominant. Since the linear rank is maximum by (iv), one can apply Theorem 2.27 to conclude that f is homaloidal. $\square$

Remark 5.7 Regarding an alternative proof above, as in Remark 4.22 (2), there is an expectation that J be of linear type. In such a case, (i) and (ii) are equivalent statements ([6, Corollary 3.2.27]).

Exercise 5.20 shows that an arbitrary codimension one symmetry preserving linear section may lack some of the properties listed in Theorem 5.6, even by dubbing a pivot entry not lying on a same row or column. This example is related to the *Hankelization* family considered in Section 10.3.

5.3 Symmetric Sections of Hollow Type

This section is based on the ideas and notation of Section 4.2.1. In order to establish a closest symmetric analog with $\mathbf{I} := \{(i, j)|1 \leq i \leq j \leq m\}$ the set of indices, one sets $\mathbf{I} = \mathcal{I} \cup \mathcal{J}$, with $\mathcal{I} = \{(1, 1), (1, 2), \ldots, (1, m)\} \cup \{(i, m + 2 - i) \,|\, 2 \leq i \leq t\}$, where $t := \lfloor m/2 \rfloor + 1$.

If no confusion arises, we denote by $\mathbf{X}$ the $m \times m$ generic symmetric matrix (a slant X was used sometimes to distinguish it from the fully generic matrix notation).

We repeat the definitions for convenience.

Definition 5.8 Let $\mathbf{L} = (\ell_{i,j})$ be a symmetric linear section of $\mathbf{X}$. Then:

1. $\mathbf{L}$ is *hollow-filled* if it is endowed with the above decomposition and $\mathbf{L}_{\mathcal{I}} = \mathbf{X}_{\mathcal{I}}$.

2. **L** is *hollow* if it is hollow-filled and strictly sparse, i.e., $\ell_{i,j} = 0$ for every $(i, j) \in \mathcal{J}$.
3. **L** is *semi-hollow of order* r, for a given integer $0 \le r \le m - 2$, if it is hollow-filled and strictly sparse along $\mathcal{J} \cap \{(i, j) | i + j \ge 2m - r + 1\}$.

5.3.1 Symmetric Hollow Sections

Consider the hollow symmetric linear section as above:

$$\mathfrak{H} = \mathfrak{H}_m := \begin{pmatrix} x_{1,1} & x_{1,2} & \cdots & x_{1,m-1} & x_{1,m} \\ x_{1,2} & 0 & \cdots & 0 & x_{2,m} \\ x_{1,3} & 0 & \cdots & x_{3,m-1} & 0 \\ \vdots & \vdots & \ddots & \vdots & \vdots \\ x_{1,m-1} & 0 & x_{3,m-1} & 0 & 0 \\ x_{1,m} & x_{2,m} & 0 & 0 & 0 \end{pmatrix}. \tag{5.18}$$

Set $\mathfrak{f} := \det \mathfrak{H}$ and keep the proviso that $\mathrm{char}(k) \ne 2$.

Letting $\tilde{\mathbf{X}}$ stand for the set of distinct entries of $\mathfrak{H}$, consider the subset

$$\mathfrak{H}_{\mathcal{I} \cup \{(2,m)\}} \subset k[\tilde{\mathbf{X}}].$$

Set:

- $\partial(\mathfrak{f})$ for the set of partial derivatives of $\mathfrak{f}$ with respect to $\mathfrak{H}_{\mathcal{I} \cup \{(2,m)\}}$.
- $\Theta(\partial(\mathfrak{f}))$ for the Jacobian matrix of $\partial(\mathfrak{f})$ with respect to $\mathfrak{H}_{\mathcal{I} \cup \{(2,m)\}}$.

Note that $\Theta(\partial(\mathfrak{f}))$ is an $(m + 1) \times (m + 1)$ submatrix of the Hessian matrix of $\mathfrak{f}$.

Proposition 5.9 ($\mathrm{char}(k) \ne 2$) *Let $\mathfrak{f}$ and t be as above. Then:*

(a) *$\mathfrak{f}$ is homaloidal.*
(b) *If m is odd, then $\mathfrak{f} = \left(\prod_{i=2}^{t} x_{i,2t+1-i} \right) q$, where*

$$q := \frac{1}{2} \det \begin{pmatrix} 2x_{1,1} & 2x_{1,2} & \cdots & 2x_{1,t-1} & 2x_{1,t} \\ 2x_{1,t+1} & 0 & \cdots & 0 & x_{t,t+1} \\ 2x_{1,t+2} & 0 & \cdots & x_{t-1,t+2} & 0 \\ \vdots & \vdots & \ddots & \vdots & \vdots \\ 2x_{1,2t-1} & x_{2,2t-1} & \cdots & 0 & 0 \end{pmatrix}.$$

If m is even, then $\mathfrak{f} = \left(\prod_{i=2}^{t-1} x_{i,2t-i} \right) q$ where

$$q = \frac{1}{2} \det \begin{pmatrix} 2x_{1,1} & 2x_{1,2} & \cdots & 2x_{1,t-1} & 2x_{1,t} \\ x_{1,t} & 0 & \cdots & 0 & x_{t,t} \\ 2x_{1,t+1} & 0 & \cdots & x_{t-1,t+1} & 0 \\ \vdots & \vdots & \ddots & \vdots & \vdots \\ 2x_{1,2t} & x_{2,2t} & \cdots & 0 & 0 \end{pmatrix}.$$

In particular, $\mathfrak{f}$ is a reduced polynomial.

(c) $\mathfrak{f}$ *is not a factor of* $\det \Theta(\partial(\mathfrak{f}))$.

Proof

(a) Write $\mathfrak{f}_{i,j} := \partial \mathfrak{f}/\partial x_{i,j}$, for $(i, j) \in \mathcal{I} \cup \mathcal{J}$ and $\Delta_{i,j}$ for the (i, j)th cofactor of $\mathfrak{H}$.
Since $\mathfrak{H}$ is a symmetric matrix, $\Delta_{i,j} = \Delta_{j,i}$ for any (i, j). By Proposition 2.10,
$\mathfrak{f}_{i,j} = \Delta_{i,j}$ possibly up to multiplication by 2 (recall that $\mathrm{char}(k) \neq 2$).
Therefore, the corresponding fields of fractions coincide. Thus, it is equivalent
to show that the rational map defined by the set $\boldsymbol{\Delta} = \{\Delta_{i,j} \,|\, (i, j) \in \mathcal{I} \cup \mathcal{J}\}$ is
birational.

Now, the adjugate matrix $\mathrm{adj}(\mathfrak{H}_m)$ has the following shape:

$$\mathrm{adj}(\mathfrak{H}_m) = \begin{pmatrix} \Delta_{1,1} & \Delta_{1,2} & \cdots & \Delta_{1,m-1} & \Delta_{1,m} \\ \Delta_{1,2} & * & \cdots & * & \Delta_{2,m} \\ \Delta_{1,3} & * & \cdots & \Delta_{3,m-1} & * \\ \vdots & \vdots & \ddots & \vdots & \vdots \\ \Delta_{1,m} & \Delta_{2,m} & \cdots & * & * \end{pmatrix}, \tag{5.19}$$

where $*$ denotes some unspecified entry. From the adjugate formula (2.1), one
gets the following linear relations among the cofactors of the set $\boldsymbol{\Delta}$:

$$x_{1j}\Delta_{1,1} + x_{j,m+2-j}\Delta_{1,m+2-j} = 0, \quad 2 \leq j \leq t$$

$$x_{1j}\Delta_{1,1} + x_{m+2-j,j}\Delta_{1,m+2-j} = 0, \quad t+1 \leq j \leq m,$$

as a result of the zero entries in $\mathfrak{f}\,\mathbb{I}_{m \times m}$, and

$$\left(\sum_{j=1}^{m} x_{1,j}\Delta_{1,j} \right) - x_{1,i}\Delta_{1,i} - x_{i,m+2-i}\Delta_{i,m+2-i} = 0, \quad 2 \leq i \leq t,$$

as the expression of equating two different terms that repeat the same entry
along the diagonal of $\mathfrak{f}\,\mathbb{I}_{m \times m}$.

These relations give rise to the following $(m+t-1) \times (m+t-2)$ submatrix
of the syzygy matrix of the set $\boldsymbol{\Delta}$:

$$\mathcal{Z} = \left(\begin{array}{cccccc|ccc}
x_{1,2} & x_{1,3} & \cdots & x_{1,t} & x_{1,t+1} & \cdots x_{1,m} & x_{1,1} & \cdots & x_{1,1} \\
0 & 0 & \cdots & 0 & 0 & \cdots x_{2,m} & 0 & \cdots & x_{1,2} \\
\vdots & \vdots & & \vdots & \vdots & \vdots & \vdots & & \vdots \\
0 & 0 & \cdots & 0 & x_{t,m+2-t} & \cdots 0 & x_{1,t} & \cdots & 0 \\
0 & 0 & \cdots x_{t,m+2-t} & 0 & & \cdots 0 & x_{1,t+1} & \cdots & x_{1,t+1} \\
\vdots & \vdots & & \vdots & \vdots & \vdots & \vdots & & \vdots \\
0 & x_{3,m-1} & \cdots & 0 & 0 & \cdots 0 & x_{1,m-1} & \cdots & x_{1,m-1} \\
x_{2,m} & 0 & \cdots & 0 & 0 & \cdots 0 & x_{1,m} & \cdots & x_{1,m} \\
\hline
0 & 0 & \cdots & 0 & 0 & \cdots 0 & -x_{2,m} & \cdots & 0 \\
\vdots & \vdots & & \vdots & \vdots & \vdots & \vdots & & \vdots \\
0 & 0 & \cdots & 0 & 0 & \cdots 0 & 0 & \cdots & -x_{t,m+2-t}
\end{array}\right).$$

It can be shown that $\mathcal{Z}$ has full rank. If we knew at this stage that the set $\boldsymbol{\Delta}$ is algebraically independent over k (e.g., if the ideal they generate is of linear type), we would conclude by Theorem 2.27.

Instead, by a quirk, we argue directly upon the submatrix of the Jacobian dual matrix ([6, Definition 3.2.18]) induced by $\mathcal{Z}$. Namely, introducing dual variables

$$\mathbf{Y} = \{y_{1,j}\ (1 \le j \le m),\ y_{i,m+2-i}\ (2 \le i \le t)\},$$

the transpose of the following matrix B is a submatrix of the Jacobian dual matrix of the set $\boldsymbol{\Delta}$

$$\left(\begin{array}{cccccc|ccc}
0 & 0 & \cdots & 0 & 0 & \cdots 0 \;\; 0 & y_{1,1} & \cdots & y_{1,1} \\
y_{1,1} & 0 & \cdots & 0 & 0 & \cdots 0 \;\; 0 & 0 & \cdots & y_{1,2} \\
0 & y_{1,1} & \cdots & 0 & 0 & \cdots 0 \;\; 0 & y_{1,3} & \cdots & y_{1,3} \\
\vdots & \vdots & & \vdots & \vdots & \vdots \;\; \vdots & \vdots & & \vdots \\
0 & 0 & \cdots y_{1,1} & 0 & \cdots 0 \;\; 0 & & y_{1,t} & \cdots & 0 \\
0 & 0 & \cdots & 0 & y_{1,1} & \cdots 0 \;\; 0 & y_{1,t+1} & \cdots & y_{1,t+1} \\
\vdots & \vdots & & \vdots & \vdots & \vdots \;\; \vdots & \vdots & & \vdots \\
0 & 0 & & 0 & 0 & \cdots y_{1,1} \;\; 0 & y_{1,m-1} & \cdots & y_{1,m-1} \\
0 & 0 & & 0 & 0 & \cdots 0 \;\; y_{1,1} & y_{1,m} & \cdots & y_{1,m} \\
\hline
y_{1,m} & 0 & \cdots & 0 & 0 & \cdots 0 \;\; y_{1,2} & -y_{2,m} & \cdots & 0 \\
0 & y_{1,m-1} & \cdots & 0 & 0 & \cdots y_{1,3} \;\; 0 & 0 & \cdots & 0 \\
\vdots & \vdots & & \vdots & \vdots & \vdots \;\; \vdots & \vdots & & \vdots \\
0 & 0 & \cdots y_{1,t+1} & y_{1,t} & \cdots & 0 \;\; 0 & 0 & \cdots & -y_{t,m+2-t}
\end{array}\right). \qquad (5.20)$$

Let B_1 be the $(m+t-2) \times (m+t-2)$ submatrix of B obtained by omitting the first row. By [6, Theorem 2.18], it suffices to show that $\det B_1$ does not vanish when the entries of B_1 are evaluated via $y_{i,j} \mapsto \Delta_{i,j}$. We have

$$\det B_1 = y_{1,1}^{m-1} \prod_{i=2}^{t} y_{i,m+2-i} + T,$$

where T is a polynomial in $k[\mathbf{y}]$ that does not depend on $y_{1,1}$. Evaluating as indicated gives the expression

$$\Delta_{1,1}^{m-1} \prod_{i=2}^{t} \Delta_{i,m+2-i} + T(\Delta_{1,2}, \ldots, \Delta_{1,m}, \Delta_{2,m}, \ldots, \Delta_{t,m+2-t}),$$

which does not vanish because, e.g., while its second summand belongs to the ideal $(x_{1,2}, \ldots, x_{1,m})^{m+t-1}$, the first does not.

(b) We will draw upon the details of the generic analogue of the hollow matrix as discussed in Proposition 4.10. To avoid a potential confusion, denote the entries of the generic hollow matrix by $z_{i,j}$ and its determinant by $\mathfrak{g}$. Then Proposition 4.10 (a) gives

$$\mathfrak{g} = z_{1,1}\widetilde{P} - \sum_{i+j=m+2} z_{1,j}z_{i,1}\frac{\widetilde{P}}{z_{i,j}}, \tag{5.21}$$

where $\widetilde{P} := \prod_{i=2}^{m} z_{i,m+2-i}$. Now specialize to the entries of the present generic symmetric hollow matrix. Namely, if m is odd, setting $P = \prod_{i=2}^{t} x_{i,m+2-i}$ and noting the specialized value of $\widetilde{P}$, we get

$$\mathfrak{f} = x_{1,1}P^2 - 2\sum_{i=2}^{t} x_{1,i}x_{1,m+2-i}\frac{P^2}{x_{i,m+2-i}}$$

$$= P\left(x_{1,1}P - 2\sum_{i=2}^{t} x_{1,i}x_{1,m+2-i}\frac{P}{x_{i,m+2-i}}\right)$$

$$= \frac{P}{2}\left(2x_{1,1}P - \sum_{i=2}^{t}(2x_{1,i})(2x_{1,m+2-i})\frac{P}{x_{i,m+2-i}}\right).$$

But once again, by the proof of Proposition 4.10,

$$\det\begin{pmatrix} 2x_{1,1} & 2x_{1,2} & \cdots & 2x_{1,t-1} & 2x_{1,t} \\ 2x_{1,t+1} & 0 & \cdots & 0 & x_{t,m+2-t} \\ 2x_{1,t+2} & 0 & \cdots x_{t-1,m+2-t+1} & 0 & 0 \\ \vdots & \vdots & \ddots & \vdots & \vdots \\ 2x_{1,2t-1} & x_{2,m} & \cdots & 0 & 0 \end{pmatrix} = 2x_{1,1}P - \sum_{i=2}^{t}(2x_{1,i})(2x_{1,m+2-i})\frac{P}{x_{i,m+2-i}}.$$

Thus, we are through in this case.

Now, if m is even, setting $Q := \prod_{i=2}^{t-1} x_{i,m+2-i}$, evaluating (5.21) yields

$$\mathfrak{f} = x_{1,1} Q^2 x_{t,m+2-t} - 2 \sum_{i=2}^{t-1} x_{1,i} x_{1,m+2-i} \frac{Q^2 x_{t,m+2-t}}{x_{i,m+2-i}} - x_{1,t} x_{1,m+2-t} Q^2$$

$$= Q \left(x_{1,1} P - 2 \sum_{i=2}^{t-1} x_{1,i} x_{1,m+2-i} \frac{P}{x_{i,m+2-i}} - x_{1,t} x_{1,m+2-t} Q \right)$$

$$= \frac{Q}{2} \left((2x_{1,1}) P - \sum_{i=2}^{t-1} (2x_{1,i})(2x_{1,m+2-i}) \frac{P}{x_{i,m+2-i}} - (2x_{1,t}) x_{1,m+2-t} Q \right).$$

Similarly, by the proof of Proposition 4.10, the factor

$$(2x_{1,1}) P - \sum_{i=2}^{t-1} (2x_{1,i})(2x_{1,m+2-i}) \frac{P}{x_{i,m+2-i}} - (2x_{1,t}) x_{1,m+2-t} Q$$

is the determinant of the matrix

$$\begin{pmatrix} 2x_{1,1} & 2x_{1,2} & \cdots & 2x_{1,t-1} & 2x_{1,t} \\ x_{1,t} & 0 & \cdots & 0 & x_{t,m+2-t} \\ 2x_{1,t+1} & 0 & \cdots x_{t-1,m+2-t+1} & 0 \\ \vdots & \vdots & \ddots & \vdots & \vdots \\ 2x_{1,2t} & x_{2,m} & \cdots & 0 & 0 \end{pmatrix}.$$

This concludes the proof.

(c) For each i, j such that $i + j = m + 2$ denote the (i, j)th entry of $\mathfrak{H}_m$ by c_{ij}. With this notation, regardless of the parity of m, Proposition 2.10 gives

$$\frac{\partial \mathfrak{f}}{\partial x_{1,1}} = \prod_{2 \leq u \leq m} c_{u,m+2-u}, \quad \text{and}$$

$$\frac{\partial \mathfrak{f}}{\partial x_{1,i}} = \pm 2 \det \begin{pmatrix} x_{1,2} & \cdots & x_{1,m+1-i} & x_{1,m+2-i} & x_{1,m+3-i} & \cdots & x_{1,m} \\ 0 & \cdots & 0 & 0 & 0 & \cdots & c_{2,m} \\ \vdots & & \vdots & \vdots & \vdots & \ddots & \vdots \\ 0 & \cdots & 0 & 0 & c_{i-1,m+3-i} & \cdots & 0 \\ 0 & \cdots c_{i+1,m+1-i} & 0 & 0 & \cdots & 0 \\ \vdots & \ddots & \vdots & \vdots & \vdots & \cdots & 0 \\ c_{m,2} & \cdots & 0 & 0 & 0 & \cdots & 0 \end{pmatrix}, \quad \text{if } 2 \leq i \leq m.$$

Barely visible, the row containing the entry $c_{i,m+2-i}$ is missing.

Then, somewhat laborious but straightforward calculation yields the following relations:

$$\frac{\partial^2 \mathfrak{f}}{\partial x_{1,j}\partial x_{1,1}} = \frac{\partial^2 \mathfrak{f}}{\partial x_{1,1}\partial x_{1,j}} = 0 \quad \text{if } 1 \leq j \leq m,$$

$$\frac{\partial^2 \mathfrak{f}}{\partial x_{2,m}\partial x_{1,1}} = \frac{\partial^2 \mathfrak{f}}{\partial x_{1,1}\partial x_{2,m}} = 2 \cdot \prod_{3 \leq u \leq m-1} c_{u,m+2-u} \quad \text{if } 2 \leq j \leq m,$$

$$\frac{\partial^2 \mathfrak{f}}{\partial x_{1,j}\partial x_{1,i}} = \frac{\partial^2 \mathfrak{f}}{\partial x_{1,i}\partial x_{1,j}} = 0 \quad \text{if } j \neq m+2-i,$$

and

$$\frac{\partial^2 \mathfrak{f}}{\partial x_{1,j}\partial x_{1,i}} = \frac{\partial^2 \mathfrak{f}}{\partial x_{1,i}\partial x_{1,j}} = \pm 2 \cdot \prod_{\substack{2 \leq u \leq m \\ u \neq i}} c_{u,m+2-u} \quad \text{if } j = m+2-i.$$

From these we get

$$\Theta(\partial(\mathfrak{f})) = \begin{pmatrix} 0 & 0 & \cdots & 0 & 2 \cdot \prod_{3 \leq u \leq m-1} c_{u,m+2-u} \\ 0 & 0 & \cdots \pm 2 \cdot \prod_{\substack{2 \leq u \leq m \\ u \neq 2}} c_{u,m+2-u} & * \\ \vdots & \vdots & \ddots & \vdots & \vdots \\ 0 & \pm 2 \cdot \prod_{\substack{2 \leq u \leq m \\ u \neq 2}} c_{u,m+2-u} \cdots & 0 & * \\ 2 \cdot \prod_{3 \leq u \leq m-1} c_{u,m+2-u} & * & \cdots & * & * \end{pmatrix}.$$

Thus, $\det\Theta(\partial(\mathfrak{f}))$ is a monomial. In particular, since $\mathfrak{f}$ is not a monomial by item (a), then $\det\Theta(\partial(\mathfrak{f})) \not\equiv 0 \mod \mathfrak{f}$. $\qquad\square$

Remark 5.10 Since $\mathfrak{f}$ is homaloidal, its partial derivatives are algebraically independent over k and so are the partial derivatives in the subset $\partial(\mathfrak{f})$. Item (c) above shows that in any characteristic $\neq 2$, the rank of the Hessian matrix of $\mathfrak{f}$ is at least $m+1$ – one expects this rank to be one less than the maximum $3m-2$ in any characteristic.

Item (c) of the above proposition also holds for the case of a hollow-filled symmetric linear section $\mathbf{L}$. Borrowing the previous notation, reset:

- $\mathcal{J} = \mathbf{I} \setminus \mathcal{I}.$
- $f := \det \mathbf{L}.$
- $\partial(f) =$ the set of partial derivatives of f with respect to $\mathfrak{H}_{\mathcal{I}\cup\{(2,m)\}}.$
- $\Theta(\partial(f) =$ the Jacobian matrix of $\partial(f)$ with respect to $\mathfrak{H}_{\mathcal{I}\cup\{(2,m)\}}.$

148   5 Symmetry Preserving Linear Sections of the Generic Symmetric Matrix

Corollary 5.11 *With the above notation,* $\det \Theta(\partial(f)) \neq 0 \bmod f$.

Proof Let $\mathfrak{f}$ be as in the previous proposition. Note that f is reduced since it specializes to the reduced $\mathfrak{f}$. Write $f = \mathfrak{f} + T$, where $\mathfrak{f}$ is as above and T belongs to the ideal $\mathfrak{n} := (\mathbf{L}_{\mathcal{J}})$. In particular,

$$\Theta(\partial(f)) \equiv \Theta(\partial(\mathfrak{f})) \bmod \mathfrak{n}.$$

Thus, it is enough to show that $\det \Theta(\partial(\mathfrak{f})) \neq 0 \bmod \mathfrak{f}$, which is the content of Proposition 5.9(c). $\qquad\square$

As in the generic case, we pose:

Conjecture 5.12 The gradient ideal of $\mathfrak{f}$ is of linear type.

5.3.2 Symmetric Semi-hollow Linear Sections

In this part we fix integers m, r with $1 \leq r \leq m - 2$ and consider the symmetric semi-hollow linear section of the $m \times m$ generic symmetric matrix, to wit:

$$\begin{pmatrix}
x_{1,1} & \cdots & x_{1,m-r} & x_{1,m-r+1} & x_{1,m-r+2} & \cdots & x_{1,m-1} & x_{1,m} \\
x_{1,2} & \cdots & x_{2,m-r} & x_{2,m-r+1} & x_{2,m-r+2} & \cdots & x_{2,m-1} & x_{2,m} \\
\vdots & \cdots & \vdots & \vdots & \vdots & \cdots & \vdots & \vdots \\
x_{1,m-r} & \cdots & x_{m-r,m-r} & x_{m-r,m-r+1} & x_{m-r,m-r+2} & \cdots & x_{m-r,m-1} & x_{m-r,m} \\
x_{1,m-r+1} & \cdots & x_{m-r,m-r+1} & x_{m-r+1,m-r+1} & x_{m-r+1,m-r+2} & \cdots & x_{m-r+1,m-1} & 0 \\
x_{1,m-r+2} & \cdots & x_{m-r,m-r+2} & x_{m-r+1,m-r+2} & x_{m-r+2,m-r+2} & \cdots & 0 & 0 \\
\vdots & \ddots & \vdots & \vdots & \vdots & \ddots & \vdots & \vdots \\
x_{1,m-1} & \cdots & x_{m-r,m-1} & x_{m-r+1,m-1} & 0 & \cdots & 0 & 0 \\
x_{1,m} & \cdots & x_{m-r,m} & 0 & 0 & \cdots & 0 & 0
\end{pmatrix}. \qquad (5.22)$$

By analogy to the notation employed in the generic case, assuming that m is fixed in the context, we denote it by $\mathbf{L}(\mathcal{S})[r]$.

Remark 5.13 If $\mathrm{char}(k) \neq 2$, by Proposition 2.10, any cofactor $\Delta_{i,j}$ of $\mathbf{L}(\mathcal{S})[r]$ such that $i + j \leq 2m - r$ is a scalar multiple of the partial derivative $\partial f / \partial x_{i,j}$ (the scalar is actually $\pm 1/2$).

Throughout we assume $\mathrm{char}(k) \neq 2$. At the other end, quite a bit of the arguments employed in the subsequent results work as well for the generic symmetric matrix (i.e., $r = 0$), a case sufficiently considered in Section 3.3. Therefore, we will assume throughout that $r \geq 1$.

Lemma 5.14 *Fix integers m, r with $1 \leq r \leq m - 2$, and let S and $\mathbf{L}(\mathcal{S})[r]$ denote, respectively, the generic $m \times m$ square matrix and its linear section as above. Then the number of distinct entries in S set to zero to get $\mathbf{L}(\mathcal{S})[r]$ is*

$$\mathfrak{o}(r) := \begin{cases} \frac{(r+1)^2}{4} & \text{if } r \text{ is odd} \\[2mm] \frac{r(r+2)}{4} & \text{if } r \text{ is even} \end{cases}. \tag{5.23}$$

In particular, the number of distinct variables appearing as entries in $\mathbf{L}(S)[r]$ *is* $\binom{m+1}{2} - \frac{(r+1)^2}{4}$ *(respectively,* $\binom{m+1}{2} - \frac{r(r+2)}{4}$*) if* r *is odd (respectively, even).*

Proof To count the distinct variables appearing as entries in the linear section sector of S, we may proceed, e.g., column-wise from right to left. If r is even, we get the summation $r + (r-2) + (r-4) + \cdots + 2$ as many such variables. In other words,

$$\sum_{l=1}^{r/2} 2l = 2 \sum_{l=1}^{r/2} l = 2\binom{r/2+1}{2} = r(r+2)/4.$$

If r is odd, get instead

$$\sum_{l=1}^{(r+1)/2} (2l-1) = 2\binom{(r+1)/2+1}{2} - (r+1)/2 = (r+1)^2/4.$$

This proves the statements. $\square$

In the notation of Section 1.5.2, $\mathfrak{o}(r) = \operatorname{codim}_S \mathbf{L}(S)[r]$.

We start with some basic properties of the submaximal minors of the matrix $\mathbf{L}(S)[r]$.

Proposition 5.15 *Let* $R = k[\mathbf{x}]$ *denote the polynomial ring in the nonzero entries of* $\mathbf{L}(S)[r]$*, with* $1 \leq r \leq m - 2$. *Set* $f := \det \mathbf{L}(S)[r]$ *and* $I = I_{m-1}(\mathbf{L}(S)[r])$. *One has:*

(a) *f is a nonzero irreducible polynomial.*
(b) *I specializes from the symmetric generic case. In particular, it is a linearly presented Cohen–Macaulay ideal of codimension 3.*
(c) *I is a prime ideal if and only if $r \leq m - 3$.*

Proof

(a) Note that for an arbitrary $1 \leq r \leq m - 2$, $f = \det \mathbf{L}(S[r])$ maps homomorphically to $\det \mathbf{L}(S[m-2])$ by an obvious map of the defining ground ring. Thus, it suffices to show that $f = \det \mathbf{L}(S[m-2])$ is irreducible and nonzero. Laplace expansion of the determinant $f = \det \mathbf{L}(S[m-2])$ along the first row yields the following $x_{1,1}$-monoid

$$f = \left(\prod_{i+j=m+2} x_{i,j} \right) x_{1,1} + H.$$

Thus, we need to show that no variable $x_{i,j}$ with $i + j = m + 2$ divides H. The latter is however clear by the following modular expression:

$$f \equiv H \equiv \prod_{i+j=m+1} x_{i,j} \not\equiv 0 \mod (x_{i,j} \mid i + j = m + 2),$$

from which also obviously follows that $f \neq 0$.

(b) This follows from Proposition 2.6.

(c) One implication is trivial as for $r = m - 2$ there is a non-prime submaximal minor.

Conversely, let $r \leq m - 3$. As we are in the range $r \geq 1$ ($r = 0$ is the generic symmetric matrix), then $m \geq 4$.

CLAIM. Let P be an associated prime of $R/I_{m-1}(\mathbf{L}(\mathcal{S})[r])$. Then, for some $1 \leq i \leq 3$, one has $x_{i,m} \notin P$.

This is clear: as otherwise $(\Delta_{m,m}, x_{1,m}, x_{2,m}, x_{3,m}) \subset P$, contradicting the fact that P has height 3 as an associated prime of the codimension three Cohen–Macaulay ring $R/I_{m-1}(\mathbf{L}(\mathcal{S})[r])$.

We induct on m. For this, consider the following symmetric matrix, obtained from $\mathbf{L}(\mathcal{S})[r]$ by permuting the first and last columns (respectively, the first and last rows):

$$\left(\begin{array}{cccccccccc}
0 & x_{1,m} & x_{2,m} & \cdots & x_{m-r,m} & 0 & 0 & \cdots & 0 & 0 \\
x_{1,m} & x_{1,1} & x_{1,2} & \cdots & x_{1,m-r} & x_{1,m-r+1} & x_{1,m-r+2} & \cdots & x_{1,m-2} & x_{1,m-1} \\
x_{2,m} & x_{1,2} & x_{2,2} & \cdots & x_{2,m-r} & x_{2,m-r+1} & x_{2,m-r+2} & \cdots & x_{2,m-2} & x_{2,m-1} \\
x_{3,m} & x_{1,3} & x_{2,3} & \cdots & x_{3,m-r} & x_{3,m-r+1} & x_{3,m-r+2} & \cdots & x_{3,m-2} & x_{3,m-1} \\
\vdots & \vdots & \vdots & \cdots & \vdots & \vdots & \vdots & \cdots & \vdots & \vdots \\
x_{m-r,m} & x_{1,m-r} & x_{2,m-r} & \cdots & x_{m-r,m-r} & x_{m-r,m-r+1} & x_{m-r,m-r+2} & \cdots & x_{m-r,m-2} & x_{m-r,m-1} \\
0 & x_{1,m-r+1} & x_{2,m-r+1} & \cdots & x_{m-r,m-r+1} & x_{m-r+1,m-r+1} & x_{m-r+1,m-r+2} & \cdots & x_{m-r+1,m-2} & 0 \\
0 & x_{1,m-r+2} & x_{2,m-r+2} & \cdots & x_{m-r,m-r+2} & x_{m-r+1,m-r+2} & x_{m-r+2,m-r+2} & \cdots & 0 & 0 \\
\vdots & \vdots & \vdots & \ddots & \vdots & \vdots & \vdots & \ddots & \vdots & \vdots \\
0 & x_{1,m-2} & x_{2,m-2} & \cdots & x_{m-r,m-1} & x_{m-r+1,m-2} & 0 & \cdots & 0 & 0 \\
0 & x_{1,m-1} & x_{2,m-1} & \cdots & x_{m-r,m-1} & 0 & 0 & \cdots & 0 & 0
\end{array} \right).$$

Since elementary operations do not change an ideal of minors, the ideal of submaximal minors of this matrix gives back $I_{m-1}(\mathbf{L}(\mathcal{S})[r])$.

In addition, since the 2×2 submatrix of the upper left corner of these matrix is invertible in $R[x_{1,m}^{-1}]$, then $I_{m-1}(\mathbf{L}(\mathcal{S})[r]) = I_{m-3}(\mathbf{L}(\widetilde{\mathcal{S}})[r - 1])$, where

$$\mathbf{L}(\tilde{\mathcal{S}})[r-1]) = \begin{pmatrix}
\tilde{x}_{2,2} & \cdots & \tilde{x}_{2,m-r} & \tilde{x}_{2,m-r+1} & \tilde{x}_{2,m-r+2} & \cdots & \tilde{x}_{2,m-2} & \tilde{x}_{2,m-1} \\
\tilde{x}_{2,3} & \cdots & \tilde{x}_{3,m-r} & \tilde{x}_{3,m-r+1} & \tilde{x}_{3,m-r+2} & \cdots & \tilde{x}_{3,m-2} & \tilde{x}_{3,m-1} \\
\vdots & \cdots & \vdots & \vdots & \vdots & \cdots & \vdots & \vdots \\
\tilde{x}_{2,m-r} & \cdots & \tilde{x}_{m-r,m-r} & \tilde{x}_{m-r,m-r+1} & \tilde{x}_{m-r,m-r+2} & \cdots & \tilde{x}_{m-r,m-2} & \tilde{x}_{m-r,m-1} \\
\tilde{x}_{2,m-r+1} & \cdots & \tilde{x}_{m-r,m-r+1} & \tilde{x}_{m-r+1,m-r+1} & \tilde{x}_{m-r+1,m-r+2} & \cdots & \tilde{x}_{m-r+1,m-2} & 0 \\
\tilde{x}_{2,m-r+2} & \cdots & \tilde{x}_{m-r,m-r+2} & \tilde{x}_{m-r+1,m-r+2} & \tilde{x}_{m-r+2,m-r+2} & \cdots & 0 & 0 \\
\vdots & \ddots & \vdots & \vdots & \vdots & \ddots & \vdots & \vdots \\
\tilde{x}_{2,m-2} & \cdots & \tilde{x}_{m-r,m-1} & \tilde{x}_{m-r+1,m-2} & 0 & \cdots & 0 & 0 \\
\tilde{x}_{2,m-1} & \cdots & \tilde{x}_{m-r,m-1} & 0 & 0 & \cdots & 0 & 0
\end{pmatrix}$$

with

$$\tilde{x}_{i,j} = \begin{cases} x_{i,j} - \dfrac{x_{i,m}(x_{1,m}x_{1,j}-x_{1,1}x_{j,m})+x_{1,i}x_{1,m}x_{i,m}}{x_{1,m}^2}, & \text{if } 2 \leq i \leq j \leq m-r \\ x_{i,j}, & \text{otherwise.} \end{cases}$$

Now, the above tilde variables are algebraically independent over k (see, e.g., [1, Lemma 2.1]).

Thus, by the inductive assumption, $R[x_{1,m}^{-1}]/I_{m-1}(\mathbf{L}(\mathcal{S})[r]))R[x_{1,m}^{-1}]$ is a domain. Hence, there is a unique associated prime P of $R/I_{m-1}(\mathbf{L}(\mathcal{S})[r])$ such that $x_{1,m} \notin P$.

If P is the unique associated prime of $R/I_{m-1}(\mathbf{L}(\mathcal{S})[r]))$, then $x_{1,m}$ is regular on $R/I_{m-1}(\mathbf{L}(\mathcal{S})[r]))$. In this case, $R/I_{m-1}(\mathbf{L}(\mathcal{S})[r]))$ is a domain as was to be shown.

Otherwise, suppose that there is another associated prime Q of $R/I_{m-1}(\mathbf{L}(\mathcal{S})[r]))$ such that $Q \neq P$. In particular, $x_{1,m} \in Q$. By the above claim, there is some $1 \leq i \leq 3$ such that $x_{i,m} \notin Q$. Since $R[x_{i,m}^{-1}]/I_{m-1}(\mathbf{L}(\mathcal{S})[r]))R[x_{i,m}^{-1}]$ is also a domain, Q is the unique associated prime of $R/I_{m-1}(\mathbf{L}(m)[r])$ such that $x_{i,m} \notin Q$. Hence, $x_{i,m} \in P$. Since

$$PR[x_{1,m}^{-1}]/I_{m-1}(\mathbf{L}(\mathcal{S})[r]))R[x_{1,m}^{-1}] = 0,$$

we have $x_{i,m}x_{1,m}^r \in I_{m-1}(\mathbf{L}(\mathcal{S})[r]))$ for some integer $r \geq 0$.

Now, consider the R-map $\varphi : R \to k$ such that

$$\varphi(x_{u,v}) = \begin{cases} 1, & \text{if } (u,v) = (1,m) \text{ or } (u,v) = (i,m) \\ 0, & \text{otherwise.} \end{cases}$$

Since φ maps the entire $I_{m-1}(\mathbf{L}(m)[r])$ to zero, this contradicts the last inclusion, and hence the assumed inequality $Q \neq P$. $\qquad\square$

The following lemma will play a role in one of the main results to follow. Keeping the previous notation, let J stand for the gradient ideal of $\det \mathbf{L}(\mathcal{S})[r]$.

Lemma 5.16 *Let*

$$t = \left(\prod_{\substack{i+j=2m-r \\ m-r \leq i \leq m-\frac{r-1}{2}}} x_{i,j} \right)^{2}, \quad \text{if } r \text{ is odd,}$$

and

$$t = \left(\prod_{\substack{i+j=2m-r \\ m-r \leq i \leq m-\frac{r}{2}}} x_{i,j} \right)^{2}, \quad \text{if } r \text{ is even.}$$

Then $t \in J : \Delta_{i,j}$, *for every* $1 \leq i \leq j \leq m$.

Proof We argue for odd r, the even case is similar with small adjustments.

Set $\delta := m - \frac{r-1}{2}$ for display convenience. The argument is a bit unsavory but will show something more, namely, for every $0 \leq l \leq (r-1)/2$, introduce the analogue

$$t_l = (x_{\delta-1,\delta} \cdot x_{\delta-2,\delta+1} \cdots x_{\delta-1-l,\delta+l})^{2}.$$

We then prove that $t_l \in J : \Delta_{i,j}$, for $1 \leq i \leq j \leq \delta + l$, where $\Delta_{i,j}$ is any entry in the following $(\delta + l) \times (\delta + l)$ submatrix of the cofactor matrix of $\mathbf{L}(\mathcal{S})[r]$:

$$\begin{pmatrix} \Delta_{1,1} & \Delta_{1,2} & \cdots & \Delta_{1,\delta+l} \\ \Delta_{2,1} & \Delta_{2,2} & \cdots & \Delta_{2,\delta+l} \\ \vdots & \vdots & \cdots & \vdots \\ \Delta_{\delta+l,1} & \Delta_{\delta+l,2} & \cdots & \Delta_{\delta+l,\delta+l} \end{pmatrix}. \tag{5.24}$$

The stated result follows by setting $l = (r-1)/2$.

We induct on l. For the initial step, observe that there is one single cofactor not belonging to J, namely, $\Delta_{\delta,\delta}$. The adjugate relation for the matrix (5.24) yields

$$x_{1,\delta-1}\Delta_{\delta,1} + \cdots + x_{\delta-1,\delta-1}\Delta_{\delta,\delta-1} + x_{\delta-1,\delta}\Delta_{\delta,\delta} = 0.$$

By Remark 5.13, any cofactor $\Delta_{i,j}$ such that $i + j < 2\delta = 2m - r + 1$ belongs to J; hence, $x_{\delta-1,\delta}\Delta_{\delta,\delta} \in J$. Therefore, certainly $x_{\delta-1,\delta}^{2}\Delta_{i,j} \in J$ for all $1 \leq i \leq j \leq \delta$.

Now suppose, by the inductive step, that $t_l \in J : \Delta_{i,j}$ for all $1 \leq i \leq j \leq \delta + l$. In other words, $t_l \in J : \Delta_{i,j}$ for all $1 \leq i, j \leq \delta + l$ because $\Delta_{i,j} = \Delta_{j,i}$, by symmetry. Consider the next submatrix

$$
\left(
\begin{array}{cccc|c}
\Delta_{1,1} & \Delta_{1,2} & \cdots & \Delta_{1,\delta+l} & \Delta_{1,\delta+l+1} \\
\Delta_{2,1} & \Delta_{2,2} & \cdots & \Delta_{2,\delta+l} & \Delta_{2,\delta+l+1} \\
\vdots & \vdots & \cdots & \vdots & \vdots \\
\Delta_{\delta+l,1} & \Delta_{\delta+l,2} & \cdots & \Delta_{\delta+l,\delta+l} & \Delta_{\delta+l,\delta+l+1} \\
\hline
\Delta_{\delta+l+1,1} & \Delta_{\delta+l+1,2} & \cdots & \Delta_{\delta+l+1,\delta+l} & \Delta_{\delta+l+1,\delta+l+1}
\end{array}
\right), \tag{5.25}
$$

obtained by the obvious bordering of the previous submatrix. Equally obviously, $t_{l+1} = t_l\, x^2_{\delta-l-2,\delta+(l+1)}$. Therefore, it suffices to show that $t_l\, x^2_{\delta-l-2,\delta+(l+1)}\Delta_{i,j} \in J$ for all $1 \le i \le j \le \delta + l + 1$.

Since $t_l\Delta_{i,j} \in J$ for all $1 \le i \le j \le \delta+l$ and $\Delta_{i,\delta+l+1} \in J$ for all $i \le \delta-l-2$, it remains to see that $t_l\, x^2_{\delta-l-2,\delta+l+1}\Delta_{i,\delta+l+1} \in J$, for $\delta - l - 1 \le i \le \delta + l + 1$.

The adjugate relation for (5.25) yields the following equality:

$$
\sum_{j=1}^{\delta-l-2} x_{j,\delta-l-2}\Delta_{i,j} + \sum_{j=\delta-l-1}^{\delta+l+1} x_{\delta-l-2,j}\Delta_{i,j} = 0, \tag{5.26}
$$

for $\delta - l - 1 \le i \le \delta + l + 1$. Multiplying through by t_l for $i \ne \delta + l + 1$ yields

$$
\sum_{j=1}^{\delta-l-2} t_l\, x_{j,\delta-l-2}\Delta_{i,j} + \sum_{j=\delta-l-1}^{\delta+l+1} t_l\, x_{\delta-l-2,j}\Delta_{i,j} = 0,
$$

for $\delta - l - 1 \le i \le \delta + l$.

Since, by hypothesis, $t_l\,\Delta_{i,j} \in J$ for all $1 \le i \le j \le \delta + l$, this expression affords $t_l \cdot x_{\delta-l-2,\delta+l+1}\Delta_{i,\delta+l+1} \in J$, for every $\delta - l - 1 \le i \le \delta + l$.

Taking (5.26) for the extreme value $i = \delta + l + 1$ and multiplying its two sides by $t_l\, x_{\delta-l-2,\delta+l+1}$ give $t_l\, x^2_{\delta-l-2,\delta+l+1}\Delta_{\delta+l+1,\delta+l+1} \in J$, which does the case $i = \delta + l + 1$ as well. ⊓

Theorem 5.17 *With the previous notation, one has:*

(i) *Let $J \subset R$ denote the gradient ideal of $f := \det \mathbf{L}(S)[r]$. Then*

$$
\mathrm{codim}(J) = \begin{cases} 2 \ \textit{if } m - r = 2 \\ 3 \ \textit{otherwise.} \end{cases}
$$

In particular, $R/(f)$ is a normal domain if and only if $m - r \ge 3$.

(ii) *The submaximal minors define a birational map $\mathbb{P}^{\binom{m+1}{2}-\mathfrak{o}(r)-1} \dashrightarrow (\mathbb{P}^{\binom{m+1}{2}-1})^{\vee}$ onto its image.*

(iii) *The homogeneous coordinate ring of the polar image $\mathfrak{P}(f) \subset \mathbb{P}^{\binom{m+1}{2}-\mathfrak{o}(r)-1}$ is a ladder (symmetric) determinantal ring of dimension $\binom{m+1}{2} - 2\mathfrak{o}(r)$; in particular, the image of the rational map in (ii) is a cone over the polar image $\mathfrak{P}(f)$ with vertex cut by $\mathfrak{o}(r)$ coordinate hyperplanes.*

Proof

(i) Since f is irreducible (radical would suffice), the gradient J has codimension at least 2. On the other hand, J is contained in the ideal of $(m-1)$-minors and the latter has codimension at most 3, hence so does J. In the case $m - r = 2$, it is quite evident that $J \subset \left(x_{1,m}, x_{2,m}\right)$. Therefore, one is done with this case.

Now assume that $m - r \geq 3$. Let $a_{i,j}$ denote the (i, j)-entry of $\mathbf{L}(S)[r]$.

By Remark 5.13, $\Delta_{1,1}, \ldots, \Delta_{1,m} \in J$. Thus, by the adjugate formula, $f = \sum_{i=1}^{m} x_{1,i} \Delta_{1,i} \in J$. Let t be the monomial as in Lemma 5.16, whose support involves only variables $x_{i,j}$ on the sub-anti-diagonal $i + j = 2m - r$. Its main role is that $t \Delta_{m,m} \in J$.

Let $\widetilde{\Delta}_{i,j}$ denote the (i, j)th cofactor of the $m \times m$ generic symmetric matrix S, and $\widetilde{f}$, the determinant of S. By Lemma 2.5,

$$\mathrm{in}(\widetilde{\Delta}_{1,1}) = \prod_{i+j=m+2} a_{i,j}, \ \mathrm{in}(\widetilde{f}) = \prod_{i+j=m+1} a_{i,j} \text{ and } \mathrm{in}(t \cdot \widetilde{\Delta}_{m,m}) = t \cdot \prod_{i+j=m} a_{i,j}.$$

Since these monomials do not involve the variables $x_{i,j}$ such that $1 \leq i \leq j \leq m$ and $i + j > 2m - r$, as in the proof of the Proposition 2.6, one has that the regular sequence $\widetilde{f}, \widetilde{\Delta}_{1,1}, t \cdot \widetilde{\Delta}_{m,m}$ specializes, that is, the sequence $f, \Delta_{1,1}, t \cdot \Delta_{m,m} \in J$ is a regular sequence. Thus, J has codimension at least three.

(ii) By Proposition 5.15 (iii), I has maximal linear rank. On the other hand, its analytic spread is maximal by Proposition 2.3. Therefore, by Theorem 2.27, the submaximal minors define a birational map onto the image.

(iii) By and large the general scope of the argument parallels that of Theorem 4.30 (a). Since the numbers are different, for the sake of clarity, we proceed ab initio.

Consider the matrix $\mathbf{L}(S)[r]$ written in the dual variables $y_{i,j}$ and set $S := k[y_{i,j}]$. Let $\mathcal{L} = \mathcal{L}(m, r)$ denote the ladder ideal depicted below, with respect to the lifted $(m - 1) \times (m - 1)$ symmetric matrix of indeterminates. As such, let $I_{m-r}(\mathcal{L})$ stand for the ideal generated by the $(m - r) \times (m - r)$ minors of $\mathbf{L}(S)[r]$ involving only the entries of $\mathcal{L}$.

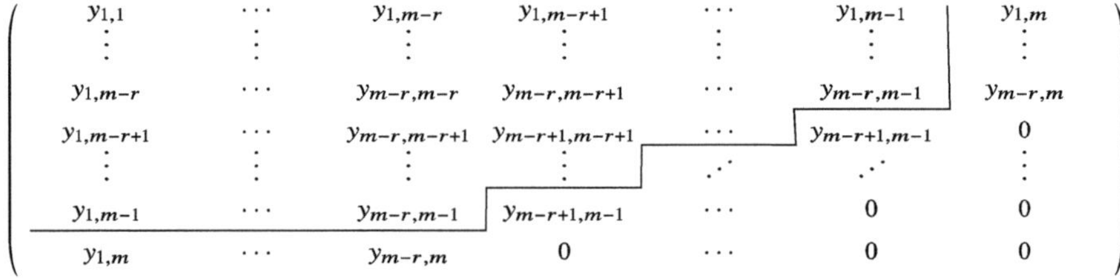

Then the ring $K[\mathcal{L}]/I_{m-r}(\mathcal{L})$ is one of the so-called symmetric ladder determinantal rings. Recall the notation where $\Delta_{j,i}$ stands for the cofactor of the (i, j)th entry of $\mathbf{L}(S)[r]$ and $f_{i,j}$ denotes the $x_{i,j}$-derivative of f.

CLAIM 1. The homogeneous defining ideal of the image of the polar map of f contains the ideal $I_{m-r}(\mathcal{L})$.

Given integers $1 \leq i_1 < i_2 < \ldots < i_{m-r} \leq m - 1$, consider the following submatrix of the adjugate matrix of $\mathbf{L}(\mathcal{S})[r]$:

$$F = \begin{pmatrix} \Delta_{1,i_1} & \Delta_{i_1,2} & \Delta_{i_1,3} & \cdots & \Delta_{i_1,m-i_{m-r}+(m-r-1)} \\ \Delta_{1,i_2} & \Delta_{i_2,2} & \Delta_{i_2,3} & \cdots & \Delta_{i_2,m-i_{m-r}+(m-r-1)} \\ \vdots & \vdots & \vdots & \cdots & \vdots \\ \Delta_{1,i_{m-r}} & \Delta_{i_{m-r},2} & \Delta_{i_{m-r},3} & \cdots & \Delta_{i_{m-r},m-i_{m-r}+(m-r-1)} \end{pmatrix}.$$

Letting

$$C = \begin{pmatrix} x_{1,i_{m-r}+1} & x_{1,i_{m-r}+2} & \cdots & x_{1,m-1} & x_{1,m} \\ \vdots & \vdots & \cdots & \vdots & \vdots \\ x_{m-r,i_{m-r}+1} & x_{m-r,i_{m-r}+2} & \cdots & x_{m-r,m-1} & x_{m-r,m} \\ x_{m-r+1,i_{m-r}+1} & x_{m-r+1,i_{m-r}+2} & \cdots & x_{m-r+1,m-1} & 0 \\ \vdots & \vdots & \cdots & \vdots & \vdots \\ x_{m-i_{m-r}+(m-r-2),i_{m-r}+1} & x_{m-i_{m-r}+(m-r-2),i_{m-r}+2} & \cdots & 0 & 0 \\ x_{m-i_{m-r}+(m-r-1),i_{m-r}+1} & 0 & \cdots & 0 & 0 \end{pmatrix},$$

the adjugate formula $\mathrm{adj}(\mathbf{L}(\mathcal{S})[r])\mathbf{L}(\mathcal{S})[r] = \det \mathbf{L}(\mathcal{S})[r]\mathbb{I}_m$ yields the relation

$$FC = 0.$$

Since the columns of C are linearly independent, it follows that the rank of F is at most $m - i_{m-r} + (m - r - 1) - (m - i_{m-r}) = (m - r) - 1$. In other words, the maximal minors of the matrix

$$Y = \begin{pmatrix} y_{1,l_1} & y_{l_1,2} & y_{l_1,3} & \cdots & y_{l_1,m-l_{m-r}+(m-r-1)} \\ y_{1,i_2} & y_{i_2,2} & y_{i_2,3} & \cdots & y_{i_2,m-i_{m-r}+(m-r-1)} \\ \vdots & \vdots & \vdots & \cdots & \vdots \\ y_{1,i_{m-r}} & y_{i_{m-r},2} & y_{i_{m-r},3} & \cdots & y_{i_{m-r},m-i_{m-r}+(m-r-1)} \end{pmatrix}$$

all vanish upon evaluation $y_{i,j} \mapsto \Delta_{i,j}$.

Since $f_{i,j}$ is a scalar multiple of $\Delta_{i,j}$ by Proposition 2.10, the maximal minors of Y vanish on the partial derivatives of f, thus proving that the homogeneous defining ideal of the image of the polar map of f contains the ideal $I_{m-r}(\mathcal{L})$.

CLAIM 2: The codimension of the ideal $I_{m-r}(\mathcal{L}(m, r))$ is at least $o(r)$.

We induct with the following inductive hypothesis: letting $1 \leq i \leq r-1$, for any $(m-i) \times (m-i)$ matrix of the form $\mathbf{L}(\mathcal{S})[r]$, the ideal $I_{(m-i)-(r-i)}(\mathcal{L}(m-i, r-i))$ has codimension at least $o(r - i)$.

Note that $(m - i) - (r - i) = m - r$; hence, the size of the inner minor does not change in the inductive step.

We descend with regard to i; thus, the induction step starts out at $i = r - 1$; hence $m - i = m - r + 1$, and we are in the situation of an $(m - r + 1) \times (m - r + 1)$ matrix of the form $\mathbf{L}(\mathcal{S})[1]$. Clearly, the ladder ideal $I_{(m-(r-1))-(r-(r-1))}\mathcal{L}(m - r + 1, r - (r - 1))$ is a principal ideal. Therefore, its codimension is $\mathfrak{o}(1) = 1$ as desired.

In order to construct a suitable inductive predecessor, let $\widetilde{\mathcal{L}}$ denote the set of entries to the left and above the stair-like polygonal depicted below

$$
\left(
\begin{array}{cccccc}
y_{1,1} & \cdots & y_{1,m-r} & \cdots & y_{1,m-3} & y_{1,m-2} \\
\vdots & \vdots & \vdots & \vdots & \vdots & \vdots \\
y_{1,m-r} & \cdots & y_{m-r,m-r} & \cdots & y_{m-r,m-3} & y_{m-r,m-2} \\
y_{1,m-r+1,} & \cdots & y_{m-r,m-r+1} & \cdots & y_{m-r+1,m-3} & \\
\vdots & \vdots & \vdots & \vdots & & \\
y_{1,m-3} & \cdots & y_{m-r,m-3} & \cdots & & \\
y_{1,m-2} & \cdots & y_{m-r,m-2} & & &
\end{array}
\right).
$$

Note that $\widetilde{\mathcal{L}}$ is a ladder ideal of the form $\mathcal{L}(m - 1, r - 1)$ on the correspondingly lifted $(m - 2) \times (m - 2)$ generic symmetric matrix, while sitting on an $(m - 1) \times (m - 1)$ matrix $\mathbf{L}(\mathcal{S})[r - 1]$. In particular $I_{m-r}(\widetilde{\mathcal{L}})$ is a Cohen–Macaulay prime ideal (see [2]). By the inductive hypothesis, the codimension of $I_{m-r}(\widetilde{\mathcal{L}})$ is at least $\mathfrak{o}(r - 1)$.

Note that $\widetilde{\mathcal{L}}$ is a subset $\mathcal{L}$; hence there is a natural ring surjection:

$$
S := \frac{k[\mathcal{L}]}{I_{m-r}(\widetilde{\mathcal{L}})k[\mathcal{L}]} = \frac{k[\widetilde{\mathcal{L}}]}{I_{m-r}(\widetilde{\mathcal{L}})}[\mathcal{L} \setminus \widetilde{\mathcal{L}}] \twoheadrightarrow \frac{k[\mathcal{L}]}{I_{m-r}(\mathcal{L})}.
$$

Since $\mathfrak{o}(r - 1) + \lceil \frac{r}{2} \rceil = \mathfrak{o}(r)$, it suffices to exhibit $\lceil \frac{r}{2} \rceil$ elements of $I_{m-r}(\mathcal{L})$ forming a regular sequence over S.

Consider the matrix

$$
\left(
\begin{array}{ccc|c}
y_{1,1} & \cdots & y_{1,m-r-1} & y_{1,m-i-1} \\
\vdots & \ddots & \vdots & \vdots \\
y_{1,m-r-1} & \cdots & y_{m-r-1,m-r-1} & y_{m-r-1,m-i-1} \\
\hline
y_{1,m-i} & \cdots & y_{m-r-1,m-i} & y_{m-r+i,m-i-1}
\end{array}
\right)
\tag{5.27}
$$

for $i = 0, 1, \ldots, \lceil \frac{r}{2} \rceil - 1$. Let $\Delta_i \in I_{m-r}(\mathcal{L})$ denote the determinant of the above matrix, for $i = 0, 1, \ldots, \lceil \frac{r}{2} \rceil - 1$.

The claim is that $\mathbf{\Delta} = \{\Delta_0, \ldots, \Delta_{\lceil \frac{r}{2} \rceil - 1}\}$ is a regular sequence on S.

Let δ denote the $(m-r-1)$-minor in the upper left corner of (5.27). Clearly, δ is a regular element on S as the defining ideal of the latter is a prime ideal generated in degree $m - r$. Therefore, it suffices to show that the localized sequence

$$\Delta_\delta = \{(\Delta_0)_\delta, \ldots, (\Delta_{\lceil \frac{r}{2} \rceil - 1})_\delta\}$$

is a regular sequence on S_δ. On the other hand, since S is Cohen–Macaulay, it suffices to show that $\dim S_\delta / \Delta_\delta S_\delta = \dim S_\delta - \lceil \frac{r}{2} \rceil$.

Write $\mathcal{Y} = \{y_{m-r,m-1}, y_{m-r+1,m-2}, \ldots, y_{m-r+(\lceil \frac{r}{2} \rceil - 1), m-(\lceil \frac{r}{2} \rceil - 1)-1}\}$. Note that, for every $i = 0, \ldots, \lceil \frac{r}{2} \rceil - 1$, one has $(\Delta_i)_\delta = y_{m-r+i,m-i-1} + (1/\delta)\Gamma_i$, with $y_{m-r+i,m-i-1} \in \mathcal{Y}$ and $\Gamma_i \in k[\mathcal{L} \setminus \mathcal{Y}]$. The association $y_{m-r+i,m-i-1} \mapsto -(1/\delta)\Gamma_i$ therefore defines a ring homomorphism

$$k[\mathcal{L}]_\delta / (\Delta_\delta) = (k[\mathcal{Y}][\mathcal{L} \setminus \mathcal{Y}])_\delta / (\Delta_\delta) \simeq k[\mathcal{L} \setminus \mathcal{Y}]_\delta.$$

This entails a ring isomorphism

$$\frac{S_\delta}{\Delta_\delta S_\delta} \simeq \frac{k[\mathcal{L} \setminus \mathcal{Y}]_\delta}{(I_{m-r}(\widetilde{\mathcal{L}})k[\mathcal{L} \setminus \mathcal{Y}])_\delta}.$$

Thus, $\dim S_\delta / \Delta_\delta S_\delta = \dim k[\mathcal{L}]_\delta - \lceil \frac{r}{2} \rceil - \mathrm{codim}\, I_{m-r}(\widetilde{\mathcal{L}})_\delta = \dim S_\delta - \lceil \frac{r}{2} \rceil$. Therefore, $\mathrm{codim}\,(I_{m-r}(\mathcal{L}))$ is at least $\mathrm{codim}\,(I_{m-r}(\widetilde{\mathcal{L}})) + \lceil \frac{r}{2} \rceil = \mathrm{o}(r)$.

In order to show that $I_{m-r}(\mathcal{L})$ is the homogeneous defining ideal $\mathfrak{J}$ of the polar image $\mathfrak{P}(f) \subset \mathbb{P}^{\binom{m+1}{2}-\mathrm{o}(r)-1}$, it suffices to show that $\mathrm{ht}\,\mathfrak{J} \le \mathrm{o}(r)$. But, as in the proof of (ii), the homogeneous defining ideal $\mathfrak{D}$ of the image of Ψ has height $\mathrm{o}(r)$. Therefore, the extended ideal of $\mathfrak{J}$, which is contained in $\mathfrak{D}$, has height at most $\mathrm{o}(r)$; hence, so does $\mathfrak{J}$ itself.

It remains to argue that the image is a cone over $\mathfrak{P}(f)$. The reason has essentially been given. In detail, recall that the k-vector space $[I]_{m-1}$ is spanned by $[J]_{m-1}$ and additional minors $\Delta_1, \ldots, \Delta_{\mathrm{o}(r)}$. Counting k-algebra dimensions, $\dim k[[I]_{m-1}] = \binom{m+1}{2} - \mathrm{o}(r)$, while $\dim k[[J]_{m-1}] = \binom{m+1}{2} - 2\mathrm{o}(r)$ (Theorem 5.17). This triggers the algebraic independence of $\Delta_1, \ldots, \Delta_{\mathrm{o}(r)}$ over k; hence, the larger algebra is a polynomial ring over the smaller one in $\mathrm{o}(r)$ indeterminates, which is the meaning of the assertion. □

Remark 5.18

(1) In contrast to the issue in Conjecture 4.37 as to whether R/J is Cohen–Macaulay when $r = m - 2$, here the picture is that R/J has an embedded component of codimension 3 (see Exercise 5.22).

(2) Some basic questions concerning the structure of the ideal of submaximal minors – such as, e.g., its primeness as a function of o – are largely open.

Exercises

(The problems below will by and large require nontrivial computer assistance.)

5.19 Consider the 3×3 case of the symmetric section $\mathbf{L}(S)$ of (5.1) and the notation of Theorem 5.6, posing $I = I_2(\mathbf{L}(S))$:

1. Show that $I = (J, \Delta_{3,3})$.
2. Show that I is the unmixed component of J (so, in particular, J has height three).
3. What is the embedded component of J?

5.20 Consider the linear section of the 3×3 generic symmetric matrix by replicating $x_{1,1}$ onto $x_{2,3}$. Changing the names of the variables for the sake of visualization, the resulting linear section is the matrix

$$\mathfrak{S} = \begin{pmatrix} x_1 & x_2 & x_3 \\ x_2 & x_4 & x_1 \\ x_3 & x_1 & x_5 \end{pmatrix}.$$

1. Show that the linear rank of the gradient ideal $J \subset k[x_1, x_2, x_3, x_4, x_5]$ of $\det \mathfrak{S}$ is 3, one below the maximum $5 - 1 = 4$. Thus, item (ii) of Theorem 5.6 fails here.
2. Show that J is an ideal of linear type; hence by [3, Proposition 3.4], $\det \mathfrak{S}$ is not homaloidal, showing that item (i) of Theorem 5.6 fails as well.

5.21 Consider the following sparse linear section of the 4×4 generic symmetric matrix:

$$\mathfrak{S} = \begin{pmatrix} 0 & x_1 & x_2 & x_3 \\ x_1 & 0 & x_4 & x_5 \\ x_2 & x_4 & 0 & x_6 \\ x_3 & x_5 & x_6 & 0 \end{pmatrix}.$$

Pretty much as in the previous problem, show:

1. The gradient ideal J of $\det \mathfrak{S}$ has linear rank 3.
2. J is a Cohen–Macaulay radical ideal.
3. J is of linear type (hence, not homaloidal).

5.22 Consider the hollow matrix of (5.18) for $m = 3$ ($\mathrm{char}(k) \neq 2$):

1. Show directly that the gradient ideal J of the determinant $\mathfrak{f}$ has at least four independent linear syzygies.

    (HINT: Show that the subideal $(\mathfrak{f}_{1,1}, \mathfrak{f}_{1,2}, \mathfrak{f}_{1,3})$ is $x_{2,3}$ times an ideal generated by a regular sequence, which gives three independent (reduced Koszul) syzygies; argue that $\{\mathfrak{f}_{1,1}, \mathfrak{f}_{1,3}, \mathfrak{f}_{2,3}\}$ have an obvious additional linear syzygy.)

2. Show that $\mathfrak{f}_{1,1}, \mathfrak{f}_{1,2}, \mathfrak{f}_{1,3}, \mathfrak{f}_{2,3}$ are algebraically independent over k.
   (HINT: Argue that the Hessian of $\mathfrak{f}$ has rank four.)
3. What are the odds against mimicking the above reasoning for the case where $m = 4$?

5.23 Consider the semi-hollow matrix of (5.22) for $r = m-2$ and let $J \subset R$ denote the gradient ideal of its determinant:

1. Justify that J is not a *syzygetic ideal* (see [6, Chapter 8, Subsection 8.4.1.2]).
   Let $m = 3$.
2. Prove that R/J is almost Cohen–Macaulay with linear minimal free resolution.
   (HINT: The details of the resolution may require computer assistance.)
3. Give the unmixed component and the embedded component of R/J.
4. Digress on the linear section by $x_{2,2} - x_{1,3}$ (cf. Chapter 6).

References

1. M. Barile, Arithmetical ranks of ideals associated to symmetric and alternating matrices. J. Algebra **176** (1995), 59–82 151
2. A. Conca, Ladder determinantal rings. J. Pure Appl. Algebra **98**, 119–134 (1995) 156
3. A. Doria, H. Hassanzadeh, A. Simis, A characteristic free criterion of birationality. Adv. Math. **230**, 390–413 (2012) 158
4. S. Goto, S. Tachibana, A complex associated with a symmetric matrix. J. Math. Kyoto Univ. **17**, 51–54 (1977) 133, 134
5. T. Jósefiak, Ideals generated by minors of a symmetric matrix. Comment. Math. Helvetici **53**, 595–607 (1978) 133
6. Z. Ramos, A. Simis, *Graded Algebras in Algebraic Geometry*. Expositions in Mathematics, vol. 70 (De Gruyter, Berlin, 2022) 141, 144, 159
7. A. Simis, *Commutative Algebra* (De Gruyter Textbook, Berlin, 2020) 133

Chapter 6
Linear Sections of the Generic Square Hankel Matrix

Abstract Hankel matrices (or their Toeplitz versions) have a classic history in Analysis and Functional Operator Theory. One of the best-known problems, which permeates the theory of orthogonal polynomials, is the Hamburger Momentum Problem. Some modern aspects of this activity are reflected in probabilities. Typical Hankel matrices in this connection are matrices with real or complex number entries, and some of the problems ask when they enjoy the property of being positive-definite. Furthermore, these matrices are allowed to be infinite. In algebraic geometry a Hankel matrix appears with functional or, more concretely, polynomial entries in an arbitrary number of indeterminates. The best-known model is when these entries are the same as the indeterminate ones. In this form, they give rise to some of the determinantal ideals studied in the book. Two examples of important algebraic varieties that can be defined using Hankel matrices are the secant manifolds of the classic rational normal curve and the normal rational scrolls. The chapter gives an overall survey of the generic square Hankel matrix with a combinatorial flavor that reflects upon the related algebraic invariants. The discussion of Hankel linear sections covers a good mileage in the chapter, with an emphasis on the section now known as a sub-Hankel matrix. For the early commutative algebra treatment of Hankel matrices we refer to Watanabe (Proc School Sci Tokai Univ 32:11–21, 1997) and Conca (Adv Math 138:263–292, 1998).

6.1 Nuts and Bolts

Given positive integers s, n such that $n = 2s - 1$, and elements $\mathbf{a} = \{a_1, \ldots, a_n\}$ of a commutative ring R, the *Hankel matrix* of order $s \times (n - s + 1)$ defined by $\mathbf{a}$ is the following matrix

$$\mathcal{H}_{s,n-s+1}(\mathbf{a}) := \begin{pmatrix} a_1 & a_2 & a_3 & \ldots & a_s & \ldots & a_{n-s} & a_{n-s+1} \\ a_2 & a_3 & a_4 & \ldots & a_{s+1} & \ldots & a_{n-s+1} & a_{n-s+2} \\ a_3 & a_4 & a_5 & \ldots & a_{s+2} & \ldots & a_{n-s+2} & a_{n-s+3} \\ \vdots & \vdots & \vdots & \ldots & \vdots & & \vdots & \vdots \\ a_s & a_{s+1} & a_{s+2} & \ldots & a_{2s-1} & \ldots & a_{n-1} & a_n \end{pmatrix}. \tag{6.1}$$

Z. Ramos, A. Simis, *Determinantal Ideals of Square Linear Matrices*,
https://doi.org/10.1007/978-3-031-55284-7_6

Popularized by the XIX Century British school under the designation *orthosymmetric* (variation: *persymmetric*), it became increasingly germane in other areas, mainly due to being the mirror image of the as much popular Toeplitz matrix.

If the elements **a** are clear from the context we usually omit them in the notation.

A notable case is the one where these elements are indeterminates over a ground ring k (typically, a field), in which case we call the above matrix the *generic Hankel matrix* of size $s \times (n - s + 1)$ over k, a terminology to be used throughout without further ado.

Under special focus here is the case where $s = n - s + 1 := m$, the square Hankel matrix $\mathcal{H}_{m,m}$, henceforth denoted $\mathcal{H}_m$ for simplicity.

The ideals of minors of a Hankel matrix have a notable property. It is known that, for any $1 \le t \le \min\{s, n - s + 1\}$, the ideal of t-minors of the generic Hankel matrix $\mathcal{H}_{s,n-s+1}$ over an algebraically closed field is prime and of codimension $n - 2t + 2$. A proof of this statement is in [9, Proposition 4.3], where the result itself is referred to earlier authors – see also [26, Theorem 1] for the case where the ground field is replaced by a Noetherian ring with a view toward generic perfection. The proof of this result in the first reference uses a distinguished trait of a Hankel matrix, first made explicit in work of Gruson and Peskine ([12, Lemme 2.3]), which for our convenience is stated in the following form:

Proposition 6.1 *For any $t \le s$, one has $I_t(\mathcal{H}_{s,n-s+1}) = I_t(\mathcal{H}_{t,n-t+1})$.*

This property allows to reduce to the case of maximal minors. In this case, to get the codimension one may use the fact that the Hankel matrix specializes to a well-known shape involving only $2m - 2t + 1$ variables – but see Lemma 6.13 later on for a more precise statement. In particular, in the case of the square Hankel matrix $\mathcal{H}_m$, its ideal of submaximal minors coincides with the ideal of maximal minors of $\mathcal{H}_{m-1,,m+1}$.

6.2 Special Results for the Square Generic Hankel Matrix

Much of the section first appeared in [15, 16], and [7].

We focus on the square generic Hankel matrix $\mathcal{H}_m$ of order m over a field k and on its determinant f. Let $R = k[\mathbf{x}] = k[x_1, \ldots, x_{2m-1}]$ stand for the corresponding ground polynomial ring on the entries of $\mathcal{H}_m$ over the field k.

6.2.1 Related Combinatorics

As before, $\mathcal{H}_{r,s}$ denotes the $r \times s$ generic Hankel matrix over a field k. When $r = s =: m$, one has $I_{m-1}(\mathcal{H}_m) = I_{m-1}(\mathcal{H}_{m-1,m+1})$ (Proposition 6.1). The advantage of this transfer is that not only one deals with maximal minors of a simple case of a non-square generic Hankel matrix but also their minimal number of generators becomes the predicted one $\binom{m+1}{2}$. Therefore, we deal in the sequel with the maximal minors of $\mathcal{H}_{m-1,m+1}$.

In the following we will repeatedly make the usual abuse of calling the determinant of a square submatrix by the word "minor."

In this part, the minor of $\mathcal{H}_{m-1,m+1}$ with columns $i_1 < \cdots < i_{m-1}$ will be denoted $[i_1, \ldots, i_{m-1}]$. Pretty much as in the case of a generic matrix [2, chapter 4], the set $\mathcal{P}_{m-1,m+1}$ of these minors is a poset by ordering the underlying indices, setting

$$[i_1, \ldots, i_{m-1}] \leq [j_1, \ldots, j_{m-1}] \quad \Leftrightarrow \quad i_1 \leq j_1, \ldots, i_{m-1} \leq j_{m-1}.$$

As an illustration, in the case of $m = 5$, one has

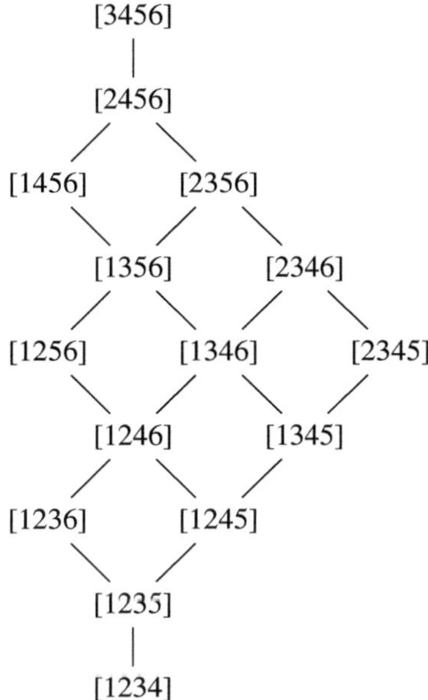

where a link to the successive upper level denotes $\leq$.

This poset has some remarkable properties:

Proposition 6.2 (Properties of $\mathcal{H}_{m-1,m+1}$).

(1) *Let* $\binom{m}{2} \leq l \leq \binom{m+2}{2} - 3$. *The lth* level *of the poset is the subset of minors* $[i_1, \ldots, i_{m-1}]$ *such that* $i_1 + \cdots + i_{m-1} = l$ *(i.e., the minors indexed by the ordered partitions of l).*

(2) *A maximal minor in the lth level admits at most two comparable maximal minors on the* $(l + 1)$*th level.*

To see this, note that a maximal minor $[i_1, \ldots, i_{m-1}]$ *is uniquely determined by the complementary columns in the matrix. Thus, one can write*

$$A = [i_1, \ldots, i_{m-1}] = [1, \ldots, i-1, \hat{i}, \ldots, j-1, \hat{j}, \ldots, m+1],$$

where $\widehat{}$ *denotes deletion. Therefore, in this notation there are at most two comparable minors with A in the successive upper level, namely*

$$[1, \ldots, \widehat{i-1}, i, \ldots, j-1, \hat{j}, \ldots, m+1] \ \text{and}$$

$$[1, \ldots, i-1, \hat{i}, \ldots, \widehat{j-1}, j, \ldots, m+1].$$

The above properties depend only on the size of the matrix. Subtler properties are as follows:

(3) (*Level elements*) *The cofactors of* $y_{t,u}$ *and* $y_{u,t}$ *coincide and the* (*determinants of the*) *maximal minors on the lth level of the poset are the distinct cofactors along the anti-diagonal* $\{y_{t,u} | t+u = 2m-l\}$. *In particular, there is a "central" level with largest possible number of elements.*

(4) (*Partial derivatives*) *Let* $f_j, 1 \le j \le 2m-1$, *denote the partial derivative of* $\det \mathcal{H}_{m,m}$ *with respect to* x_j. *Then* f_{2m-l} *is a* (*possibly vanishing in positive characteristic*) $\mathbb{Z}$-*linear combination of the maximal minors of* $\mathcal{H}_{m-1,m+1}$ *in the lth level of the poset.*

   (Note that this is reminiscent of the result in Proposition 2.10).

(5) (*Plücker relations*) *In the generic case it is a classical result that the so-called* Plücker relations *generate the defining ideal of the maximal minors. A Plücker relation in the case of a generic matrix of arbitrary size* $p \times q$ *has a somewhat cumbersome expression, with many indices floating around (see* [2, Lemma 4.4]). *Luckily, in the present environment, where* $p = m-1, q = m+1$, *assuming that the characteristic of the ground field is zero, the expression becomes simpler and these simpler expressions can be shown to generate the ideal of relations. This is based on the elementary observation to the effect that two maximal minors have in common at least* $m-3$ *indices. Thus, the intersection*

$$\{i_1, \ldots, i_{m-1}\} \cap (\{1, \ldots, m+1\} \setminus \{k_1, \ldots, k_{m-1}\})$$

has at most two elements. Consequently, the typical Plücker relation has at most 3 terms, each a product of two minors, while the relevant ones correspond to the case where the above intersection has exactly 2 elements, with the cases of 0 and 1 element yielding, respectively, an empty equation or an identity. Thus, the shape can be described up to certain signs by assuming that, e.g., $k_1 = i_1, \ldots, k_{m-3} = i_{m-3}$ *and the above intersection is the set* $\{i_{m-2}, i_{m-1}\}$, *which affords the relation*

$$[i_1, \ldots, i_{m-1}] [k_1, \ldots, k_{m-1}] = [i_1, \ldots, i_{m-3}, i_{m-2}, i_{m-1}][i_1, \ldots, i_{m-3}, k_{m-2}, k_{m-1}]$$

$$= [i_1, \ldots, i_{m-3}, \widehat{i_{m-2}}, i_{m-1}, k_{m-2}][i_{m-2}, i_1, \ldots, i_{m-3}, \widehat{k_{m-2}}, k_{m-1}]$$

$$+ [i_1, \ldots, i_{m-3}, i_{m-2}, \widehat{i_{m-1}}, k_{m-2}][i_{m-1}, i_1, \ldots, i_{m-3}, \widehat{k_{m-2}}, k_{m-1}].$$

Up to signs and reordering of indices, this is the shape of a Plücker relation that intervenes in the proof of Theorem 6.5 in the next section.

Perhaps more surprising than property (5) above is the following result, which for the best of our knowledge first appeared in [7]:

Proposition 6.3 *Let k be a field and let*

$$\mathcal{G} = (y_{u,v})_{\substack{1 \le u \le m-1 \\ 1 \le v \le m+1}} \text{ and } \mathcal{H} = (x_{i+j-1})_{1 \le i \le m-1, 1 \le j \le m+1}$$

denote, respectively, the $(m-1) \times (m+1)$ generic matrix over k and the generic Hankel matrix of the same size over k. Then the respective special fiber algebras of the ideals $I_{m-1}(\mathcal{G})$ and $I_{m-1}(\mathcal{H})$ of maximal minors are isomorphic as graded k-algebras.

Proof Consider the natural k-algebra specialization map

$$k[y_{u,v}] \twoheadrightarrow k[x_1, \ldots, x_{2m-1}], \quad y_{u,v} \mapsto x_{u+v-1}, \quad 2 \le u + v \le 2m \tag{6.2}$$

of the corresponding ground polynomial rings. Since each of the ideals in question is equigenerated, its fiber algebra is graded k-isomorphic to the respective k-subalgebra of the polynomial ring generated by the maximal minors, which one denotes here by $k[\Delta(\mathcal{G})]$ and $k[\Delta(\mathcal{H})]$, respectively.

Clearly, the map (6.2) restricts to a k-algebra surjection

$$k[\Delta(\mathcal{G})] \twoheadrightarrow k[\Delta(\mathcal{H})]. \tag{6.3}$$

But both algebras have Krull dimension $2m - 1$.

Indeed, $k[\Delta(\mathcal{G})]$ is, up to normalization, graded isomorphic to the homogeneous coordinate ring of the Grassmannian $G(m-1, m+1)$, while the latter has dimension $2(m-1) = 2m - 2$ (Proposition 3.7).

The fact that also $\dim k[\Delta(\mathcal{H})] - 2m - 1$ follows from identifying the ideal of maximal minors $I_{m-1}(\mathcal{H})$ with the ideal of submaximal minors of the associated $m \times m$ Hankel matrix (see Proposition 6.1). After this is done the dimension comes out immediately off Proposition 2.3.

To conclude, the kernel of the surjection (6.3) is a prime ideal of height zero in a domain, hence must be the zero ideal. $\qquad\square$

Remark 6.4 A more geometric type of argument is devised in [20, Proposition 4.4]. For arbitrary size of a Hankel matrix this turns out to be false, simply because the two dimensions are different, but perhaps more crucial as regards the present material, since minimal quadratic relations other than Plücker relations come in the picture as is explained in [4] and [5]. As will be noted at the end of the next part, Plücker relations are not the right tool to deal with the semi-hollow sections $\mathcal{H}[r]$.

6.2.2 The Generic Polar Map

In this part, unless otherwise stated, we assume that $\mathrm{char}(k) = 0$. Recall from Section 3.1 the notion of the polar map of a form f in a polynomial ring over k and its homaloidness.

Our interest is in the case where $f = \det \mathcal{H}_m$. As before, let $\mathbf{f} = \{f_1, \ldots, f_{2m-1}\}$ denote the partial derivatives of f as elements of the polynomial ring $R = k[x_1, \ldots, x_{2m-1}]$. As is well-known, f is homaloidal if and only if the homogeneous inclusion $k[\mathbf{f}] \subset k[R_{m-1}]$ induces an isomorphism of the respective fields of fractions. In addition, let $\boldsymbol{\Delta} \subset R$ denote the set of the $(m-1)$-minors of $\mathcal{H}_m$. By Proposition 2.10, one has yet another homogeneous inclusion $k[\mathbf{f}] \subset k[\boldsymbol{\Delta}]$.

The following fundamental result has been essentially obtained in [15, Proposition 3.37] (see also [16, Proposition 3.13]).

Theorem 6.5 $(\mathrm{char}(k) = 0)$ $\sqrt{(\mathbf{f})k[\boldsymbol{\Delta}]} = (\boldsymbol{\Delta})k[\boldsymbol{\Delta}]$.

Proof It suffices to show that any $\Delta \in \boldsymbol{\Delta}$ has a power in $(\mathbf{f})k[\boldsymbol{\Delta}]$.

By property (4) of Proposition 6.2, every partial derivative $f_j \in R$ is a sum of maximal minors of $\mathcal{H}_{m-1,m+1}$, while by (3) of that proposition, every Δ is a summand of some f_j.

We will then argue, by descending induction on $j = 1, \ldots, 2m - 1$, that any maximal minor $\Delta \in \boldsymbol{\Delta}$, which is a summand of f_j, has a power in $(\mathbf{f})k[\boldsymbol{\Delta}]$.

For $j = 2m - 1, 2m - 2$ this is trivial since f_{2m-1}, f_{2m-2} are themselves (signed) minors up to a coefficient. For clarity in the inductive step, we display the details of one additional step – although this step is really embedded in the general inductive step. Namely, one deals now with the minor $\Delta = [1, \ldots, n-3, n-1, n]$, a summand of the partial derivative f_{2m-3}, whose complementary summands as a consequence of properties (3) and (4) of Proposition 6.2 boil down to the sole minor $\Delta' = [1, \ldots, n-2, n+1]$. Write $f_{2m-3} = \lambda \Delta + \Delta'$.

Next, consider the following Plücker relation:

$$\Delta \Delta' = 1/2[1, \ldots, n-3, n-1, n+1]f_{2m-2} - [1, \ldots, n-3, n, n+1]f_{2m-1}. \tag{6.4}$$

Then, $\Delta \Delta' \in (\mathbf{f})k[\boldsymbol{\Delta}]$.

On the other hand, one has the obvious relation:

$$\Delta^2 - 1/\lambda \Delta (\lambda \Delta + \Delta') + 1/\lambda \Delta \Delta' = 0. \tag{6.5}$$

From this follows immediately that $\Delta^2 \in (\mathbf{f})k[\boldsymbol{\Delta}]$.

Now, for the general inductive step, fix $j < 2m - 3$ and assume that, for any t such that $j < t \leq 2m - 1$, any minor appearing as a summand of f_t admits a power in $(\mathbf{f})k[\boldsymbol{\Delta}]$. Then one wishes to prove that the same holds for any minor appearing as a summand of f_j.

The argument is similar, by showing that any such Δ satisfies an equation of integral dependence of the form

$$\Delta^2 + a\Delta f_j + bg = 0, \tag{6.6}$$

for suitable $a, b \in k$, where g is a (signed) sum of products of two maximal minors, one of which always appears as a summand of some f_t, with $t > j$. (This step requires again the use of a suitable Plücker relation.)

By the inductive hypothesis the latter minors have a power in $(\mathbf{f})k[\Delta]$, and therefore so does every such g. Finally, (6.6) implies that some power of Δ belongs to $(\mathbf{f})k[\Delta]$ as well.

For the explicit details of the required Plücker relation we refer to [16, The proof of Proposition 3.13]. $\square$

Corollary 6.6 (char(k) $=$ 0) *The ring inclusion $k[\mathbf{f}] \subset k[\Delta]$ is an integral extension. In particular, $\dim k[\mathbf{f}] = \dim k[\Delta]$ and the partial derivatives of f are algebraically independent over k.*

Proof Fairly general, let $A = k[A_1] \subset B = k[B_1]$ denote a homogeneous inclusion of standard graded algebras over k. Then $B_1 \subset \sqrt{A_1 B}$ implies that B is finitely generated as an A-module. In fact, letting $B_1^N \subset (A_1 B)_N = A_1^N + A_1 B^{N-1}$ for certain $N \geq 1$, then it is clear that B is generated by $\{B_1, B_1^2, \ldots, B_1^{N-1}\}$ as an A-module.

The assertion about the algebraic independence follows since the ideal $I_{m-1}(\mathcal{H}_m) = (\Delta)$ of submaximal minors has maximal analytic spread by Proposition 2.3. $\square$

One also retrieves the result of [15, Proposition 3.3.11]:

Corollary 6.7 (char(k) $=$ 0) *The Hessian of $f := \det \mathcal{H}_m$ does not vanish.*

The following theorem solves affirmatively some of the questions in [16, Conjecture 3.16].

Theorem 6.8 *Let $\mathcal{H}_m$ denote the $m \times m$ ($m \geq 3$) generic Hankel matrix and let $k[\mathbf{f}] \subset k[\Delta]$ be as above. Then:*

(a) *The k-algebra $k[\Delta]$ is a Gorenstein unique factorization domain whose regularity over the polynomial ring R is $m - 2$.*
    Suppose, moreover, that char(k) $=$ 0. Then:
(b) *The extension $k[\mathbf{f}] \subset k[\Delta]$ is a Noether normalization.*
(c) *The polar map of $f = \det \mathcal{H}_m$ is not homaloidal.*
(d) *The gradient ideal $J := (\mathbf{f}) \subset R$ is a reduction of the ideal of minors $I_{m-1}(\mathcal{H}_m)$ with reduction number $m - 2$.*

Proof

(a) It is classical that the homogeneous coordinate ring of a Grassmannian is Gorenstein, as it follows from a result of [17], by using that it is Cohen–Macaulay [13] and a unique factorization domain [22].

As for the regularity, one knows that the degree of the Hilbert series (as a rational function) of a Cohen–Macaulay standard graded k-algebra C coincides with its a-invariant $a(C)$. On the other hand, under the same hypothesis, since $k[\Delta]$ coincides with the special fiber of the ideal $(\Delta) \subset R$, by Proposition 2.15) and the comment right after it, the regularity of C coincides with the degree of the polynomial in the numerator of its Hilbert series. Since the denominator has degree $\dim C$, it follows that the regularity in the case of the Grassmannian (with the notation of Proposition 6.3) is $2m - 1 - a(k[\Delta(\mathcal{G})])$. On the other hand, $a(k[\Delta(\mathcal{G})]) = -(m + 1)$ ([1, Corollary 1.4]).

Therefore, the regularity of $k[\Delta(\mathcal{G})]$ is $2m - 1 - (m + 1) = m - 2$.

To conclude, one uses Proposition 6.3.

(b) By Corollary 6.6 the extension $k[\mathbf{f}] \subset k[\Delta]$ is integral and $k[\mathbf{f}]$ is a polynomial ring over k.

(c) Since $k[\mathbf{f}]$ is a polynomial ring it is integrally closed in its field of fractions. If the extension $k[\mathbf{f}] \subset k[\Delta]$ is birational then Corollary 6.6 implies that $k[\mathbf{f}] = k[\Delta]$. But this is impossible since for $m \geq 3$ there is some cofactor which is not in the k-linear span of $k[\mathbf{f}]$ (actually the k-linear span of the $(m - 1)$-minors has dimension $\binom{m+1}{2} > 2m - 1$ for $m \geq 3$).

(d) The first of the two statements is a well-known consequence of (b). To see it, note that the k-algebra $k[\Delta]$ is graded isomorphic to the special fiber of the ideal $I_{m-1}(\mathcal{H}_m) = (\Delta) \subset R$. Then the result is explained in the proof of [25, Theorem 1.77]

For the reduction number, since $k[\Delta]$ is Cohen–Macaulay, by Proposition 2.15 the reduction number of J is the degree of the polynomial in the numerator of its Hilbert series. But this is also the regularity by the facts in the proof of (a). Therefore, one gets the required value. $\square$

Remark 6.9 The result of item (c) has also been obtained in [18] by a geometric argument.

6.3 Linear Sections Preserving the Hankel Structure

In this section we focus on linear sections of the generic Hankel matrix that preserve the Hankel structure. Such matrices are, in particular, symmetry preserving sections of the generic symmetric matrix.

6.3.1 Hollow Like Version

Consider the following Hankel hollow section of the symmetric generic matrix $\mathbf{X}$ as in Section 5.8, where $\mathbf{X}_{\mathcal{I}} = \{x_{1,1}, x_{1,2}, \ldots, x_{1,m}\} \cup \{x_{2,m}\}$.

$$\mathfrak{H}(\mathcal{H}_m) = \begin{pmatrix} x_{1,1} & x_{1,2} & x_{1,3} & \cdots & x_{1,m-1} & x_{1,m} \\ x_{1,2} & 0 & 0 & \cdots & 0 & x_{2,m} \\ x_{1,3} & 0 & 0 & \cdots & x_{2,m} & 0 \\ \vdots & \vdots & \vdots & \ddots & \vdots & \vdots \\ x_{1,m-1} & 0 & x_{2,m} & 0 & \cdots & 0 \\ x_{1,m} & x_{2,m} & 0 & \cdots & 0 & 0 \end{pmatrix}. \tag{6.7}$$

Proposition 6.10 (char(k) $= 0$) *Let* $\mathfrak{f} := \det \mathfrak{H}(\mathcal{H}_m)$. *Then:*

(i) $\mathfrak{f} = x_{2,m}^{m-2}\,\widetilde{q}$, *where* $\widetilde{q} = \frac{1}{x_{2,m}^{m-3}}\frac{\partial \mathfrak{f}}{\partial x_{2,m}} + cx_{1,1}x_{2,m}$ *for a suitable nonzero scalar* $c \in k$.

(ii) $\mathfrak{f}$ *is homaloidal.*

(iii) *The Hessian of* $\mathfrak{f}$ *is* $d\,x_{2,m}^m$ *for some nonzero* $d \in k$. *In particular,* $\mathfrak{f}$ *is not a factor of its Hessian.*

Proof

(i) One can consider $\mathfrak{H}(\mathcal{H}_m)$ as a specialization of the hollow linear section of the $m \times m$ generic symmetric matrix of (5.18). As such $\mathfrak{f}$ is the image of its determinant via the map of k-algebras

$$\zeta : k[x_{1,i}\ (1 \le i \le m),\ x_{i,m-i+2}\ (1 \le i \le \lfloor m/2 \rfloor + 1)] \to k[x_{1,i}\ (1 \le i \le m), x_{2,m}]$$

such that $x_{1,i} \mapsto x_{1,i}\ (1 \le i \le m)$ and $x_{i,m-i+2} \mapsto x_{2,m}\ (1 \le i \le \lfloor m/2 \rfloor + 1)$. Drawing upon the notation and details of Proposition 5.9 (b), we get:

$$\mathfrak{f} = \frac{\zeta(P)}{2}\left(2x_{1,1}\zeta(P) - \sum_{i=2}^{t}(2x_{1,i})(2x_{1,m+2-i})\frac{\zeta(P)}{\zeta(x_{i,m+2-i})}\right).$$

$$= x_{2,m}^{2t-3}\left(x_{1,1}x_{2,m} - 2\sum_{i=2}^{t}x_{1,i}x_{1,m+2-i}\right)$$

$$= x_{2,m}^{m-2}\left(x_{1,1}x_{2,m} - 2\sum_{i=2}^{t}x_{1,i}x_{1,m+2-i}\right) \tag{6.8}$$

if m is odd, and

$$\mathfrak{f} = \frac{\zeta(Q)}{2}\left((2x_{1,1})\zeta(P) - \sum_{i=2}^{t-1}(2x_{1,i})(2x_{1,m+2-i})\frac{\zeta(P)}{\zeta(x_{i,m+2-i})}\right.$$

$$-(2x_{1,t})x_{1,m+2-t}\zeta(Q)\Bigg)$$

$$= x_{2,m}^{2t-4}\left(x_{1,1}x_{2,m} - 2\sum_{i=2}^{t-1} x_{1,i}x_{1,m+2-i} - x_{1,t}x_{1,m+2-t}\right)$$

$$= x_{2,m}^{m-2}\left(x_{1,1}x_{2,m} - 2\sum_{i=2}^{t-1} x_{1,i}x_{1,m+2-i} - x_{1,t}x_{1,m+2-t}\right)$$

if m is even.

A moment reflection let us retrieve the stated shape in the item.

(ii) Since the derivatives of $\mathfrak{f}$ and $x_{2,m}\widetilde{q}$ define the same rational map, it suffices to deal with the latter.

By inspection, the Jacobian ideal J of $x_{2,m}\widetilde{q}$ is generated by the quadrics

$$\mathfrak{Q} = \{x_{2,m}^2, x_{1,2}x_{2,m}, \ldots, x_{1,m}x_{2,m}, x_{1,1}x_{2,m} + Q\}$$

for a suitable quadric Q in other variables than $x_{1,1}$. But, the rational map defined by $\mathfrak{Q}$ is a generalized de Jonquières transformation with identity support (see [19, Subsection 7.2.1]). Thus, $x_{2,m}\widetilde{q}$ is homaloidal.

(iii) This follows immediately from the shape of the partial derivatives in (ii). $\square$

6.3.2 Semi-Hollow Like Version

In this part we fix integers m, r with $0 \leq r \leq m - 2$ and consider the Hankel semi-hollow like linear section, i.e., the Hankel version of the semi-hollow symmetric section considered in Section 5.3.2. Such matrices have recently been discussed in connection with the rational theory of Taylor polynomials [6].

Precisely, set:

$$\mathcal{H}_m[r] = \begin{pmatrix} x_{1,1} & x_{1,2} & \cdots & x_{1,m-r} & x_{1,m-r+1} & \cdots & x_{1,m-1} & x_{1,m} \\ x_{1,2} & x_{1,3} & \cdots & x_{1,m-r+1} & x_{1,m-r+2} & \cdots & x_{1,m} & x_{2,m} \\ \vdots & \vdots & \vdots & \vdots & & \vdots & \vdots & \vdots \\ x_{1,m-r} & x_{1,m-r+1} & \cdots & x_{1,m-r+2} & x_{1,m-r+3} & \cdots & x_{m-r-1,m} & x_{m-r,m} \\ x_{1,m-r+1} & x_{1,m-r+2} & \cdots & x_{1,m-r+3} & x_{1,m-r+4} & \cdots & x_{m-r,m} & 0 \\ x_{1,m-r+2} & x_{1,m-r+3} & \cdots & x_{1,m-r+4} & x_{1,m-r+5} & \cdots & 0 & 0 \\ \vdots & \vdots & \vdots & \vdots & & & & \vdots \\ x_{1,m} & x_{2,m} & \cdots & x_{m-r,m} & 0 & \cdots & 0 & 0 \end{pmatrix}. \tag{6.9}$$

One retrieves the fully generic case as $\mathcal{H}_m[0]$, while $\mathcal{H}_m[m-2]$ was dubbed *sub-Hankel* in [3] and will be treated in Section 6.4. This notation will also be used when the Hankel matrix is not necessarily square, namely, $\mathcal{H}_{s,n-s+1}[r]$.

The ground ring is the polynomial ring $k[x_{1,1}, \ldots, x_{1,m}, x_{2,m}, \ldots, x_{m-r,m}]$. In the sequel it will be convenient to rename these variables as

$$x_1 := x_{1,1}, \ldots, x_m := x_{1,m}, x_{m+1} := x_{2,m}, \ldots, x_{2m-r-1} := x_{m-r,m}.$$

If no confusion arises, when r is fixed in the discussion, we will denote this ring simply by R.

The following elementary result will be used throughout without further ado.

Proposition 6.11 *Let $\mathcal{H}_m[r]$ denote the linear section as in (6.9) with $0 \le r \le m-2$ and let $R = k[x_1, \ldots, x_{2m-1-r}]$. Then*

 (i) $\det \mathcal{H}_m[r] \ne 0$.
 (ii) $\det \mathcal{H}_m[r] \in R$ *is irreducible.*
 (iii) *The ideal $I_{m-1}(\mathcal{H}_m[r]) \subset R$ has maximal analytic spread.*

Proof

 (i) There are many elementary ways of verifying the non-vanishing of $f := \det \mathcal{H}_m[r]$. Perhaps an easy one is to see that f has a unique nonzero pure term in x_m, namely, the product of the entries along the main anti-diagonal.
 (ii) For any $0 \le r \le m-2$, $\det \mathcal{H}_m[r]$ maps homomorphically onto $\det \mathcal{H}_m[m-2]$ by an obvious map of the ground ring. Thus, it suffices to deal with the latter. Here, one has $\det \mathcal{H}_m[m-2] = (-1)^{m-1} x_1 x_{m+1}^{m+1} + g$, where $g \in k[x_2, \ldots, x_{m+1}]$ has a pure power of x_m as a nonzero term. Thus, $\det \mathcal{H}_m[m-2]$ is monic in x_1 and x_{m+1} is not a factor of g, i.e., it is a primitive polynomial, hence irreducible.
 (iii) With (i), it follows immediately from Proposition 2.3. $\qquad\qquad\square$

Remark 6.12 Since f is homogeneous, a rather convoluted alternative argument for the case $r \le m-3$ consists in showing that $R/(f)$ is normal. Since $R/(f)$ is a hypersurface ring, it suffices to prove that it is locally regular in codimension one. By Proposition 6.19 below, proved independently, the gradient ideal J has codimension $3 = 1 + 2$ provided $r \le m-3$. This proves that f is irreducible when $r \le m-3$. This sort of argument repeats itself over and over when enough is known about the gradient ideal.

6.3.3 The Lower Minors of $\mathcal{H}_m[r]$

Ideals of minors of a (not necessarily square) Hankel matrix over a Noetherian ring, having maximal possible grade R are perfect, a result established in [26] in the footsteps of the classical results of Eagon–Hochster–Northcott. In particular, in the generic case, if R is a domain, these ideals are prime and similar results hold concerning normality.

In this subsection we establish what can be recovered in the case of the linear section $\mathcal{H}_m[r]$ over a field.

As a preliminary we need a result about the lower minors of a not necessarily square generic Hankel matrix.

Let $\mathcal{H} := \mathcal{H}_{a,b}$ denote the generic Hankel matrix of order $a \times b$, with $a \le b$ and entries $x_1, \ldots, x_{a+b-1}$ over a field k. Set $R = k[x_1, \ldots, x_{a+b-1}]$ for the ground polynomial ring on these entries. Then the ideal $I_a(\mathcal{H})$ of maximal minors has the expected codimension $b - a + 1$, and moreover, $R/I_a(\mathcal{H})$ is a Cohen–Macaulay ring (Theorem 1.24, via [11, Proposition 6.3]).

Lemma 6.13 *With the above notation, an explicit system of parameters of $R/I_a(\mathcal{H})$ is given by the residues of the elements $x_1, \ldots, x_{a-1}, x_{b+1}, \ldots, x_{a+b-1}$.*

Proof Set $\mathbf{X} = (x_{i,j})$ for the fully generic $a \times b$ matrix $\mathbf{X} = (x_{i,j})$, for which it is well-known that $I_a(\mathbf{X}) \subset S := k[x_{i,j}]$ has codimension $b - a + 1$. Consider the usual $(a-1)(b-1)$ independent binomial linear forms in the $x_{i,j}$'s that define $\mathcal{H}$ and let $D \subset S$ denote the ideal generated by these forms. The isomorphism $S/D \simeq R$ induces an isomorphism $S/I_a(\mathbf{X}) \,(\mathrm{mod}\,D) \simeq R/I_a(\mathcal{H})$.

Note $\dim S/I_a(\mathbf{X}) = ab - (b - a + 1) = (b + 1)(a - 1)$ and $\dim R/I_a(\mathcal{H}) = a + b - 1 - (b - a + 1) = 2(a - 1)$, hence, the difference of these dimensions is $(a - 1)(b - 1)$. Since both rings are Cohen–Macaulay, it follows that D is a regular sequence on $S/I_a(\mathbf{X})$. On the other hand, by [24, Theorem 1], the initial ideal of $I_a(\mathbf{X})$ in a convenient order is generated by the products of the entries along the anti-diagonals of the a-minors. Therefore, it follows from this that the stated set $\{x_1, \ldots, x_{a-1}, x_{b+1}, \ldots, x_{a+b-1}\}$ is a regular sequence on $R/I_a(\mathcal{H})$, hence a system of parameters as it has the right length. $\qquad\square$

For the next result we set $R := k[x_1, \ldots, x_{2m-1}]$ and $\bar{R} := k[x_1, \ldots, x_{2m-r-1}]$ for the respective ground polynomial rings of $\mathcal{H}_m$ and of $\mathcal{H}_m[r]$. We can also view the latter as the residue ring of the first by the ideal $(x_{2m-r}, \ldots, x_{2m-1})$.

Proposition 6.14 *Assume that $0 \le r \le m - 2$ and let $1 \le t \le m$. Then*

(i) $\operatorname{codim} I_t(\mathcal{H}_m[r]) = \min\{2(m - t) + 1, 2m - t - r\}$ *and* $\bar{R}/I_t(\mathcal{H}_m[r])$ *is a Cohen–Macaulay ring.*

(ii) $I_t(\mathcal{H}_m[r])$ *is a prime ideal if and only if either $t = 1$ or else $t \ge r + 2$.*

Proof

(i) By Proposition 6.1, the ideal $I_t(\mathcal{H}_m[r]) \subset \bar{R}$ is generated by the maximal minors of $\mathcal{H}_{t,2m-t}[r]$. Therefore its codimension is at most $2m - t - t + 1 = 2(m - t) + 1$.

We analyze the two cases separately:

(1) $t \ge r + 2$.

Let $\mathcal{H}$ denote the uniquely defined generic Hankel matrix of size $t \times (2m - t)$ over the ground ring R. By Proposition 6.1, one has $I_t(\mathcal{H}_m) = I_t(\mathcal{H})$. Setting

$A := R/I_t(\mathcal{H})$, one has $R/(x_{2m-r}, \ldots, x_{2m-1}) \simeq \bar{R}$ and

$$A/(x_{2m-r}, \ldots, x_{2m-1})A \simeq \bar{R}/I_t(\mathcal{H}_m[r]).$$

Since the strand $\{x_{2m-r}, \ldots, x_{2m-1}\}$ is a subset of the system of parameters determined in Lemma 6.13, as applied with $a = t, b = 2m - t$, it is a regular sequence over A. Therefore, $\bar{R}/I_t(\mathcal{H}_m[r])$ is Cohen–Macaulay.

Moreover, since $\dim(\bar{R}/I_t(\mathcal{H}_m[r])) = \dim(A) - r$, one gets codim $I_t(\mathcal{H}_m[r])$ $= 2(m - t) + 1$.

(2) $t \leq r + 1$.

In this situation, we get $2m - t - (2m - r - 1) = r - t + 1$ null columns at the right end of the matrix $\mathcal{H}_{t,2m-t}[r]$, blurring the original requirement for the nature of the degeneration. Yet, removing these additional null columns yields a matrix of the shape $\mathcal{H}_{t,2m-r-1}[t-1]$ such that $I_t(\mathcal{H}_{t,2m-r-1}[t-1]) = I_t(\mathcal{H}_m[r])$.

Applying [9, Proposition 4.3] to the generic Hankel matrix $\mathcal{H}$ of size $t \times (2m - r - 1)$ over the ring $R = k[x_1, \ldots, x_{2m-r+t-2}]$ yields that $R/I_t(\mathcal{H})$ has dimension $2(t - 1)$. By the same token as before, the strand $\{x_{2m-r}, \ldots, x_{2m-r+t-2}\}$ is a subset of the system of parameters determined in Lemma 6.13, hence it is a regular sequence over $A = R/I_t(\mathcal{H})$. Therefore, $\bar{R}/I_t(\mathcal{H}_m[r])$ is Cohen–Macaulay. Doing the arithmetic once more, one gets codim $I_t(\mathcal{H}_m[r]) = 2m - t - r$.

(ii) Clearly, $I_1(\mathcal{H}_m[r])$ is prime. Now, suppose that $t \leq r + 1$. If $t = r + 1$, the t-minor on the bottom-right corner of the matrix contains a product of variables, hence is not prime. Clearly, this implies that no ideal of t-minors is prime either, for $t \leq r + 1$.

For the reverse implication we proceed as follows. Note that the hypothesis puts us in the first case of (i), in which the ideal $I_t(\mathcal{H}_m[r])$ coincides with the ideal of maximal minors of the matrix $\mathcal{H}_{t,2m-t}[r]$ and has codimension $2(m - t) + 1$. Since the generic Hankel matrix is 1-generic (see [11, Proposition 6.3]) then by Theorem 1.24 the ideal $(x_{i_1}, \ldots, x_{i_s}, I_t(\mathcal{H}_{t,2m-t}))$ is prime provided $s \leq t - 2$, where $\mathcal{H}_{t,2m-t}$ denotes the corresponding generic Hankel matrix. But since $r \leq t - 2$ by hypothesis, we are through.   □

Remark 6.15 Note that a particular feature of the above result and its method of proof is that, for any given $1 \leq t \leq m$ and any $0 \leq r \leq m - 2$, the dimension of the k-linear span of the t-minors of $\mathcal{H}_m[r]$ is the same as that of $\mathcal{H}_m$.

For convenience, we isolate two special cases of the above proposition.

Corollary 6.16 *Assume that* $0 \leq r \leq m - 2$. *One has:*

(i) *If* $m \geq 3$, $I_{m-1}(\mathcal{H}_m[r])$ *has codimension 3 and is a prime ideal if and only if* $r \leq m - 3$.

(ii) *If* $m \geq 4$, $I_{m-2}(\mathcal{H}_m[r])$ *has codimension 5 and is a prime ideal if and only if* $r \leq m - 4$.

Remark 6.17 Note that, when $r \geq 1$, even for $r \leq m - 3$, the ring $R/I_{m-1}(\mathcal{H}_m[r])$ is not always normal, a property that may require $m - r \gg 3$.

It may be interesting to compare the generic, generic symmetric and the Hankel situations as regards invariants of the semi-hollow linear section. Thus, let the size be $m \times m$ with $0 \leq r \leq m - 2$. Let PI and CM be short for polar image and Cohen–Macaulay, respectively. The overall picture looks as follows:

	Rank of Hessian	Polar image (PI)	Primality of I_{m-1}	Special fiber of I_{m-1}
Generic	$m^2 - r(r+1)$	Gorenstein ladder	if $r \leq m - 3$	Cone over PI
Symmetric	$\binom{m+1}{2} - 2\mathfrak{o}(r)$	CM ladder	?	Cone over PI
Hankel	$2m - r - 1$ (maximum)	Trivial	if $r \leq m - 3$	?

For the generic and symmetric generic cases see Sections 4.5 and 5.3.2, respectively. In regard to the question mark at the end of the third row, by Proposition 6.1, one can reduce to the case where I_{m-1} is replaced by the maximal minors of an $(m - 1) \times (m + 1)$ semi-hollow like Hankel matrix. In the case where $r = 0$ (generic Hankel), one knows that the defining ideal of the special fiber is generated by Plücker relations ([5, Theorem 4.7]) – more precisely, the fiber is isomorphic as k-algebra to the homogeneous coordinate ring of the Grassmannian $G(m-1, m+1)$ as proved in Proposition 6.3.

Yet, already for $m = 4, r = 1$, there is a minimal defining relation of degree 3 besides the quadratic ones. Curiously, in this case the defining ideal of the special fiber is a Gorenstein ideal generated by five quadrics and one cubic, where the quadrics generate a Gorenstein ideal of codimension three and generic structural Pfaffian

$$
\begin{pmatrix}
0 & -y_1 & -y_2 & -y_3 & -y_4 \\
y_1 & 0 & -y_5 & -y_6 & -y_7 \\
y_2 & y_5 & 0 & -y_8 & -y_9 \\
y_3 & y_6 & y_8 & 0 & -y_{10} \\
y_4 & y_7 & y_9 & y_{10} & 0
\end{pmatrix} .
$$

However, this behavior is not a pattern throughout.

Conjecture 6.18 Let $1 \leq r \leq m - 2$.

1. The special fiber of $I_{m-1}(\mathcal{H}_m[r])$ is Cohen–Macaulay.
2. The Rees algebra of $I_{m-1}(\mathcal{H}_m[r])$ is Cohen–Macaulay and of fiber type.
3. The defining ideal of the special fiber of $I_{m-1}(\mathcal{H}_m[r])$ is minimally generated by quadrics and cubics.

These questions are specially intriguing because they actually refer to the maximal minors of a Hankel matrix $\mathcal{H}_{m-1,m+1}[r]$, with $r \leq m - 1$. For low values of m, computer assistance calculation confirms all three. In the case $r = m - 2$ the defining ideal of the special fiber has even minimal binomial generators with non squarefree terms.

6.3.4   The Gradient Ideal of $\mathcal{H}_m[r]$

6.3.4.1   The Codimension

The codimension of the gradient ideal J of $\det \mathcal{H}_m[r]$ comes in as one of the basic ingredients in this subsection. The result below is a neat consequence of the nature of J in its relation to the cofactors of $\mathcal{H}_m[r]$ and of the adjugate relation.

Proposition 6.19 (char$(k) \neq 2$) *Let $J \subset R = k[x_1, \ldots, x_{2m-r-1}]$ denote the gradient ideal of* $\det \mathcal{H}_m[r]$, *with* $0 \leq r \leq m - 2$. *Then*

$$\operatorname{codim}(J) = \begin{cases} 2 & \text{if } m - r = 2 \\ 3 & \text{otherwise.} \end{cases}$$

Proof By Proposition 2.10, J is contained in $I_{m-1}(\mathcal{H}_m[r])$. The latter has codimension 3 by Corollary 6.16. Therefore, J has codimension at most 3.

By Proposition 6.11, $\det \mathcal{H}_m[r]$ is irreducible, hence codim $J \geq 2$.

For $r = m - 2$ it is easily checked that the support of every nonzero term of $\det \mathcal{H}_m[r]$ is of degree ≥ 2 in x_m, x_{m+1}. This implies that $J \subset (x_m, x_{m+1})$.

Thus, we assume that $m - r \geq 3$ and induct on $m - r$. The argument, though slightly long, conveys quite a bit of the role of the hollow like status.

(INITIAL STEP) If $m - r = 3$ then $\operatorname{codim}(J) = 3$.

More precisely, given a prime ideal P containing J, we will prove that $P \supset I_{m-1}(\mathcal{H}_m[r])$ or $P \supset (x_m, x_{m+1}, x_{m+2})$. We split the argument in two cases.

CASE 1: $x_{m+2} \in P$.

Note that $f_1 = x_{m+1}^{m-1} + H$ with $H \in (x_{m+2})$. Thus, since $(x_{m+2}, f_1) \subset P$ then $x_{m+1} \in P$. By a similar token, $f_m - x_m^{m-1} + G$ with $G \in (x_{m+1}, x_{m+2})$. This way, since $(x_{m+1}, x_{m+2}, f_m) \subset P$, we see that $x_m \in P$. Therefore, $(x_m, x_{m+1}, x_{m+2}) \subset P$.

CASE 2: $x_{m+2} \notin P$.

Set $\operatorname{adj}(\mathcal{H}_m[m - 3]) := (\Delta_{i,j})$, with $\Delta_{i,j}$ standing for the cofactor of the (j, i)-entry of $\mathcal{H}_m[m - 3]$.

CLAIM. For every $2 \leq k \leq m$, the entries of its $k \times k$ submatrix

$$\begin{pmatrix} \Delta_{1,1} & \cdots & \Delta_{1,k} \\ \vdots & \cdots & \vdots \\ \Delta_{k,1} & \cdots & \Delta_{k,k} \end{pmatrix}$$

belong to P.

As a consequence, taking $k = m$ we have $I_{m-1}(\mathcal{H}_m[r]) \subset P$, as was to be shown.

To prove this claim we induct on k.

Let $k = 2$. The last row of $\mathcal{H}_m[m-3]$ is $[x_m \, x_{m+1} \, x_{m+2} \, 0 \cdots 0]$. Equating the respective $(m,1)$-entries in the two sides of the adjugate relation $\mathcal{H}_m[m-3]\,\mathrm{adj}(\mathcal{H}_m[m-3]) = f\,\mathbb{I}_m$ then yields

$$x_m \Delta_{1,1} + x_{m+1}\Delta_{2,1} + x_{m+2}\Delta_{3,1} = 0. \tag{6.10}$$

Since $f_1 = \Delta_{1,1}$ and $f_2 = (1/2)\Delta_{2,1}$, it follows from (6.10) that $\Delta_{3,1} \in P$. On the other hand, since the entries of $\mathcal{H}_m[m-3]$ in the slots (doubly) indexed by $(1,3), (2,2), (3,1)$ are all equal to x_3, it follows from Proposition 2.10 that $f_3 = \Delta_{2,2} + 2\Delta_{3,1}$. Therefore, $\Delta_{2,2} \in P$ as well. Hence, for $k = 2$ the statement is true.

Now, assuming the statement valid for some $k \geq 2$, we show that it holds for $k + 1$. Consider the block partition

$$\left(\begin{array}{ccc|c} \Delta_{1,1} & \cdots & \Delta_{1,k} & \Delta_{1,k+1} \\ \vdots & \ddots & \vdots & \vdots \\ \Delta_{k,1} & \cdots & \Delta_{k,k} & \Delta_{k,k+1} \\ \hline \Delta_{k+1,1} & \cdots & \Delta_{k+1,k} & \Delta_{k+1,k+1} \end{array}\right).$$

The entries in the $(m-k+2)$th row of $\mathcal{H}_m[m-3])$ are

$$x_{m-k+2}, \; x_{m-k+3}, \; \cdots, \; x_m, \; x_{m+1}, \; x_{m+2}, \; 0, \; \cdots, \; 0.$$

By a similar token as before, equating the respective $(m-k+2, j)$-entries on both sides of the adjugate relation, for $1 \leq j \leq k+1$, yields the following relations:

$$x_{m-k+2}\Delta_{1,j} + \cdots + x_{m+1}\Delta_{k,j} + x_{m+2}\Delta_{k+1,j} = 0, \text{ for all } 1 \leq j \leq k+1. \tag{6.11}$$

These equalities, along with the inductive hypothesis, yield

$$\Delta_{j,k+1} \in P \text{ and } \Delta_{k+1,j} \in P, \tag{6.12}$$

for every $1 \leq j \leq k$. Finally, for $j = k+1$ one has

$$x_{m-k+2}\Delta_{1,k+1} + \cdots + x_{m+1}\Delta_{k,k+1} + x_{m+2}\Delta_{k+1,k+1} = 0.$$

From this and (6.12), it follows that $\Delta_{k+1,k+1} \in P$. This concludes the inductive step in the proof of the claim.

Thus, we are through with the initial step $m - r = 3$ of the main induction.

(INDUCTIVE STEP) From $m - r$ to $m - r + 1 = m - (r-1)$ corresponds to a descending induction step from r to $r - 1$. Thus, we are given the matrix $\mathcal{H}_m[r-1]$, with $r - 1 \leq m - 4$, and the corresponding gradient ideal $J \subset R = k[x_1, \ldots, x_{2m-r}]$ and assume by induction that the corresponding gradient ideal $J' \subset R' = k[x_1, \ldots, x_{2m-r-1}]$ of $\det \mathcal{H}_m[r]$ has codimension 3. Since the latter matrix is a linear section of the former by trading 0 for x_{2m-r}, clearly $(J', x_{2m-r}) \subset (J, x_{2m-r})$ as ideals in the larger ring R, hence the codimension of J is at least 3 since (J', x_{2m-r}) has codimension 4. $\square$

Remark 6.20

(1) The proposition will be a consequence of the more precise Theorem 6.24 (a) below. Indeed, the proofs bear some similarity but are mutually independent.
(2) It has been seen in the proof of Proposition 6.14 (i) that, for arbitrary $r \leq m-3$, the entry subset $\{x_{2m-r}, x_{2m-(r-1)}, \ldots, x_{2m-1}\}$ of the fully generic $m \times m$ Hankel matrix $\mathcal{H}_m$ is a regular sequence modulo the submaximal minors of $\mathcal{H}_m$. *It would seem* that this sequence is equally regular modulo the gradient ideal of $\det \mathcal{H}_m$ (see Conjecture 6.27 below). Thus, the specialization still has codimension 3 in the entry ring of $\mathcal{H}_m[r]$. Unfortunately, the specialized ideal is larger than the gradient ideal of $\det \mathcal{H}_m[r]$, hence one resorts to the inductive argument above to bypass this apparent obstruction.

6.3.4.2 The Associated Primes

In this part we assume throughout that $r \leq m - 3$. The case where $r = m - 2$ will be thoroughly dissected in Section 6.4, bearing on the results of [3] and [16].

We first develop a few aspects of lower ideals of minors as related to the adjugate formula, some of which were scratched in the previous section. The findings are a bit technical, so we refer to the main aftermath as contained in Theorem 6.24.

For any integer $1 \leq j \leq m - 2$, we consider the following block-decompositions

$$\mathcal{H}_m[r] = \begin{bmatrix} U_{m-j} \\ D_j \end{bmatrix} \tag{6.13}$$

where U_{m-j} is an $(m - j)$-rowed submatrix, and

$$\mathrm{adj}(\mathcal{H}_m[r]) = \begin{bmatrix} A_{m-j} & B_{m-j} \\ B'_j & C_j \end{bmatrix}, \qquad B'_j = B^t_{m-j}, \tag{6.14}$$

where A_{m-j} is a square $(m - j)$-rowed matrix.

The notation is such that the subscript of any of the blocks denotes its number of rows.

Block multiplication and the adjugate formula (2.1) yield

$$f\, \mathbb{I}_m = \mathrm{adj}(\mathcal{H}_m[r])\, \mathcal{H}_m[r] = \begin{bmatrix} A_{m-j}U_{m-j} + B_{m-j}D_j \\ B^t_{m-j}U_{m-j} + C_jD_j \end{bmatrix}, \tag{6.15}$$

where $f = \det \mathcal{H}_m[r]$.

Lemma 6.21 (char(k) does not divide m) *Let $I \subset R$ be an ideal such that $J \subset I$. Then, with the notation of 6.15, one has:*

$$I_1(A_{m-j}U_{m-j}) \subset I \Rightarrow I_1(B_{m-j})I_j(D_j) \subset I$$

and, similarly,

$$I_1(B^t_{m-j}U_{m-j}) \subset I \Rightarrow I_1(C_j)I_j(D_j) \subset I.$$

Proof It suffices to consider the first of the above implications. Since char(k) does not divide m, Euler's identity implies that $f \in J \subset R = k[x_1, \ldots, x_{2m-r-1}]$, and hence the entries on the rightmost matrix belong to J as well. Therefore, $I_1(A_{m-j}U_{m-j} + B_{m-j}D_j) \subset I$ and, since $I_1(A_{m-j}U_{m-j}) \subset I$ by hypothesis, then $I_1(B_{m-j}D_j) \subset I$.

Note that $I_j(D_j)$ is the ideal of maximal minors of D_j. Thus, letting M denote an arbitrary $j \times j$ submatrix of D_j, one has $I_1(B_{m-j}M) \subset I$. Thus, $I_1(B_{m-j}\,M\,\mathrm{adj}(M)) = I_1(\det M\, B_{m-j}) \subset I$. Therefore, $I_j(D_j)I_1(B_{m-j}) \subset I$, as required. $\qquad\square$

Note that Lemma 6.21 still holds by replacing the block matrices of (6.14) by their respective ith rows, for an arbitrary i.

As an additional notation, given a matrix $\mathcal{M}$ and an index i, $L_i(\mathcal{M})$ will stand for its ith row.

Write $[L_1(A_{m-j})\ L_1(B_{m-j})]$ for the first row of the adjugate matrix $\mathrm{adj}(\mathcal{H}_m[r])$ as decomposed in (6.14). Below $\Delta_{i,j}$ denotes the cofactor of the (j, i)-entry of $\mathcal{H}_m[r]$.

The following consequence highlights the case where $j = m - 2$.

Corollary 6.22 (char(k) $\neq 2$ and does not divide m) *With the above notational convention, one has $I_{m-2}(D_{m-2})I_1([L_1(A_2)\ L_1(B_2)]) \subset J$.*

Proof Of course, $L_1(A_2) = [\Delta_{11}\ \Delta_{1,2}]$, hence its entries belong to J since $\partial f/\partial x_1 = \Delta_{1,1}$ and $\partial f/\partial x_2 = 2\Delta_{1,2}$. Clearly, then $I_1(L_1(A_2)U_2) \subset J$. By Lemma 6.21 one deduces that $I_{m-2}(D_{m-2})I_1(L_1(B_2)) \subset J$. $\qquad\square$

So far, j was fixed. Now we make it vary in a certain range in order to decide when a given prime ideal containing J contains the ideal $I_{m-1}(\mathcal{H}_m[r])$. For example, the prime ideal $Q = (x_m, x_{m+1}, \ldots, x_{2m-r-1})$ contains J but not $I_{m-1}(\mathcal{H}_m[r])$, while adding the entry x_{m-1} to Q gives a prime ideal containing the submaximal minors. Precisely, one has:

Proposition 6.23 (char(k) $\neq 2$ and does not divide m) *Suppose that $r \leq m - 4$. Let $P \supset J$ be a prime ideal. If there exists an index j in the range $r + 2 \leq j \leq m - 2$ such that $I_j(D_j) \subset P$ and $I_{j-1}(D_{j-1}) \not\subset P$, then $I_{m-1}(\mathcal{H}_m[r]) \subset P$.*

Proof Since $I_j(D_j) \subset P$, (6.15) implies that the entries on the row $[A_{m-j}\ B_{m-j}]$ (resp., $[A_{m-j}\ B_{m-j}]^t$) belong to P. On the other hand, since $\mathcal{H}_m[r]$ is a coordinate linear section of the generic $\mathcal{H}_m$, Proposition 2.10 is applicable. But note that the cofactors of two entries symmetrically positioned with respect to the main diagonal are equal. Therefore, for every $r + 2 \leq j \leq m - 2$ one gets

$$\Delta_{m-j+1,m-j+1} = \partial f / \partial x_{2(m-j+1)-1} - 2 \sum_{i=\max\{1,m-2j+2\}}^{m-j} \Delta_{2(m-j+1)-i,i}.$$

But since the entries on the row $[A_{m-j} \; B_{m-j}]$ (resp., $[A_{m-j} \; B_{m-j}]^t$) belong to P, we have $\Delta_{m-j+1,m-j+1} \in P$.

Therefore, the entries of $L_{m-j+1}(A_{m-j+1}) = [\Delta_{m-j+1,1} \cdots \Delta_{m-j+1,m-j+1}]$ belong to P. It follows that $I_1(L_{m-j+1}(A_{m-j+1})U_{m-j+1}) \subset P$. Then the upshot from Lemma 6.21 is that

$$I_{j-1}(D_{j-1})I_1(L_{m-j+1}(B_{m-j+1})) \subset P.$$

But since $I_{j-1}(D_{j-1}) \not\subset P$, then

$$I_1(L_{m-j+1}(B_{m-j+1})) = (\Delta_{m-j+1,m-j+2}, \ldots, \Delta_{m-j+1,m}) \subset P.$$

This in turn tells us that

$$I_1([A_{m-j+1} \; B_{m-j+1}]) \subset P \qquad (\text{resp.,} \;\; I_1([A_{m-j+1} \; B_{m-j+1}]^t) \subset P). \qquad (6.16)$$

Thus, $I_1(B^t_{m-j+1}U_{m-j+1}) \subset P$ and once again Lemma 6.21 implies that

$$I_{j-1}(D_{j-1})I_1(C_{j-1}) \subset P.$$

It follows that

$$I_1(C_{j-1}) \subset P. \qquad (6.17)$$

A moment reflection will convince us that (6.16) and (6.17) imply the inclusion $I_{m-1}(\mathcal{H}_m[r]) \subset P$, as was to be shown. $\square$

Theorem 6.24 (char$(k) \neq 2$ and does not divide m) *Let $J \subset R$ denote the gradient ideal of* $\det \mathcal{H}_m[r]$, *with* $1 \leq r \leq m - 3$, *and let Q denote the ideal generated by the $m - r$ nonzero variables of the right most column of $\mathcal{H}_m[r]$. Then:*

(a) *The minimal primes of R/J are Q and $P := I_{m-1}(\mathcal{H}_m[r])$.*
(b) *J is not a reduction of P.*
(c) *The unmixed and minimal components of J coincide if and only if $r = m - 3$.*

Proof

(a) It suffices to prove that any prime ideal P containing J necessarily contains either $Q = (x_m, x_{m+1}, \ldots, x_{2m-r-1})$ or $I_{m-1}(\mathcal{H}_m[r])$. With the notation of (6.13), we split the argument in two cases:

Case 1: $I_{r+1}(D_{r+1}) \subset P$.

Since $x_{2m-r-1}^{r+1} \in I_{r+1}(D_{r+1})$ then $x_{2m-r-1} \in P$. More: for any $m \le u \le 2m - r - 2$ there exists an $(r + 1)$-minor in $I_{r+1}(D_{r+1})$ of the form

$$x_u^{r+1} + x_{u+1}g_{u+1} + \cdots + x_{2m-r-1}g_{2m-r-1},$$

for certain forms $g_{u+1}, \ldots, g_{2m-r-1} \in R$. Decreasing induction on u then wraps up the argument for the inclusion $(x_m, \ldots, x_{2m-r-1}) \subset P$.

Case 2: $I_{r+1}(D_{r+1}) \not\subset P$.

If there exists an index j in the range $r+2 \le j \le m-2$ such that $I_j(D_j) \subset P$ then Proposition 6.23 implies that $I_{m-1}(\mathcal{H}_m[r]) \subset P$.

Thus, assume that $I_j(D_j) \not\subset P$ for every $r + 2 \le j \le m - 2$. In particular, $I_{m-2}(D_{m-2}) \not\subset P$. By Corollary 6.22,

$$I_{m-2}(D_{m-2})I_1([L_1(A_2) \ L_1(B_2)]) \subset P$$

and since $I_{m-2}(D_{m-2}) \not\subset P$, then

$$I_1([L_1(A_2) \ L_1(B_2)]) \subset P. \tag{6.18}$$

Thus, $\Delta_{1,3} \in P$. But $\Delta_{2,2} = \partial f / \partial x_3 - 2\Delta_{1,3} \in J \subset P$. Therefore, the entries $L_2(A_2) = [\Delta_{2,1} \ \Delta_{2,2}]$ belong to P. Then, from Lemma 6.21 one has

$$I_{m-2}(D_{m-2})I_1([L_2(A_2) \ L_2(B_2)]) \subset P$$

hence,

$$I_1([L_2(A_2) \ L_2(B_2)]) \subset P. \tag{6.19}$$

From (6.18) and (6.19) one has

$$I_1([A_2 \ B_2]) \subset P \qquad (\text{resp.,} \ \ I_1([A_2 \ B_2]^t) \subset P). \tag{6.20}$$

From these inclusions and Lemma 6.21 it obtains $I_{m-2}(D_{m-2})I_1(C_{m-2}) \subset P$, and hence,

$$I_1(C_{m-2}) \subset P. \tag{6.21}$$

Clearly, (6.20) and (6.21) imply that $I_{m-1}(\mathcal{H}_m[r]) \subset P$.

(b) If J is a reduction of P, at least $\sqrt{J} = P$, which would contradict the result in (a).

(c) This is an immediate consequence of (a). $\square$

Remark 6.25

(i) Computational evidence suggests that the Q-primary component of J is generated by $\binom{m-1}{r}$ forms of degree r (thus coinciding with Q when $r = 1$).
(ii) The structure of the embedded associated primes of R/J is quite involved. The following two primes seem to be candidates in general: (x_{m-1}, Q) (in codimension $m - r + 1$) and the radical of $I_{m-2}(\mathcal{H}_m[r])$ (in codimension 5).

6.3.4.3   Linear Behavior

In order to refrain from repeating the usual forbidden values of the characteristic, we assume once for all that the ground field has characteristic zero.

For the generic Hankel matrix $\mathcal{H}_m$ it was proved that the linear rank of J is 3 ([15, Theorem 3.3.5]). At the other end of the spectrum, for the linear section $\mathcal{H}_m[m - 2]$ we will show in the next section that the linear rank of J is maximal possible (Proposition 6.35) and that J is in addition of linear type (Theorem 6.43).

Though not obvious at all, one expects that, for $1 \leq r \leq m - 3$, the linear rank of the gradient ideal $J_m[r]$ of $\det \mathcal{H}_m[r]$ be squeezed in-between. In this regard, we state:

Conjecture 6.26 Let $1 \leq r \leq m - 3$. Then the linear rank of $J_m[r]$ is 2.

We would like to base ourselves on the following paragon, itself conjectural as well:

Conjecture 6.27 If J is the gradient ideal of the fully generic $m \times m$ Hankel matrix, then the sequence $\{x_{m+3}, \ldots, x_{2m-1}\}$ is regular modulo J.

Proposition 6.28 *Conjecture* 6.27 *implies Conjecture* 6.26.

Proof Set $R := k[x_1, \ldots, x_{2m-1}]$, $\bar{R} := R/(x_{2m-r}, \ldots, x_{2m-1})$, $\bar{J} := J\bar{R} \subset \bar{R}$, so

$$\bar{J} = (J, x_{2m-r}, \ldots, x_{2m-1})/(x_{2m-r}, \ldots, x_{2m-1}) \subset k[x_1, \ldots, x_{2m-r-1}].$$

One asserts that the module of linear syzygies of $\bar{J}$ is a free $\bar{R}$-module of rank 3.

To see this, note that since $r \leq m - 3$ then $2m - r \geq m + 3$, Conjecture 6.27 implies that $\{x_{2m-r}, \ldots, x_{2m-1}\}$ is a regular sequence modulo J.

The minimal graded resolution of R/J specializes to that of $\bar{R}/\bar{J}$. In particular, the module of linear syzygies of $\bar{J}$ has the same structure as that of J. But, in the fully generic case, this module has been shown to be free of rank 3 ([15, Theorem 3.3.5 and its proof]).

Now, the generators of $J_m[r]$ are part of a minimal set of generators of $\bar{J}$ in the natural order of the partial derivatives, as follows from Proposition 2.10. In particular, any linear syzygy of $J_m[r]$ gives one of $\bar{J}$ by filling down enough zeros. This implies that the module of syzygies of $J_m[r]$ is a submodule of that of $\bar{J}$ and is free of rank ≤ 3.

On the other hand, according to [15, Chapter 3, Section 3.3.2, p.22], the submodule of linear syzygies of the fully generic J has the matrix form

$$\begin{pmatrix} 0 & \beta_1 x_2 & \gamma_1 x_1 \\ \alpha_2 x_1 & \beta_2 x_3 & \gamma_2 x_2 \\ \vdots & \vdots & \vdots \\ \alpha_{2m-2} x_{2m-3} & \beta_{2m-1} x_{2m-1} & \gamma_{2m-2} x_{2m-2} \\ \alpha_{2m-1} x_{2m-2} & 0 & \gamma_{2m-1} x_{2m-1} \end{pmatrix} \tag{6.22}$$

with $\alpha_i, \beta_j, \gamma_l$ elements of k, where $\alpha_i \neq 0$, for $2 \leq i \leq 2m - 1$. Note that there is no k-elementary operation that kills the last coordinate of the above syzygies. Since the module specializes to the module of linear syzygies of $\bar{J}$, the latter is obtained by setting to zero the entries $x_{2m-r}, \ldots, x_{2m-1}$. Thus, there is a linear syzygy of $\bar{J}$ whose $(2m - r)$th coordinate $\alpha_{2m-r} x_{2m-r-1}$ does not vanish. Since $f_m[r]$ has only $2m - r - 1$ derivatives, this syzygy cannot be a syzygy of $J_m[r]$. Therefore, the module of syzygies of $J_m[r]$ is free of rank at most 2.

But, by the same token, the β and γ version of the syzygies of $\bar{J}$ are syzygies of $J_m[r]$. We conclude that the module of syzygies of $J_m[r]$ is free of rank exactly 2; in particular, it has linear rank 2. $\qquad\square$

Remark 6.29

(1) Conjecture 6.26 fails in positive characteristic. For example, in characteristic 3 the linear rank of $J_4[1]$ is 3.
(2) If $r = m - 2$ a similar argument crumbles down because the element x_{m+2} is not regular on $R/(J, x_{2m-1}, \ldots, x_{m+3})$, and neither on $J_m[m - 2]$ for that matter. And, in fact, we know that $J_m[m - 2]$ has a whole batch of new linear syzygies forcing maximal linear rank.
(3) What are the odds against the auxiliary conjecture? First, since one is in a homogeneous environment, one can just as well consider the sequence $\{x_{2m-1}, \ldots, x_{m+3}\}$ in reverse order. At each step this way one is actually asking about the associated primes of $\bar{J}$, not those of $J_m[r]$ exactly. Thus, it is urgent to compare both, a task that can be carried under severe hypothesis. In any case, it would seem that the conjecture could use information about the associated primes of $J_m[r]$, a problem that has been poorly accessed so far (Remark 6.25).

Conjecture 6.30 $(\operatorname{char}(k) = 0)$ Let $0 \leq r \leq m - 2(m \geq 3)$ and $J = J_f$, where $f = \det \mathcal{H}_m[r]$. Then:

(1) J satisfies condition (F_1).
(2) J is of linear type.

Knowingly, (2) implies (1), while the latter is a front-runner feature of an ideal of linear type. Another front runner of (2) is the k-linear independence of the partial derivatives of f, equivalently, the maximal value of the analytic spread of J. This property holds true as a consequence of the following basic result:

Theorem 6.31 (char(k) $= 0$) *Let* $f = \det \mathcal{H}_m[r]$. *For* $0 \leq r \leq m - 2$, *the Hessian* $h(f)$ *does not vanish.*

Due to its length, the proof is given in the Appendix. Shorter proofs are available in the cases $r = 0$ (Corollary 6.7) and $r = m - 2$ (Section 6.4).

A weak piece of (2) is known in the fully generic case, namely, that J is generically a complete intersection (Theorem 6.8 (d)).

Conversely, proving affirmatively (2) would give a short argument for both Theorem 6.8 (c) and Conjecture 6.32 below.

Recall, as mentioned before, that $f := \det \mathcal{H}_m[r]$ is homaloidal for $r = m - 2$ ([3]).

We state:

Conjecture 6.32 Assume that $0 \leq r \leq m - 2$. If $f := \det \mathcal{H}_m[r]$ is homaloidal then $r = m - 2$.

We now add a few guideline-like comments on some of the main questions of the section.

1. Elementary obstructions in the case $r < m - 2$ are: (i) the ground polynomial ring R has dimension at least $m + 2$ (while the dimension is $m + 1$ in the case $r = m-2$); (ii) $J \subset R$ – and hence $(J, \Delta) \subset R$ too – is an ideal of codimension 3 (while J has codimension 2 and (J, Δ) has codimension 3 in the case $r = m-2$).
2. The reason one cannot extend the method of proof of Theorem 6.8 (a) to the semi-hollow like environment is that the extension $k[\mathbf{f}] \subset k[\Delta]$ is not integral anymore. In fact, at least if $r \leq m - 3$, then $P \not\subset \sqrt{J}$ since R/J admits minimal primes other than P. From the point of view of the actual argument of Theorem 6.8 (a), although one still has a poset of maximal minors, the existing Plücker relations do not necessarily come up with at least one factor belonging to $k[\mathbf{f}]$.
3. Note that, since $k[\Delta]$ has maximal dimension, for a linear section $\mathcal{H}_m[r]$ one still has $\dim k[\mathbf{f}] = \dim k[\Delta]$ in characteristic zero due to Theorem 6.31 – of course, this equality is equivalent to the non-vanishing of the Hessian, so an a priori proof of the dimension equality would give a more elegant proof of the latter. Anyway, if one assumes homaloidness for $r \leq m - 3$, then the corresponding field extension is trivial and since $k[\mathbf{f}]$ is a polynomial ring, by [23, Proposition 3.7 and Proposition 6.1 (a)] one has the equality

$$e(k[\Delta]) = e(k[\mathbf{f}], k[\Delta]) + 1,$$

where $e(k[\mathbf{f}], k[\Delta])$ is the relative multiplicity introduced in [23]. In order to derive a contradiction one needs an a priori knowledge of a sufficiently large lower bound for $e(k[\Delta])$, a presently unseen goal.

Note that the latter relative multiplicity can be computed as the ordinary multiplicity of the graded algebra

$$G/0 :_G (\Delta)G,$$

where G stands for the associated graded ring of the ideal $(\mathbf{f})k[\Delta]$ of the k-algebra $k[\Delta]$.

4. Picking up from a slightly different angle, still assuming that f is homaloidal, letting $\Delta \in \mathbf{\Delta}$ denote the cofactor of the (m, m) entry, the extension $A := k[\mathbf{f}] \subset B := k[\mathbf{f}, \Delta]$ is birational for even more reason. Since Δ has same degree as the elements of $\mathbf{f}$, it must be an element of the field of fractions $k(\mathbf{f})$ of the form $G(\mathbf{f})/F(\mathbf{f})$, with $F(\mathbf{f}), G(\mathbf{f}) \in k[\mathbf{f}]$ homogeneous of degrees, say, $s - 1, s$, respectively. Since $k[\mathbf{f}]$ is a polynomial ring over k, this forces a presentation of B as a k-algebra over a polynomial ring $k[\mathbf{t}, u]$, with $\mathbf{t} = \{t_1, \ldots, t_{2m-r-1}\}$, as follows:

$$B \simeq k[\mathbf{t}, u]/(uF(\mathbf{t}) + G(\mathbf{t})),$$

where $t_i \mapsto f_i, u \mapsto \Delta$, and F, G are forms in $\mathbf{t}$ of degrees, $s - 1, s$, respectively.

The proof along these steps would then claim that this is impossible unless $r = m - 2$. In other words, one would show that such a presentation is only possible if $f_1 = \partial f/\partial x_1$ is a pure power, in which case necessarily $f_1 = x_{2m-r-1}^{m-1} = x_{m+1}^{m-1}$.

5. It is equivalent to show that if $r < m - 2$, the defining equation e of B for a presentation with set of generators $\{\mathbf{f}, \Delta\}$ has u-degree ≥ 2 (note that e has no pure power term in u since u is not integral).

6. A computer calculation for the simplest situation ($m = 4, r = 1$) gives that e has degree 15 and u-degree 5, well beyond what is needed.

Note that either $F(\mathbf{t})$ or $G(\mathbf{t})$ must involve t_1 because the set $\{f_2, \ldots, f_{2m-r-1}, \Delta\}$ is algebraically independent over k. Suppose that $F(\mathbf{t})$ effectively involves t_1. Then by a k-linear change of variables fixing t_1 – i.e., of the form $t_1 \mapsto t_1, t_i \mapsto t_1 + \alpha_i t_i$, for $2 \leq i \leq 2m - r - 1$ – one can assume that $F(\mathbf{t})$ has a non-vanishing pure power term αt_1^{s-1}. Then we would have $\Delta f_1^{s-1} \in J^s$. In this case it would suffice to prove that for no exponent l one has

$$\Delta \in J^l : f_1^{l-1}.$$

The feeling is that an early obstruction is the initial degree of this colon ideal which may turn out to be $> m - 1$. Note that the colon ideal contains the ideal $J^l : J^{l-1}$, but one cannot derive anything from this as J is seemingly Ratliff–Rush closed – at least if one takes for granted that the associated graded ring G of J has positive depth given the expectation that J is actually of linear type (not just analytically independent) and G is Cohen–Macaulay.

As a slight confirmation, if $r = m - 2$ is the case then a presentation equation as above can actually be written in the form $uF(\mathbf{t}) + G(\mathbf{t})$ with $F = t_1^{e(B)-1}$, that is, $F(\mathbf{f}) = f_1^{e(B)-1} = x_{m+1}^{(m-1)(e(B)-1)}$.

6.4   Special Results for the Sub-Hankel Matrix

For given $m \geq 2$, consider the following special case of an $m \times m$ semi-hollow like Hankel matrix

$$
\mathcal{H}_{x_0,\ldots,x_m}[m-2] =
\begin{pmatrix}
x_0 & x_1 & x_2 & x_3 & \cdots & x_{m-2} & x_{m-1} \\
x_1 & x_2 & x_3 & x_4 & \cdots & x_{m-1} & x_m \\
x_2 & x_3 & x_4 & x_5 & \cdots & x_m & 0 \\
\vdots & \vdots & \vdots & \vdots & \cdots & \vdots & \vdots \\
x_{m-2} & x_{m-1} & x_m & 0 & \cdots & 0 & 0 \\
x_{m-1} & x_m & 0 & 0 & \cdots & 0 & 0
\end{pmatrix}.
$$

It has been dubbed a *sub–Hankel matrix* of order m (see [3]). Note that the subscript m of the previous sections has been replaced by the more informative set of variables. In addition, we kept the indexing of the variables as originally in [3] for reference convenience. As usual in the Hankel case, we assume at least that $\mathrm{char}(k) \neq 2$.

Due to its relevance from both the algebraic and the geometric viewpoint, this section will be very detailed and lengthy. Thus, one may experiment focusing on the subsections of one's most interest.

6.4.1   Properties and Syzygies of the Partial Derivatives

Set $R := k[x_0, \ldots, x_m]$. Let $f^{(m)}(x_0, \ldots, x_m)$ denote the determinant of the semi-hollow linear section $\mathcal{H}_{x_0,\ldots,x_m}[m-2]$. Consistently with this notation, we let $f^{(j)}(x_{m-j}, \ldots, x_m)$ denote the determinant of the $j \times j$ submatrix $\mathcal{H}_{x_{m-j},\ldots,x_m}[j-2]$ of $\mathcal{H}_{x_0,\ldots,x_m}[m-2]$ obtained by omitting the first $m-j$ columns from the right and the last $m-j$ rows from bottom. Set $f^{(0)} = 1$.

We need the following algebraic structural lemmas about the partial derivatives of $f^{(m)}(x_0, \ldots, x_m)$. We stress the use throughout of Proposition 2.10, which is applicable to the matrices in question due to the fact that they are coordinate linear sections of the generic matrix of the same size.

Lemma 6.33 *Let $m \geq 2$. Then:*

$$
\frac{\partial f^{(m)}(x_0, \ldots, x_m)}{\partial x_i} = (-1)^m x_m \frac{\partial f^{(m-1)}(x_1, \ldots, x_m)}{\partial x_{i+1}}, \quad 0 \leq i \leq m-2. \tag{6.23}
$$

Proof The relation is clear for $i = 0$: by Proposition 2.10, the derivative $\partial f^{(m)}/\partial x_0$ is the cofactor of x_0, a (signed) $(m-1)$th power of x_m. On the other hand, $\partial f^{(m-1)}/\partial x_1$ is itself a (signed) $(m-2)$th power of x_m.

For $1 \leq i \leq m - 2$, shifting the indices $i \mapsto i - 1$ along the entries of the matrix $\mathcal{H}_{x_1,\ldots,x_m}[m - 3]$ yields the $(m - 1) \times (m - 1)$ matrix $\mathcal{H}_{x_0,\ldots,x_{m-1}}[m - 3]$, of the same shape as $\mathcal{H}_{x_0,\ldots,x_m}[m - 2]$. Therefore, the first $m - 1$ partial derivatives of each have the same shape, except for a variable shift. Now apply to the partial derivatives of $\mathcal{H}_{x_1,\ldots,x_m}[m - 3]$ the reversed indices shifting $j \mapsto j + 1$, for $1 \leq j \leq m - 1$, and multiply each by x_m, thus retrieving the partial derivatives of $\mathcal{H}_{x_0,\ldots,x_m}[m - 2]$ of order $1 \leq i \leq m - 2$. $\qquad\square$

In the sequel, if the ground variables $x_0, \ldots, x_m$ are self-understood, we denote $f^{(m)}(x_0, \ldots, x_m)$ simply by $f^{(m)}$.

Lemma 6.34 *For $0 \leq i \leq m - 1$, one has*

$$\frac{\partial f^{(m)}}{\partial x_0}, \ldots, \frac{\partial f^{(m)}}{\partial x_i} \in k\,[x_{m-i}, \ldots, x_m] \qquad (6.24)$$

and the g.c.d. of these partial derivatives is x_m^{m-i-1}.

Proof A close scrutiny of these partial derivatives, in terms of the cofactors via Proposition 2.10, show that (6.24) holds. The assertion about the *gcd* follows by iterating (6.23). $\qquad\square$

Proposition 6.35 (Linear Syzygies) *The gradient ideal of $f^{(m)}$ has maximal linear rank.*

Proof Set $f = f^{(m)}$ and $f_l = \frac{\partial f}{\partial x_l}, l = 0, \ldots, m$, for lighter reading.
By Proposition 2.10, one has

$$f_l = d_l \Delta_{l-\lfloor l/2 \rfloor, \lfloor l/2 \rfloor} + \sum_{u=0}^{\lfloor l/2 \rfloor - 1} 2\Delta_{l-u,u} \quad (l = 0, \ldots, m), \qquad (6.25)$$

where

$$d_l = \begin{cases} 1 & \text{if } l \text{ is odd} \\ 2 & \text{otherwise.} \end{cases}$$

Given $1 \leq l \leq m$, let L_l and C_l denote the lth row of $\mathrm{adj}(\mathcal{H}_{x_0,\ldots,x_m}[m - 2])$ and the lth column of $\mathcal{H}_{x_0,\ldots,x_m}[m - 2]$, respectively. Then, for any i in the range $1 \leq i \leq m - 1$, set

$$F_i := \sum_{l=0}^{i-1} 2(i - l)L_{l+1}\,C_{m-i+l+1}.$$

Disentangling, one has:

$$F_i = \left(\sum_{u=0}^{i} x_{m-i+u}\Delta_{u,0}\right) + \cdots + \left(\sum_{u=0}^{i-l} x_{m-i+u+l}\Delta_{u,l}\right) + \cdots$$

$$+ \left(\sum_{u=0}^{1} x_{m-i+u+i-1}\Delta_{u,i-1}\right)$$

$$= 2i\,x_{m-i}\Delta_{0,0} + x_{m-i+1}\sum_{u=0}^{1}2(i-u)\Delta_{1-u,u} + \cdots$$

$$+x_{m-i+l}\sum_{u=0}^{l}2(i-u)\Delta_{l-u,u} + \cdots$$

$$= 2i\,x_{m-i}\Delta_{0,0} + (2i-1)x_{m-i+1}2\Delta_{0,1} + \cdots$$

$$+ (2i-l)x_{m-i+l}\left(\sum_{u=0}^{\lfloor l/2\rfloor-1}2\Delta_{l-u,u} + d_l\Delta_{l-\lfloor l/2\rfloor,\lfloor l/2\rfloor}\right) + \cdots$$

$$= 2i\,x_{m-i}f_0 + (2i-1)x_{m-i+1}f_1 + \cdots + (2i-l)x_{m-i+l}f_l + \cdots + ix_m f_i,$$

where the last equality follows from (6.25).

But, by the adjugate formula (2.1), one has $F_i = 0$, hence we obtain the following $m-1$ independent linear relations of the partial derivatives:

$$2i\,x_{m-i}\,f_0+(2i-1)x_{m-i+1}f_1+\cdots+(2i-l)x_{m-i+l}f_l+\cdots+ix_m f_i = 0. \qquad (6.26)$$

In addition, by a similar token, there is the following linear syzygy involving the derivative with respect to x_m as well:

$$\sum_{l=0}^{m-2}(m-l-1)\,x_l\,\frac{\partial f^{(m)}}{\partial x_l} - x_m\,\frac{\partial f^{(m)}}{\partial x_m} = 0. \qquad (6.27)$$

Thus, we have accounted for m linear syzygies. It suffices to prove that the rank of the submatrix of the syzygy matrix consisting of these m syzygies has rank m. But this follows from the ladder-like shape of this submatrix according to (6.26) and (6.27) (see Proposition 6.37 for a more precise assertion). $\qquad\square$

Remark 6.36 From (6.27), via (6.23), we derive an expression of the mth partial derivative in terms of the partial derivatives of $\det \mathcal{H}_{x_1,\dots,x_m}[m-3]$; and vice versa. Thus, if we could derive the latter expression first we would retrieve the linear syzygy (6.27). Alas, this expression is not so obvious a priori.

Fixing the integer $m \geq 2$ and the ground variables $x_0, \dots, x_m$, from now on we set $f = f^{(m)}$ and denote the gradient ideal of f by J. In addition, given $0 \leq i \leq m-1$, we let $J_i \subset k[x_{m-i},\dots,x_m]$ denote the ideal generated by the partial derivatives $\partial f/\partial x_0, \ldots, \partial f/\partial x_i$ upon factoring out their gcd. This ideal should not be confused with the ith graded part of J, the latter being zero for $i \leq m - 2$.

Proposition 6.37 *With the above notation, for every $0 \leq i \leq m - 1$, the ideal J_i is a codimension two perfect ideal of linear type with linear presentation.*

Proof The assertion is readily checked for $m = 2$ since $f = x_0 x_2 - x_1^2$. Thus, we assume henceforth that $m \geq 3$.

The assertion is trivial for $i = 0$, so assume that $1 \leq i \leq m - 1$. Note that the gcd of the partial derivatives of f with respect to $x_0, \ldots, x_i$ is x_m^{m-i-1}. Set $R^{[i]} = k[x_{m-i}, \ldots, x_m]$.

We claim that the presentation matrix of J_i is an $(i + 1) \times i$ recurrent matrix having the form

$$\Phi^{[i]} = \left(\begin{array}{c|c} \begin{array}{c} 2\,x_{m-i} \\ \frac{2i-1}{i}\,x_{m-i+1} \\ \vdots \\ \frac{i+1}{i}\,x_{m-1} \\ x_m \end{array} & \begin{array}{c} \\ \Phi^{[i-1]} \\ \\ \mathbf{0} \end{array} \end{array} \right), \qquad \mathbf{0} = \underbrace{(0, \ldots, 0)}_{i-1}, \tag{6.28}$$

where the first column comes from (6.26) by dividing the field coefficients by i throughout. By induction on i, one has that the last $i-1$ columns of $\Phi^{[i]}$ are relations of J_i, hence the full matrix $\Phi^{[i]}$ is a matrix of relations of J_i and, moreover, its linear part has maximal rank $(= i)$.

On the other hand, by a well-known acyclicity criterion in this case, it suffices to check that the columns of $\Phi^{[i]}$ are relations of the generators of J_i and the determinantal ideal $I_i(\Phi^{[i]})$ has codimension at least 2. Thus, we are left with finding two i-minors of $\Phi^{[i]}$ without nontrivial common factor. Let δ_1 (respectively, δ_2) denote the minor obtained by deleting the first (respectively, the last) row of $\Phi^{[i]}$. By induction on i, $\delta_1 = \pm x_m^i$ and δ_2 admits a summand of the form $\pm(i + 1)x_{m-1}^i$ that results from multiplying the entries along the anti-diagonal of the first i rows – indeed, by (6.26), taking $k = i - 1$, the coefficient of the $(i, i - 1)$ entry is $(i + 1)/i$, hence the product is $(i + 1/i)(i/i - 1) \cdots (3/2)(2/1) = i + 1$. It follows that δ_1 is not divisible by x_m. Therefore, $I_i(\Phi^{[i]})$ has codimension at least 2, as required.

Thus, J_i has a free resolution of length one, and since it has codimension at least 2 then it is a codimension two perfect ideal. In order to conclude that it is of linear type it suffices to show that it satisfies the condition G_∞ (Proposition 2.22). For this, since $\mathrm{codim}(I_i(\Phi^{[i]})) \geq 2 = i - i + 2 = \mathrm{rank}(\Phi^{[i]}) - i + 2$, we only have to check that

$$\mathrm{codim}(I_t(\Phi^{[i]})) \geq i - t + 2, \quad \text{for} \quad 1 \leq t \leq i - 1. \tag{6.29}$$

We proceed by induction on i, so $\mathrm{codim}(I_t(\Phi^{[i-1]})) \geq i - 1 - t + 2 = i - t + 1$ holds true in the range $1 \leq t \leq i - 2$. Therefore, one needs, for every t in the range $1 \leq t \leq i - 1$, an additional t-minor of $\Phi^{[i]}$ which is a nonzero divisor modulo the ideal $I_t(\Phi^{[i-1]})$. Since $\Phi^{[i-1]}$ has entries in the polynomial ring $R^{[i-1]} = k[x_{m-i+1}, \ldots, x_m]$, it suffices to show that there exists such a minor effectively involving the variable $x_{m-i} \notin R^{[i-1]}$. Supposing this were not the case, the full matrix of relations $\Phi^{[i]}$ would have entries entirely contained in $R^{[i-1]}$ and, since the generators of J_i are the maximal minors of $\Phi^{[i]}$, they would all belong to $R^{[i-1]}$, which is not the case.

This finishes the proof of the last statement. $\square$

6.4.2 *Homology of the Gradient Ideal*

We now focus on additional homological properties of the gradient ideal $J \subset R = k[x_0, \ldots, x_m]$ of the sub-Hankel determinant. Set $P = (x_{m-1}, x_m) \subset R$ throughout – note that this is against our notation in previous parts, where P used to denote the ideal of submaximal minors.

Proposition 6.38 *With the above notation and the one of the previous subsection, one has*:

(i) *P is the radical of J and all the ideals J_i, $1 \le i \le m-1$ are P-primary.*
(ii) *The R_P-module $R_P/(J_i)_P$ has length $\binom{i+1}{2}$.*
(iii) *The ideal of $R_P/(J_{m-1})_P$ generated by the forms x_m, x_{m-1}^{m-1} has length $\binom{m-1}{2}$.*

Proof

(i) This is clear from the form of these ideals: any prime ideal containing any of them has to contain P, which is then the unique minimal prime thereof. By Proposition 6.37, for $i \le m-1$, the ideal J_i is perfect, hence R/J_i is Cohen–Macaulay, thus implying that P is the only associated prime of J_i.
(ii) By the primary case of the associativity formula for multiplicities, one has

$$e(R/J_i) = \ell(R_P/(J_i)_P)\, e(R/P) = \ell(R_P/(J_i)_P),$$

since P is generated by linear forms. On the other hand, from Proposition 6.37 one has the graded free resolution

$$0 \to R(-(i+1))^i \to R(-i)^{i+1} \to R \to R/J_i \to 0. \tag{6.30}$$

Applying the multiplicity formula for Cohen–Macaulay rings with pure resolution as obtained in [14, Theorem 1.2], one derives in this case $e(R/J_i) = \binom{i+1}{2}$, as required.
(iii) Close inspection of the shape of the partial derivatives gives $(x_m, J_{m-1}) = (x_m, x_{m-1}^{m-1})$. On the other hand,

$$(x_m, J_{m-1})/J_{m-1} \simeq (x_m)/(x_m) \cap J_{m-1} \simeq R/(J_{m-1} : x_m)(1) = R/J_{m-2}(1),$$

where the equality $(J_{m-1} : x_m) = J_{m-2}$ follows from Proposition 6.37. Therefore

$$l\left((x_m, x_{m-1}^{m-1})_P/(J_{m-1})_P\right) = l(R/J_{m-2})_P = \binom{m-1}{2},$$

by part (ii). $\square$

Lemma 6.39 *With the same notation as above, one has*

$$J/J_{m-1} \simeq \frac{R}{(x_m, x_{m-1}^{m-1})}(-(m-1)).$$

Proof The following isomorphisms of R-graded modules are immediate:

$$\frac{J}{J_{m-1}} = \frac{\left(J_{m-1}, \frac{\partial f}{\partial x_m}\right)}{J_{m-1}} \simeq \frac{\left(\frac{\partial f}{\partial x_m}\right)}{J_{m-1} \cap \left(\frac{\partial f}{\partial x_m}\right)} \simeq \frac{R}{\left(J_{m-1} : \frac{\partial f}{\partial x_m}\right)}(m-1). \qquad (6.31)$$

We claim that $\left(J_{m-1} : \frac{\partial f}{\partial x_m}\right) = (x_m, J_{m-1})$. Once this is proved, we will have $\left(J_{m-1} : \frac{\partial f}{\partial x_m}\right) = (x_m, x_{m-1}^{m-1})$ as has been noted in the proof of Proposition 6.38 (iii).

Now, by (6.27), $(x_m, J_{m-1}) \subset \left(J_{m-1} : \frac{\partial f}{\partial x_m}\right)$ as trivially $J_{m-1} \subset \left(J_{m-1} : \frac{\partial f}{\partial x_m}\right)$. For the reverse inclusion, we proceed as follows.

Let $r \in (J_{m-1} : \partial f/\partial x_m)$ and $P = (x_{m-1}, x_m)$ as before. Since J_{m-1} is a P-primary ideal and $\partial f/\partial x_m \notin J_{m-1}$ then $r \in P$. Write $r = r(x_0, \ldots, x_m)$ as

$$r = x_m h(x_0, \ldots, x_m) + r'(x_0, \ldots, x_{m-1}) \in P,$$

where x_m divides no term on the second summand. Then $r'(x_0, \ldots, x_{m-1}) \in (x_{m-1})$, so let $l \in \mathbb{N}$ be such that $r'(x_0, \ldots, x_{m-1}) = x_{m-1}^l r''(x_0, \ldots, x_{m-1})$ and x_{m-1} does not divide $r''(x_0, \ldots, x_{m-1})$. It follows that $r''(x_0, \ldots, x_{m-1}) \notin P$.

Because $(x_m, J_{m-1}) = (x_m, x_{m-1}^{m-1})$ and J_{m-1} is P–primary, $r'' x_{m-1}^l \partial f/\partial x_m \in J_{m-1}$ implies $x_{m-1}^l \partial f/\partial x_m \in J_{m-1}$. Thus, for the required reverse inclusion it is enough to prove:

CLAIM. $l \geq m - 1$.

To see this, let, say,

$$x_{m-1}^l \frac{\partial f}{\partial x_m} = \sum_{i=0}^{m-1} r_i \frac{\partial f}{\partial x_i}, \qquad (6.32)$$

where $r_i \in R$.

Writing $r_i = x_m g_i(x_0, \ldots, x_m) + g_i'(x_0, \ldots, x_{m-1})$ and drawing upon (6.26) yields

$$
x_{m-1}^l \frac{\partial f}{\partial x_m} = \sum_{i=0}^{m-1} r_i \frac{\partial f}{\partial x_i} = \sum_{i=0}^{m-1} (x_m g_i(x_0, \ldots, x_m) + g_i'(x_0, \ldots, x_{m-1})) \frac{\partial f}{\partial x_i}
$$

$$
= \sum_{i=0}^{m-1} g_i(x_0, \ldots, x_m) \left(-\sum_{k=0}^{i-1} \frac{2i - k}{i} x_{m-i+k} \frac{\partial f}{\partial x_k} \right)
$$

$$
+ \sum_{i=0}^{m-1} g_i'(x_0, \ldots, x_{m-1}) \frac{\partial f}{\partial x_i}.
$$

By repeating this process for g_i and so forth, we can assume that, after finitely many steps, the coefficients of $\partial f / \partial x_i$ in (6.32) do not involve x_m.

Multiplying both sides of (6.32) by x_m and using (6.27) yields a syzygy of the ideal $J_{m-1} = (\partial f / \partial x_0, \ldots, \partial f / \partial x_{m-1})$:

$$
((m - 1)x_{m-1}^l x_0 - x_m r_0, \ \ldots, \ x_{m-2} x_{m-1}^l - x_m r_{m-2}, \ -x_m r_{m-1})^t. \tag{6.33}
$$

Thinking of this syzygy as a column vector K, we can write

$$
K = \alpha_1 C_1 + \cdots + \alpha_{m-1} C_{m-1} \tag{6.34}
$$

for suitable $\alpha_i \in k[x_0, \cdots, x_m]$, where C_i denotes the ith column of the syzygy matrix of J_{m-1} as in Proposition 6.37:

$$
\begin{pmatrix}
2x_1 & 2x_2 & 2x_3 & \ldots & 2x_{m-1} \\
\frac{2m-3}{m-1} x_2 & \frac{2m-5}{m-2} x_3 & \frac{2m-7}{m-3} x_4 & \ldots & x_m \\
\vdots & \vdots & \vdots & \ldots & \vdots \\
\frac{m+2}{m-1} x_{m-3} & \frac{m}{m-2} x_{m-2} & \frac{m-2}{m-3} x_{m-1} & \ldots & 0 \\
\frac{m+1}{m-1} x_{m-2} & \frac{m-1}{m-2} x_{m-1} & x_m & \ldots & 0 \\
\frac{m}{m-1} x_{m-1} & x_m & 0 & \ldots & 0 \\
x_m & 0 & 0 & \ldots & 0
\end{pmatrix}. \tag{6.35}
$$

Focusing on the last two rows, this affords the following relations:

$$
x_{m-2} x_{m-1}^l - x_m r_{m-2} = -r_{m-1} \frac{m}{m-1} x_{m-1} + \alpha_2 x_m \quad \text{and} \quad \alpha_1 = -r_{m-1}.
$$

As we assumed that $r_i \in k[x_0, \ldots, x_{m-1}]$ for all i, then $\alpha_2 = -r_{m-2}$. Therefore

$$- r_{m-1} \frac{m}{m-1} x_{m-1} = x_{m-2} x_{m-1}^l.$$

In particular $l \geq 1$ and hence $r_{m-1} = -\frac{m-1}{m} x_{m-2} x_{m-1}^{l-1}$.

The $(m-2)$th row yields:

$$2 x_{m-1}^l x_{m-3} - x_m r_{m-3} = -\frac{m+1}{m-1} r_{m-1} x_{m-2} - \frac{m-1}{m-2} r_{m-2} x_{m-1} + \alpha_3 x_m.$$

Since both r_{m-1} and r_{m-2} belong to $k[x_0, \ldots, x_{m-1}]$ then $\alpha_3 = -r_{m-3}$. Substituting for r_{m-1} it obtains

$$2 x_{m-1}^l x_{m-3} = -\frac{m+1}{m-1} r_{m-1} x_{m-2} - \frac{m-1}{m-2} r_{m-2} x_{m-1}$$

$$= \frac{m+1}{m} x_{m-2}^2 x_{m-1}^{l-1} - \frac{m-1}{m-2} r_{m-2} x_{m-1}.$$

Then, necessarily $l \geq 2$ and furthermore

$$r_{m-2} = \frac{m-2}{m-1} \left(-2 x_{m-1}^{l-1} x_{m-3} + \frac{m+1}{m} x_{m-2}^2 x_{m-1}^{l-2} \right) = x_{m-1}^{l-2} s_{m-2}, \tag{6.36}$$

for some $s_{m-i} \in k[x_0, \ldots, x_{m-1}]$.

Since every entry of the last nonzero subdiagonal of (6.35) is x_m, an inductive argument yields that, for every $1 \leq i \leq m-2$, one has $\alpha_i = -r_{m-i}$ and $r_{m-i} = x_{m-1}^{l-i} s_{m-i}$ with $s_{m-i} \in k[x_0, \ldots, x_{m-1}]$ and $l \geq i$. In particular, $l \geq m-2$ and $r_2 = x_{m-1}^{l-n+2} s_2$.

Finally, from the first row of K, we have:

$$(m-1) x_0 x_{m-1}^l - x_m r_0 = -2 x_1 r_{m-1} - 2 x_2 r_{m-2} - \ldots - 2 x_{m-2} r_2 - 2 x_{m-1} r_1.$$

Hence $r_0 = 0$. Rearranging yields

$$- 2 x_1 r_{m-1} - 2 x_2 r_{m-2} - \ldots - 2 x_{m-1} r_1 - (m-1) x_0 x_{m-1}^l$$

$$= 2 x_{m-2} r_2 = 2 x_{m-2} x_{m-1}^{l-n+2} s_2. \tag{6.37}$$

Since the left-hand side is divisible by x_{m-1} so is the right-hand side. Thus $l - m + 2 > 0$. In other words, $l \geq m-1$, as required in the stated claim. $\square$

Proposition 6.40 *Let J as before denote the gradient ideal of $f^{(m)} = \det \mathcal{H}_{\mathbf{x}}[m-2]$ $\in R = k[x_0, \ldots, x_m]$. Then the minimal graded free resolution of R/J has the form*

$$0 \to R(-(2m-1)) \to R(-m)^m \oplus R(-2(m-1)) \xrightarrow{\varphi} R^{m+1}(-(m-1)) \to R.$$

In particular, R/J is strictly almost Cohen–Macaulay.

Proof By Lemma 6.39 we have a minimal free resolution

$$C:\ 0 \to R(-(2m-1))\to R(-m)\oplus R(-2(m-1))\to R(-(m-1))\to J/J_{m-1}\to 0.$$

On the other hand, (6.30) gives a resolution

$$\mathcal{J}_{m-1}:\quad 0 \to R(-m)^{m-1} \to R(-(m-1))^m \to R \to R/J_{m-1} \to 0.$$

Since the inclusion $J/J_{m-1} \subset R/J_{m-1}$ induces a map of complexes $C \to \mathcal{J}_{m-1}$, the resulting mapping cone is a resolution of R/J ([10, Exercise A3.30]):

$$0 \to R(-(2m-1)) \to R(-m)^m \oplus R(-2(m-1)) \overset{\varphi}{\to} R^{m+1}(-(m-1))$$
$$\to R \to R/J \to 0, \tag{6.38}$$

(where the right end tail $R \oplus J/J_{m-1} \to R/J_{m-1} \to 0$ has been replaced by $R \to R/J \to 0$). Moreover, since for every relevant index i, the shifts of $(\mathcal{J}_{m-1})_i$ are strictly smaller than those of $(C)_i$, it follows by [loc. cit.] that (6.38) is a minimal resolution. In particular, one reads from it that the linear part φ_1 has rank n, and hence is maximal. $\qquad\square$

Corollary 6.41 *Let* $P = \sqrt{J} = (x_{m-1}, x_m)$ *as in* Proposition 6.38. *Then the P-primary component of J is J_{m-2}.*

Proof Since J_{m-2} is a P-primary ideal (Proposition 6.38 (i)), it is equivalent to show the equality $J_P = (J_{m-2})_P$. Since $J \subset J_{m-2}$ we will be done by showing the equality of lengths $l(R_P/J_P) = l(R_P/(J_{m-2})_P)$.

Now, with the present data, by the associativity formula one has $l(R_P/J_P) = e(R/J)$ – the multiplicity of R/J. To compute the latter we deploy the numerator of the Hilbert series of R/J in terms of the graded Betti numbers of R/J as in (6.38); it obtains

$$S(t) := 1 - (m + 1)t^{m-1} + t^{2n-2} + mt^n - t^{2n-1}.$$

Taking second derivatives, evaluating at $t = 1$, etc., finally gives $e(R/J) = (m - 1)(m - 2)/2 = \binom{m-1}{2}$. But the latter coincides with $l(R_P/(J_{m-2})_P)$ by Proposition 6.38 (ii). $\qquad\square$

Let us delve a little more on the details of the mapping cone in the previous proof. It is of the form

$$
\begin{array}{ccccccccc}
0 & \longrightarrow & R^{m-1} & \overset{\varphi_{m-1}}{\longrightarrow} & R^n & \xrightarrow{(\frac{\partial f}{\partial x_0},\ldots,\frac{\partial f}{\partial x_{m-1}})} & R & \longrightarrow & \dfrac{R}{J_{m-1}} & \longrightarrow & 0 \\
& & \big\uparrow{\scriptstyle g_4} & & \big\uparrow{\scriptstyle g_3} & & \big\uparrow{\scriptstyle g_2} & & \big\uparrow{\scriptstyle g_1} & & \\
0 & \longrightarrow & R & \xrightarrow{(x_{m-1}^{m-1},-x_m)^t} & R^2 & \xrightarrow{(x_m,x_{m-1}^{m-1})} & R & \longrightarrow & J/J_{m-1} & \longrightarrow & 0
\end{array}
$$

Note that g_2 is multiplication by $\partial f/\partial x_m$ and the induced map g_3 is given by the following $n \times 2$ matrix:

$$g_3 = \begin{pmatrix} (m-1)x_0 & r_0 \\ (m-2)x_1 & r_1 \\ & \vdots \\ x_{m-2} & r_{m-2} \\ 0 & r_{m-1} \end{pmatrix}.$$

We next find out the entries of g_4. By the commutativity of the mapping cone diagram, we have

$$\varphi_{m-1}g_4 = g_3 \begin{pmatrix} x_{m-1}^{m-1} \\ -x_m \end{pmatrix} = \begin{pmatrix} (m-1)x_0 & r_0 \\ (m-2)x_1 & r_1 \\ & \vdots \\ x_{m-2} & r_{m-2} \\ 0 & r_{m-1} \end{pmatrix} \begin{pmatrix} x_{m-1}^{m-1} \\ -x_m \end{pmatrix}$$

$$= \begin{pmatrix} (m-1)x_{m-1}^{m-1}x_0 - x_m r_0 \\ (m-2)x_{m-1}^{m-1}x_1 - x_m r_1 \\ \vdots \\ x_{m-2}x_{m-1}^{m-1} - x_m r_{m-2} \\ -x_m r_{m-1} \end{pmatrix}$$

where the rightmost matrix is the syzygy K in (6.33), viewed as a column vector, where $l = n - 1$. By reasoning as in the argument that ensues (6.34), one gets that the ith entry of g_4 is $-r_{m-i}$, $1 \leq i \leq m - 1$. As a result, the leftmost map in (6.38) is $\psi := (x_{m-1}^{m-1}, -x_m, g_4^t) = (x_{m-1}^{m-1}, -x_m, -r_{m-1}, \ldots, -r_1)$.

We make use of this in the following result, where J is as before.

Theorem 6.42 *The associated primes of R/J are (x_{m-1}, x_m) and (x_{m-2}, x_{m-1}, x_m).*

Proof Clearly, $P := (x_{m-1}, x_m)$ is a minimal prime thereof. Since P is also the radical of J (Proposition 6.38 (i)) then there are no additional minimal primes.

To argue for the embedded primes we proceed as follows. Let as above ψ denote the tail map of (6.38). Since R/J has homological dimension 3, any $Q \in \mathrm{Ass}(R/J)$ has codimension at most 3 and a prime Q of codimension 3 containing J is an associated prime of R/J if and only if $Q \supset I_1(\psi)$ (see, e.g., [10, Corollary 20.14(a)] for the last part). As seen above, $I_1(\psi) = (x_{m-1}^{m-1}, x_m, -r_{m-1}, \ldots, -r_1)$. On the other hand, we know from the proof of Proposition 6.39 that $r_{m-i} = x_{m-1}^{m-i-1}s_{m-i}$, where $s_{m-i} \in k[x_0, \ldots, x_{m-1}]$ for all $1 \leq i \leq m - 1$. Additionally, the inductive argument in this proof and relation (6.36) also show that, for $1 \leq i \leq m - 2$, $s_{m-i} = x_{m-1}t_{m-i} + x_{m-i}^i$, for some $t_{m-i} \in k[x_0, \ldots, x_{m-1}]$.

Therefore the condition for such a prime Q to be an associated prime of R/J is that it contain the ideal $(x_m, x_{m-1}, x_{m-2}s_2) = (x_m, x_{m-1}, x_{m-2}(x_{m-1}t_2 + x_{m-2}^{m-2})) = (x_m, x_{m-1}, x_{m-2}^{m-1})$. It follows that $Q = (x_{m-2}, x_{m-1}, x_m)$, as stated. $\qquad\square$

Quite a bit harder is the following result:

Theorem 6.43 *The gradient ideal J of the sub-Hankel determinant is of linear type.*

Proof By definition, we have to show that the natural surjective R-homomorphism $\mathcal{S}_R(J) \twoheadrightarrow \mathcal{R}_R(J)$ from the symmetric algebra of J to its Rees algebra is injective. Since the kernel of this map is the R-torsion of $\mathcal{S}_R(J)$, this is the case if and only if $\mathcal{S}_R(J)$ is a domain.

Set $\mathcal{S}_R(J) \simeq R[y_0, \cdots, y_m]/\mathcal{L}$, where $\mathcal{L}$ is the ideal generated by the 1-forms coming from the syzygies of J. We will argue as follows: since x_m belongs to the radical of J, J_{x_m} is the unit ideal in R_{x_m}. Now, suppose one shows that x_m is a nonzero divisor modulo $\mathcal{L}$. Then $\mathcal{L}_{x_m}$ is the defining ideal of the symmetric algebra of $J_{x_m} = R_{x_m}$, hence it is the zero ideal in a polynomial ring over a domain. In particular, it is a prime ideal, hence so must be $\mathcal{L}$. Therefore $\mathcal{S}_R(J)$ is a domain.

By a quirk, it will be easier to show first that y_m is nonzero divisor modulo $\mathcal{L}$ and then that (y_m, x_m) is a regular sequence modulo $\mathcal{L}$. In this case, since $\mathcal{S}_R(J)$ is a positively graded ring any permutation of a regular sequence is a regular sequence, hence (x_m, y_m) is a regular sequence as well, in particular x_m is a nonzero divisor over $\mathcal{S}_R(J)$.

Step 1. y_m is nonzero divisor modulo $\mathcal{L}$.

Let $h = h(\mathbf{x}, \mathbf{y}) \in R[y_0, \ldots, y_m]$ be such that $y_m h(\mathbf{x}, \mathbf{y}) \in \mathcal{L}$. Letting r_i be as in (6.33) and drawing upon the shape of the syzygies in (6.38), one can write

$$y_m h = \sum_{i=1}^{m+1} h_i g_i,$$

for suitable $h_i = h_i(\mathbf{x}, \mathbf{y}) \in R[y_0, \ldots, y_m]$, where

$$g_i = 2x_{m-i}y_0 + \frac{2i-1}{i}x_{m-i+1}y_1 + \cdots + x_m y_i \quad (1 \le i \le m-1)$$

and

$$g_m = (m-1)x_0 y_0 + (m-2)x_1 y_1 + \cdots + x_m y_m, \quad g_{m+1} = \sum_{j=1}^{m-1} r_j y_j + x_{m-1}^{m-1} y_m$$

generate $\mathcal{L}$.

Decompose further $h_i = h_i' + y_m h_i''$, with $h_i' \in R[y_0, \ldots, y_{m-1}]$. Then

$$y_m h(\mathbf{x}, \mathbf{y}) = y_m \sum_{i=1}^{m-1} h_i'' g_i + y_m h_m'' g_m + h_m' x_m y_m + y_m h_{m+1}'' g_{m+1} + h_{m+1}' x_{m-1}^{m-1} y_m.$$

Since in the right-hand side the terms not divisible by y_m must vanish, we get

$$h = \left(\sum_{i=1}^{m-1} h_i'' g_i + h_m'' g_m + h_{m+1}'' g_{m+1} \right) + h_m' x_m + h_{m+1}' x_{m-1}^{m-1}.$$

To show that $h \in \mathcal{L}$ it is then enough to check that

$$h_m' x_m + h_{m+1}' x_{m-1}^{m-1} \in \mathcal{L}.$$

Now a form in $\mathcal{L}$ must vanish when evaluated at $y_i \mapsto \partial f / \partial x_i$ – the generators of J. Letting $\partial \mathbf{f}$ denote these partial derivatives we have $\frac{\partial f}{\partial x_m} h(\mathbf{x}, \partial \mathbf{f}) = 0$, hence $h(\mathbf{x}, \partial \mathbf{f}) = 0$. Retrieving in terms of the expression of h implies that

$$h_m'(\mathbf{x}, \partial \mathbf{f}) \, x_m + h_{m+1}'(\mathbf{x}, \partial \mathbf{f}) \, x_{m-1}^{m-1} = 0.$$

Since $h_m', h_{m+1}' \in R[y_0, \ldots, y_{m-1}]$, the form $h_m' x_m + h_{m+1}' x_{m-1}^{m-1}$ belongs to the defining ideal of $\mathcal{R}_R(J_{m-1})$. By Proposition 6.37, J_{m-1} is of linear type, hence $h_m' x_m + h_{m+1}' x_{m-1}^{m-1}$ belongs to the defining ideal of $\mathcal{S}_R(J_{m-1})$, which is a subideal of $\mathcal{L}$ by the very proof of Proposition 6.37. This shows the contention.

Step 2. x_m is nonzero divisor modulo $(\mathcal{L}, y_m)$.

One has $(\mathcal{L}, y_m) = (g_1, \ldots, g_{m-1}, g, h, y_m)$, where $g = (m-1)x_0 y_0 + (m-2)x_1 y_1 + \cdots + x_{m-2} y_{m-2}$ and $h = \sum_{j=1}^{m-1} r_j y_j$. Then

$$\frac{R[y_0, \ldots, y_m]}{(\mathcal{L}, y_m)} \simeq \frac{R[y_0, \ldots, y_{m-1}]}{(g_1, \ldots, g_{m-1}, g, h)}$$

hence we are to show that x_m is a nonzero divisor on the rightmost ring. Let then $\kappa = \kappa(\mathbf{x}, \mathbf{y} \setminus y_m)$ be a form in $R[y_0, \ldots, y_{m-1}]$ such that $x_m \kappa \in (g_1, \ldots, g_{m-1}, g, h)$. Write $x_m \kappa = \sum_{j=1}^{m-1} \mu_j g_j + \mu g + \nu h$, for suitable forms $\mu_j, \mu, \nu \in R[y_0, \ldots, y_{m-1}]$. Evaluating $y_i \mapsto f_{x_i} := \partial f / \partial x_i$ for $i = 0, \cdots, n-1$, and taking in account the shape of g and h, we have

$$x_m \kappa(\mathbf{x}, f_{x_0}, \ldots, f_{x_{m-1}}) = \mu(\mathbf{x}, f_{x_0}, \ldots, f_{x_{m-1}}) g(\mathbf{x}, f_{x_0}, \ldots, f_{x_{m-1}})$$

$$+ \nu(\mathbf{x}, f_{x_0}, \ldots, f_{x_{m-1}}) h(\mathbf{x}, f_{x_0}, \ldots, f_{x_{m-1}})$$

$$= -\mu(\mathbf{x}, f_{x_0}, \ldots, f_{x_{m-1}}) x_m f_{x_m} - \nu(\mathbf{x}, f_{x_0}, \ldots, f_{x_{m-1}}) x_{m-1}^{m-1} f_{x_m}.$$

On the other hand, by the shape of f_{x_m}, one has $\gcd(x_m, f_{x_m}) = 1$. It follows that

$$\kappa(\mathbf{x}, f_{x_0}, \ldots, f_{x_{m-1}}) = \delta \, f_{x_m},$$

for some $\delta \in R$. Pulling back to the **y**-variables tells us that $\kappa - \delta\, y_m$ vanishes on the partial derivatives and hence belongs to the affine ideal $\tilde{\mathcal{J}}$ of all polynomials vanishing on the partial derivatives. To argue that this ideal is prime, we think of $\partial \mathbf{f} := (f_{x_0}, \ldots, f_{x_m})$ as a point in K^{m+1}, where K denotes the field of fractions of R. Then, consider the ideal of $K[\mathbf{y}]$ vanishing on $\partial \mathbf{f}$ and contract it to $R[\mathbf{y}]$. Note, moreover, that the defining ideal $\mathcal{J}$ of the Rees algebra is the largest homogeneous ideal contained in $\tilde{\mathcal{J}}$.

Now, multiplying $\kappa - \delta\, y_m$ by x_m and using that $x_m \kappa \in \mathcal{L} \subset \mathcal{J} \subset \tilde{\mathcal{J}}$, it follows that $x_m\, \delta\, y_m \in \tilde{\mathcal{J}}$. Since this element is (trivially) homogeneous in **y**, it must belong to the Rees ideal $\mathcal{J}$, hence $\delta = 0$. Therefore, $\kappa \in \tilde{\mathcal{J}}$. But, since κ is assumed to be homogeneous, it belongs to the Rees ideal $\mathcal{J}$. Thus, we have $\kappa \in \mathcal{J} \cap R[y_0, \ldots, y_{m-1}]$. But this means that κ belongs to the Rees ideal $\mathcal{J}'$ of J_{m-1}. Since the latter is of linear type we conclude as above that $\kappa \in \mathcal{L}$, as was to be shown. $\qquad\square$

An important feature of the sub-Hankel case is the next property. The first part has been proved in [3] with different tools, based on an earlier criterion established in [21, Proposition 4.1]. The present proof is based on [8] and other parts of [21].

Theorem 6.44 *Let $f \in R := k[x_0, \ldots, x_m]$ stand for the determinant of the $m \times m$ sub-Hankel matrix.*

(i) *f is homaloidal.*
(ii) *The base ideal of the inverse map is a codimension two perfect ideal generated in degree $m - 1$.*

Proof

(i) By Proposition 6.35, the gradient ideal $J = J_f$ has maximal linear rank and by Proposition 6.43 it is of linear type. Therefore, by [19, Corollary 3.2.27], the polar map of f is homaloidal.
(ii) First we claim that the linear syzygies of J span a free submodule of R^{m+1} of rank m. This follows from the maximal linear rank property and the structure of the syzygies in the graded free resolution of J as in Proposition 6.40. Then the result follows from the details of [19, Section 3.2] or, even earlier, from [21, Corollary 3.8], then a conjecture but now a true result by [19, Corollary 3.2.27]. $\qquad\square$

Exercises

The following problems will require computer assistance.

6.45 In regard to Conjecture 6.18:

1. Confirm all three statements for $m = 3$.
2. How far out values of m and r can you reach to test for the three statements?

3. The Simis–Vasconcelos conjecture states that, given an equigenerated ideal I of a polynomial ring over a field, if the Rees algebra of I is Cohen–Macaulay then so is the special fiber of I. Can you see this into play in the above environment, i.e., that the Cohen–Macaulay part of statement (2) implies statement (1)? (The authors would appreciate any insight here.)

6.46 In regard to Remark 6.17:

1. Show that $R/I_3(\mathcal{H}_4[1])$ is not a normal ring, where $R = k[x_1, \ldots, x_6]$.
2. Show that $R/I_4(\mathcal{H}_5[1])$ is a normal ring, where $R = k[x_1, \ldots, x_8]$.
3. Is it true that, for any m, r, with $m - r \geq 3$, the normality of $R'/I_{m-1}(\mathcal{H}_m[r+1])$ implies that of $R/I_{m-1}(\mathcal{H}_m[r])$, where $R = R'[x_{2m-r-1}]$?
4. Can you guess an optimum lower bound for $m - r$ such that $R/I_{m-1}(\mathcal{H}_m[r])$ is normal?

6.47 Consider the semi-hollow like section $\mathcal{H}_4[1]$ and the gradient ideal J of its determinant. Prove directly the statements of Theorem 6.24.

6.48 Discuss the gradient ideal J of $\det \mathcal{H}_4[1]$ with respect to the content of Remark 6.25:

1. Is J generically a complete intersection?
   (HINT: in the notation of loc. cit., consider the minimal prime Q)
2. What are the embedded associated primes of J?

6.49 Confirm the statements of conjectures 6.27, 6.26 and 6.30 for as large m as possible.

References

1. W. Bruns, J. Herzog, On the computation of a-invariants. Manuscripta Math. **77**, 201–213 (1992) 168
2. W. Bruns, U. Vetter, *Determinantal Rings*. Lecture Notes in Mathematics, vol. 1327 (Springer, Berlin, 1988) 163, 164
3. C. Ciliberto, F. Russo, A. Simis, Homaloidal hypersurfaces and hypersurfaces with vanishing Hessian. Adv. Math. **218**, 1759–1805 (2008) 171, 177, 183, 185, 197
4. A. Conca, J. Herzog, G. Valla, Sagbi bases with applications to blow-up algebras. J. Reine Angew. Math. **474**, 113–138 (1996) 165
5. A. Conca, Straightening law and powers of determinantal ideals of Hankel matrices. Adv. Math. **138**, 263–292 (1998) 165, 174
6. A. Conca, S. Naldi, G. Ottaviani, B. Sturmfels, Taylor polynomials of rational functions. Acta Math. Vietnamica (2024). To appear 170
7. R. Cunha, M. Mostafazadehfard, Z. Ramos, A. Simis, Coordinate sections of generic Hankel matrices. J. Algebra **611**, 285–319 (2022) 162, 165
8. A. Doria, H. Hassanzadeh, A. Simis, A characteristic free criterion of birationality. Adv. Math. **230**, 390–413 (2012) 197
9. D. Eisenbud, Linear sections of determinantal varieties. Amer. J. Math. **110**, 541–575 (1988) 162, 173

10. D. Eisenbud, Commutative Algebra with a View Toward Algebraic Geometry (Springer, Berlin, Heidelberg, New York, 1995) 193, 194

11. D. Eisenbud, *The Geometry of Syzygies: A Second Course in Algebraic Geometry and Commutative Algebra*. GTM No. (Springer, Berlin, 2005) 172, 173

12. L. Gruson, C. Peskine, Courbes de L'Espace Projectif: Variétés de Sécantes 1–32, in *Enumerative Geometry and Classical Algebraic Geometry*, ed. by P. Le Barz, Y. Hervier. Progress in Mathematics, vol. 24 (Birkhäuser, Boston, 1982) 162

13. M. Hochster, Grassmannians and their Schubert subvarieties are arithmetically Cohen–Macaulay. J. Algebra **25**, 40–57 (1973) 167

14. C. Huneke, M. Miller, A note on the multiplicity of Cohen-Macaulay algebras with pure resolution. Can. J. Math. **37**, 1149–1162 (1985) 189

15. M. Mostafazadehfard, Hankel and sub-Hankel determinants – a detailed study of their polar ideals, PhD Thesis, Universidade Federal de Pernambuco (Recife, Brazil), 2014 162, 166, 167, 181, 182

16. M. Mostafazadehfard, A. Simis, Homaloidal determinants. J. Algebra **450**, 59–101 (2016) 162, 166, 167, 177

17. M.P. Murthy, A note on factorial rings. Arch. Math. **15**, 418–420 (1964) 167

18. J.R. Nogueira, *Classes caracterí sticas e secantes de curvas racionais normais* (Portuguese), Ph.D. Thesis, Universidade Federal Fluminense, RJ, Brazil 168

19. Z. Ramos, A. Simis, *Graded Algebras in Algebraic Geometry*. Expositions in Mathematics, vol. 70 (De Gruyter, Berlin, 2022) 170, 197

20. F. Russo, A. Simis, On birational maps and Jacobian matrices. Comp. Math. **126**, 335–358 (2001) 165

21. A. Simis, Cremona transformations and some related algebras. J. Algebra **280**, 162–179 (2004) 197

22. P. Samuel, *Lectures on Unique Factorization Domains*. Math. Ser. vol. 30 (Tata Institute of Fundamental Research, Bombay, 1964) 167

23. A. Simis, B. Ulrich, W.V. Vasconcelos, Codimension, multiplicities and integral extensions. Math. Proc. Camb. Philos. Soc. **130**, 237–257 (2001) 183

24. B. Sturmfels, Gröbner bases and Stanley decompositions of determinantal rings. Math. Z. **205**, 137–144 (1990) 172

25. W. Vasconcelos, *Integral closure. Rees algebras, multiplicities, algorithms*. Springer Monographs in Mathematics (Springer, Berlin, 2005) 168

26. J. Watanabe, Hankel matrices and Hankel ideals. Proc. School Sci. Tokai Univ. **32**, 11–21 (1997) 162, 171

Chapter 7
Hankel Like Catalecticants

Abstract The meaning of *catalecticant* here does not refer to the classical (generic) catalecticant from invariant theory, a notion that only retrieves the present one in the special case of a Hankel matrix. The classical one is a symmetric matrix in the square case, representing a linear map attached to a given form in the dual space, induced by the Macaulay inverse action. The matrices in this chapter constitute a natural extension of the Hankel logistic, consisting in fixing an integer $r \geq 1$ and starting the second row $r + 1$ steps off (instead of two steps). In the square case it is only symmetric for $r = 1$, which retrieves the usual Hankel matrix. In the case these matrices have only two rows, up to column permutations they may be written as concatenations of Hankel matrices, hence, define rational normal scrolls. Here one focuses on generic such matrices, moving away from the nature of its linear sections, emphasizing instead a method of relating to such a matrix another matrix mimicking a famous construction of Gruson and Peskine. Such a method has been used before, but here one expands on the related algebraic invariants. In the square case a discussion of the dual variety to the determinant is spelled, with a handle to the problem of parabolism stated in previous chapters.

7.1 Terminology

A Hankel matrix is a particular symmetry-preserving linear section of the symmetric matrix. Another way of envisaging a Hankel matrix is as the extremal member of the family of certain r-leap catalecticant matrices. Actually, the only common member of these different families is the Hankel matrix.

© The Author(s), under exclusive license to Springer Nature Switzerland AG 2024   201
Z. Ramos, A. Simis, *Determinantal Ideals of Square Linear Matrices*,
https://doi.org/10.1007/978-3-031-55284-7_7

In this chapter we deal with the second of these families.

Definition 7.1 For given numbers m, n and $1 \leq r \leq n$, the *r-leap $m \times n$ generic catalecticant* is the matrix

$$
C(m, n, r) := \begin{pmatrix}
x_1 & x_2 & x_3 & \cdots & x_n \\
x_{r+1} & x_{r+2} & x_{r+3} & \cdots & x_{r+n} \\
x_{2r+1} & x_{2r+2} & x_{2r+3} & \cdots & x_{2r+n} \\
\vdots & \vdots & \vdots & \ddots & \vdots \\
x_{(m-1)r+1} & x_{(m-1)r+2} & x_{(m-1)r+3} & \cdots & x_{(m-1)r+n}
\end{pmatrix},
\tag{7.1}
$$

whose entries are variables in the polynomial ring $R = k[x_1, \ldots, x_{(m-1)r+n}]$.

Note that for the extreme values $r = 1$ and $r = n$ one gets the Hankel matrix and the generic matrix, respectively.

Remark 7.2 A word on the terminology seems in place. The one here does not refer to the classical (generic) catalecticant from invariant theory, a notion that only retrieves the present one in the special case of a 1-leap catalecticant, i.e., a Hankel matrix. The latter is a symmetric matrix in the square case – defined as representing a linear map attached to a given form in the dual space, induced by the Macaulay inverse action – while the one here in the square case is only symmetric for $r = 1$. Alternative terminologies, such as *extended Hankel matrix* or *shifted Hankel matrix*, have been suggested by some authors. However, one should parsimoniously abuse adjectives which are feebly mnemonic or do not express in which sense are being used.

These matrices are interesting also from the view point of homaloidal theory, i.e., the theory of Cremona transformations ([8] is a good classical source).

7.2 *r*-Leap Catalecticants Are 1-Generic

They share a typical property of the generic Hankel case.

Proposition 7.3 *The r-leap generic catalecticant is 1-generic.*

Proof We show that $C(m, n, r)$ is a submatrix of a suitable generic Hankel matrix $\mathcal{H}$. This is done by stretching the matrix to allow the "missing" Hankel blocks:

$$
\mathcal{H} =
\begin{pmatrix}
x_1 & x_2 & x_3 & \cdots & x_n \\
x_{r+1} & x_{r+2} & x_{r+3} & \cdots & x_{r+n} \\
x_{2r+1} & x_{2r+2} & x_{2r+3} & \cdots & x_{2r+n} \\
\vdots & \vdots & \vdots & \ddots & \vdots \\
x_{(m-1)r+1} & x_{(m-1)r+2} & x_{(m-1)r+3} & \cdots & x_{(m-1)r+n} \\
\hline
x_2 & x_3 & x_4 & \cdots & x_{n+1} \\
x_3 & x_4 & x_5 & \cdots & x_{n+2} \\
\vdots & \vdots & \vdots & \ddots & \vdots \\
x_r & x_{r+1} & x_{r+2} & \cdots & x_{r+n-1} \\
\hline
x_{r+2} & x_{r+3} & x_{r+4} & \cdots & x_{r+n+1} \\
x_{r+3} & x_{r+4} & x_5 & \cdots & x_{r+n+2} \\
\vdots & \vdots & \vdots & \ddots & \vdots \\
x_{2r} & x_{2r+1} & x_{2r+2} & \cdots & x_{2r+n-1} \\
\hline
\vdots & \vdots & \vdots & \ddots & \vdots
\end{pmatrix}
$$

The result follows from the 1-genericity of the generic Hankel matrix. $\qquad\square$

Corollary 7.4 *Let $C(m, n, r)$ be as in (7.1), with $m \leq n$. Then $I_m(C(m, n, r))$ is a Cohen–Macaulay prime ideal of codimension $n - m + 1$.*

Proof It follows from Proposition 1.24. $\qquad\square$

7.3 The Gruson–Peskine Paragon

It has been shown in [2, Lemme 2.3] that the ideal of minors of a Hankel matrix over a commutative ring depends only on the sum of its number of rows and its number of columns. That is to say, in the notation of Chapter 6, for any pairs of positive integers indices (r, s) and (r', s') such that $r \leq s, r' \leq s'$ and $r + s = r' + s'$, and any integer $1 \leq t \leq \min\{r, r'\}$, one has $I_t(\mathcal{H}_{r,s}) = I_t(\mathcal{H}_{r',s'})$ (see Proposition 6.1).

It has long been observed by the senior author (unpublished) that a simulacrum of this result would hold for r-leap catalecticant matrices – special cases of this paradigm appeared later in [5]. In-between and totally independently, this paradigm was obtained in all generality in [6], by reworking a proof of the Gruson–Peskine lemma by Conca [1].

In this part we wish to apply these ideas to clarify the structure of the submaximal minors of the square r-leap generic catalecticant

$$
C(m, m, r) = \begin{pmatrix}
x_1 & x_2 & \cdots & x_{m-r} & x_{m-r+1} & \cdots & x_m \\
x_{r+1} & x_{r+2} & \cdots & x_m & x_{1+m} & \cdots & x_{r+m} \\
x_{2r+1} & x_{2r+2} & \cdots & x_{m+r} & x_{1+m+r} & \cdots & x_{2r+m} \\
\vdots & \vdots & \vdots & \vdots & \vdots & \ddots & \vdots \\
x_{(m-2)r+1} & x_{(m-2)r+2} & \cdots & x_{(m-2)r+m-r} & x_{(m-2)r+1} & \cdots & x_{(m-2)r+m} \\
x_{(m-1)r+1} & x_{(m-1)r+2} & \cdots & x_{(m-1)r+m-r} & x_{(m-1)r+1} & \cdots & x_{(m-1)r+m}
\end{pmatrix}. \tag{7.2}
$$

The vertical separator is a mnemonic for the next definition.

We assume throughout that $m \geq 3$ and write $C(m, m, r) = C(m, r)$ for short. Note that the present matrix is the square case of the matrix in (7.1), and the latter will also be called upon in an argument below.

Definition 7.5 The *GP-associated matrix* to $C(m, r)$ is the r-leap $(m - 1) \times (m + r)$ generic catalecticant

$$
GP(C(m, r)) = \begin{pmatrix}
x_1 & x_2 & \cdots & x_m & x_{1+m} & \cdots & x_{r+m} \\
x_{r+1} & x_{r+2} & \cdots & x_{r+m} & x_{1+r+m} & \cdots & x_{2r+m} \\
\vdots & \vdots & \vdots & \ddots & \vdots & \ddots & \vdots \\
x_{(m-2)r+1} & x_{(m-2)r+2} & \cdots & x_{(m-2)r+m} & x_{1+(m-2)r+m} & \cdots & x_{(m-1)r+m}
\end{pmatrix}, \tag{7.3}
$$

where the first block is the $(m - 1) \times m$ submatrix of $C(m, r)$ omitting the last row, while the second block is the $(m - 1) \times r$ submatrix of $C(m, r)$ omitting the first row and the first $m - r$ columns.

The reader will have no difficulty in recognizing this construction in the terminology of [6].

Note, as a clear feature, that the respective k-vector spaces of entries in both matrices coincide. We let R denote the corresponding polynomial ring on the distinct entries over k. The following non-Hankel example is a simple illustration of the construction:

$$
C(4, 2) = \begin{pmatrix}
x_1 & x_2 & x_3 & x_4 \\
x_3 & x_4 & x_5 & x_6 \\
x_5 & x_6 & x_7 & x_8 \\
x_7 & x_8 & x_9 & x_{10}
\end{pmatrix}
\qquad
GP(C(4, 2)) = \begin{pmatrix}
x_1 & x_2 & x_3 & x_4 & x_5 & x_6 \\
x_3 & x_4 & x_5 & x_6 & x_7 & x_8 \\
x_5 & x_6 & x_7 & x_8 & x_9 & x_{10}
\end{pmatrix}.
$$

The next theorem wraps up the essential features of this construction.

Theorem 7.6 *Let $C(m, r)$ and $GP(C(m, r))$ be as above. Then*

$$
I_{m-1}(C(m, r)) \subset I_{m-1}(GP(C(m, r))).
$$

(Note that the right-hand side are maximal minors).

Moreover:

(i) *(GP, Hankel case) Let $r = 1$. Then $I_{m-1}(C(m, 1)) = I_{m-1}(GP(C(m, 1)))$ is a Cohen–Macaulay prime ideal of codimension 3.*

(ii) *If $r \geq 2$, then $I_{m-1}(C(m, r))$ is a Gorenstein proper subideal of codimension 4 of the codimension $r + 2$ Cohen–Macaulay prime ideal $I_{m-1}(GP(C(m, r)))$.*

(iii) *Let $r = 2$. Then $I_{m-1}(C(m, 2))$ is reduced, with prime decomposition*

$$I_{m-1}(C(m, 2)) = I_{m-1}(GP(C(m, 2)) \cap I_{m-2}(C(m + 1, m - 2, 2)).$$

(iv) *(Conjectured) If $r \geq 3$, then $I_{m-1}(C(m, r))$ is a prime ideal.*

Proof The opening assertion is [6, Corollary 3.2].

(i) This follows from [2, Lemme 2.3] and Corollary 7.4.

We observe for the record that, since $I_{m-1}(C(m, 1))$ has codimension ≥ 3 (hence, $= 3$) by direct inspection of its initial ideal in lexicographic order, for this assertion it would suffice to give an independent proof that $I_{m-1}(C(m, 1))$ is a prime ideal. Unfortunately, the known proofs draw directly on the Gruson–Peskine paragon.

(ii) The assertion on $I_{m-1}(GP(C(m, r)))$ is Corollary 7.4.

To see that $I_{m-1}(C(m, r))$ has codimension 4, we claim that, if $r \geq 2$, then the initial ideal of $I_{m-1}(C(m, r))$ in a convenient order has codimension ≥ 4. Since the codimension of $I_{m-1}(C(m, r))$ is at most 4 as $C(m, r)$ is a specialization of the $m \times m$ fully generic matrix, we will be done.

Let $[i_1, \ldots, i_u \mid j_1, \ldots, j_v]$ stand for the minor of (7.2) with rows $i_1, \ldots, i_u$ and columns $j_1, \ldots, i_v$. Consider the $(m - 1)$-minors

$$[1, \ldots, m - 1 \mid 1, \ldots, m - 1], \ [1, \ldots, m - 1 \mid 2, \ldots, m], \ [2, \ldots, m \mid 1, \ldots, m - 1].$$

Take the lexicographic order induced from the natural order of the variables

$$x_1 > x_2 > \cdots > x_m > x_{1+m} > \cdots > x_{r+m} > x_{1+m+r} > \cdots > x_{2r+m} > \cdots$$

$$> x_{1+(m-2)r+m} > \cdots > x_{(m-1)r+m}.$$

CLAIM The initial terms of the above minors are

$$i_1 = x_1 x_{r+2} x_{2r+3} \cdots x_{(m-2)r+m-1},$$

$$i_2 = x_2 x_{r+3} x_{2r+4} \cdots x_{(m-2)r+m},$$

and

$$i_3 = x_{r+1} x_{2r+2} x_{3r+3} \cdots x_{(m-1)r+m-1},$$

the respective products of the entries along the three main diagonals of (7.2).

For the proof, since the matrices of these minors are again s-leap generic catalecticants for some $s \geq 1$, it suffices to argue that the initial term of the determinant of an arbitrary s-leap generic catalecticant is the product of the entries along the diagonal. In this case it is clear by a Laplace expansion along the first row, where the first entry is the most expensive in the chosen order. The rest is induction on the size of the matrix, pretty much as in [11, Lemma 5] up to the chosen order of variables respecting rows and columns. The inductive process stays within the class of the square s-leap generic catalecticants.

So much for the claim.

No two among i_i, $i = 1, 2, 3$, have any common variable in their supports. Therefore, the three generate an ideal of height 3. Note that the three $(m - 1)$-minors do not involve the last entry $x_{(m-1)r+m}$. Thus, we consider the $(m - 1)$-minors effectively involving this entry. Their matrices are no longer strictly s-leap catalecticant for some s. Still, a similar argument yields that the corresponding initial terms are the products along the main diagonal hence have the form

$$j_t := x_1 x_{r+2} x_{2r+3} \cdots \widehat{x_{tr+t+1}} \cdots x_{(m-2)r+m-1} x_{(m-1)r+m},$$

where the hat denotes omission and $0 \le t \le m - 2$. Set

$$j := \frac{1}{x_{(m-1)r+m}} \sum_{t=0}^{m-2} j_t.$$

Now, since (i_1, i_2, i_3) has height 3, any of its associated primes is generated by three variables belonging each to the support of one of i_i, $i = 1, 2, 3$. Close inspection, based on this knowledge, shows that j is not an element of any of these linear prime ideals.

It follows that $I_{m-1}(C(m, r))$ is a Gorenstein ideal [9, Corollary 6.4.10].

(iii) We first prove

CLAIM $I_{m-1}(C(m, 2)) \subset I_{m-2}(C(m + 1, m - 2, 2))$.

Write down the two matrices explicitly:

$$C(m, 2) = \left(\begin{array}{cccc|cc} x_1 & x_2 & \cdots & x_{m-2} & x_{m-1} & x_m \\ x_3 & x_4 & \cdots & x_m & x_{m+1} & x_{m+2} \\ \vdots & \vdots & \vdots & \vdots & \vdots & \vdots \\ x_{2m-3} & x_{2m-2} & \cdots & x_{3(m-2)} & x_{3(m-2)+1} & x_{3(m-2)+2} \\ x_{2m-1} & x_{2m} & \cdots & x_{3(m-2)+2} & x_{3(m-2)+3} & x_{3(m-2)+4} \end{array} \right) \tag{7.4}$$

and

$$C(m + 1, m - 2, 2) = \left(\begin{array}{cccc} x_1 & x_2 & \cdots & x_{m-3} & x_{m-2} \\ x_3 & x_4 & \cdots & x_{m-1} & x_m \\ x_5 & x_6 & \cdots & x_{m+1} & x_{m+2} \\ \vdots & \vdots & \cdots & \vdots \\ x_{2m-1} & x_{2m} & \cdots & x_{3(m-2)+1} & x_{3(m-2)+2} \\ x_{2m+1} & x_{2m+2} & \cdots & x_{3(m-2)+3} & x_{3(m-2)+4} \end{array} \right). \tag{7.5}$$

The second matrix is a simulacrum of the idea of the GP paragon, this time around lowering the size of the maximal minors. In particular, its entries are still indeterminates of the same polynomial ring R, and in fact, the respective k-vector spaces of entries in both matrices coincide. It is important to note that this matrix is very particular for $r = 2$ as for higher values of r a similar construct will have a zero entry.

We argue, equivalently, with the cofactors of $C(m, 2)$. Thus, let $\Delta_{i,j}$ be the (i, j)th cofactor of $C(m, 2)$. The $(m - 2) \times m$ submatrix of $C(m, 2)$ formed by the $m - 2$ first columns is also a submatrix of $C(m + 1, m - 2, 2)$. In the same way, the $(m - 2) \times m$ submatrix of $C(m, 2)$ formed by the last $m - 2$ columns is also a submatrix of $C(m + 1, m - 2, 2)$. Thus, by a Laplace expansion, one gets

$$\Delta_{i,j} \in I_{m-2}(C(m+1, m-2, 2)), \quad 1 \le i \le m \text{ and } j \in \{1, 2, m-1, m\}. \tag{7.6}$$

It remains to deal with the cofactors $\Delta_{i,j}$ of $C(m, 2)$ such that $1 \le i \le m$ and $3 \le j \le m - 2$.

By the adjugate formula of (2.1),

$$C(m, 2) \operatorname{adj}(C(m, 2)) = \det C(m, 2)\, \mathbb{I}_m.$$

In particular, since the respective k-vector spaces of entries in both matrices coincide, and hence, $\det C(m, 2) \in I_{m-2}(C(m + 1, m - 2, 2))$, one has as well

$$C(m, 2) \operatorname{adj}(C(m, 2)) \equiv \mathbf{0} \bmod I_{m-2}(C(m + 1, m - 2, 2)).$$

Along with (7.6) it implies

$$\underbrace{\begin{pmatrix} x_3 & \cdots & x_{m-2} \\ x_5 & \cdots & x_m \\ \vdots & \ddots & \vdots \\ x_{2m+3} & \cdots & x_{3(m-2)+2} \end{pmatrix}}_{=:\mathcal{A}} \begin{pmatrix} \Delta_{1,3} & \cdots & \Delta_{m,3} \\ \vdots & \ddots & \vdots \\ \Delta_{1,m-2} & \cdots & \Delta_{m,m-2} \end{pmatrix}$$

$$\equiv \mathbf{0} \bmod I_{m-2}(C(m + 1, m - 2, 2)).$$

Recall that $I_{m-2}(C(m + 1, m - 2, 2))$ is a prime ideal by Corollary 7.4. Since $\mathcal{A}$ has maximal rank over the domain $R/I_{m-2}(C(m + 1, m - 2, 2))$, one deduces

$$\begin{pmatrix} \Delta_{1,3} & \cdots & \Delta_{m,3} \\ \vdots & \ddots & \vdots \\ \Delta_{1,m-2} & \cdots & \Delta_{m,m-2} \end{pmatrix} \equiv \mathbf{0} \bmod I_{m-2}(C(m + 1, m - 2, 2)).$$

From this $\Delta_{i,j} \in I_{m-2}(C(m+1, m-2, 2))$ for every $1 \le i \le m$ and $3 \le j \le m-2$ as desired

This takes care of the claim.

Now, by item (ii), both $I_{m-1}(C(m, 2))$ and $I_{m-1}(GP(C(m, 2)))$ are specializations of the respective generic ancestors, hence their multiplicities are well-known (see, e.g., [3, Example, p. 17]):

$$e(R/I_{m-1}(C(m, 2))) = \frac{1}{12}m^2(m - 1)(m + 1)$$

and

$$e(R/I_{m-1}(GP(C(m, 2)))) = \binom{m + 2}{m - 2}.$$

As for $I_{m-2}(C(m + 1, m - 2, 2))$, by the above Claim it has codimension at least four, hence exactly four. Thus, by the same token one has

$$e(R/I_{m-2}(C(m + 1, m - 2, 2))) = \binom{m + 1}{m - 3}.$$

Adding the last two gives the first. Since both ideals in question are minimal primes of $I_{m-1}(C(m, 2))$, the required result follows by the associative formula of multiplicities.

(iv) (Modicum) It is enticing to ask whether $R/I_{m-1}(C(m, r))$ actually satisfies property (R_1) of Serre's – hence, would be normal and, consequently, a domain. $\square$

7.4   Properties at Large

7.4.1   Parabolism and the Dual Variety

Square Hankel like catalecticants provide a good testing ground for the parabolic principle in Section 3.2. Let $m \geq 2$ and $1 \leq r \leq m$ be given integers. Let $R = k[x_0, \ldots, x_n]$ be a polynomial ring in $n+1 = (m-1)(r+1)+1$ variables (note that, for geometric convenience, the ground variables are indexed starting from zero).

The r-leap $m \times m$ generic catalecticant has the shape

$$C(m, r) = \begin{pmatrix} x_0 & x_1 & x_2 & \cdots & x_{m-1} \\ x_r & x_{r+1} & x_{r+2} & \cdots & x_{m+r-1} \\ x_{2r} & x_{2r+1} & x_{2r+2} & \cdots & x_{m+2r-1} \\ \vdots & \vdots & \vdots & \ddots & \vdots \\ x_{(m-1)r} & x_{(m-1)r+1} & x_{(m-1)r+2} & \cdots & x_{(m-1)r+(m-1)} \end{pmatrix}. \tag{7.7}$$

Note the change in the entry indices in order to comply with matters related to parabolism. The extreme values $r = 1$ and $r = m$ yield, respectively, the ordinary Hankel matrix and the generic matrix.

By Proposition 7.3, $C(m, r)$ is 1-generic, hence f is irreducible (Theorem 1.24).

Conjecture 7.7 Let f denote the determinant of the $m \times m$ r-leap catalecticant (7.7), with $1 \leq r \leq m$. Then f is parabolic and it has the expected multiplicity as a factor of $h(f)$.

The following examples may give some evidence to the conjecture, while simultaneously showing the varying nature of the complementary factor of the power of f as well as the dual variety $V(f)^*$. All calculations were done with computer assistance, in a large characteristic environment, those of the dual variety $V(f)^*$ with special significance.

Start with the lowest value $r = 1$.

Example 7.8 ($m = 3$) This is the 3×3 Hankel matrix

$$C(3, 1) = \begin{pmatrix} x_0 & x_1 & x_2 \\ x_1 & x_2 & x_3 \\ x_2 & x_3 & x_4 \end{pmatrix}.$$

The factorization of the Hessian determinant is $h(f) = 8f \frac{\partial f}{\partial x_2}$.

As to the dual variety, a later argument (Proposition 8.8) shows that dim $V(f)^* \geq m - 1 = 2$. That the dimension is exactly 2 follows by showing that $V(f)^*$ is a well-known codimension 2 subvariety of $(\mathbb{P}^4)^*$ defined by 7 cubics, obtained as a suitable projection of the 2-Veronese in $(\mathbb{P}^5)^*$. Alternatively, $V(f)^*$ is the image of the rational map $\mathbb{P}^2 \dashrightarrow (\mathbb{P}^4)^*$ defined by the Pfaffians of a suitable linear 5×5 skew-symmetric matrix in 3 variables. The following figure captures the two ways:

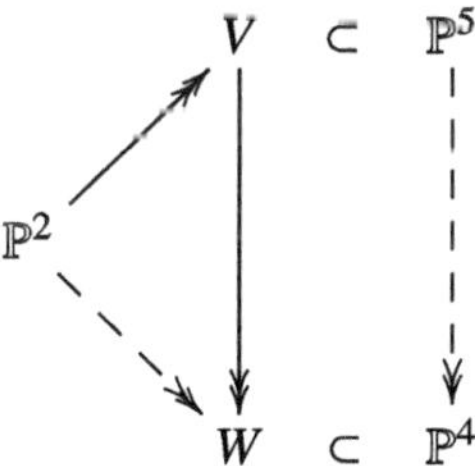

The easiest is to project from the point $(0 : 1 : 0 : 0 : 0 : -1)$ which is external to the Veronese. Then W is the (closure of the) image of the rational map defined by the quadrics $\{x^2, xy + z^2, xz, y^2, yz\}$, whose base ideal is the Gorenstein ideal generated by the maximal Pfaffians of the following skew-symmetric matrix

$$\begin{pmatrix} 0 & x & 0 & 0 & z \\ -x & 0 & y & z & 0 \\ 0 & -y & 0 & x & 0 \\ 0 & -z & -x & 0 & y \\ -z & 0 & 0 & -y & 0 \end{pmatrix}.$$

Thus, the expected multiplicity is $n - 1 - \dim V(f)^* = 3 - 2 = 1$, clearly coinciding with the effective multiplicity above.

The next value of m is computationally harder.

Example 7.9 ($m = 4$) The Hankel matrix

$$C(4, 1) = \begin{pmatrix} x_0 & x_1 & x_2 & x_3 \\ x_1 & x_2 & x_3 & x_4 \\ x_2 & x_3 & x_4 & x_5 \\ x_3 & x_4 & x_5 & x_6 \end{pmatrix}.$$

One finds that the effective multiplicity of f is 2. A reasonable guess is that the complementary factor is the square of $\partial f/\partial x_3$ up to an appropriate projective change of coordinates, in which case the prime factorization would be $H(f) = f^2(\partial f/\partial x_3)^2$ up to coordinate change.

Once more, by Proposition 8.8 $\dim V(f)^* \geq m - 1 = 3$. That it is exactly of this dimension is conjectured in general in the later Conjecture 8.9 and turns out to be true here. Moreover it is defined by quartics, which is also subsumed in the above conjecture. As it comes out, the expected multiplicity $5 - 3 = 2$ coincides with the effective multiplicity.

Example 7.10 ($m = 3$) Consider the 2-leap catalecticant

$$C(3, 2) = \begin{pmatrix} x_0 & x_1 & x_2 \\ x_2 & x_3 & x_4 \\ x_4 & x_5 & x_6 \end{pmatrix}.$$

Here one has $H(f) = fg$, where g is (up to a projective change of coordinates) the equation of the dual surface to the twisted cubic in the variables x_0, x_2, x_4, x_6. Observe that the degrees match as the equation of the dual surface to the twisted cubic has degree 4 and $H(f)$ has degree 7.

The dual variety $V(f)^* \subset (\mathbb{P}^6)^*$ is a determinantal scheme defined by the maximal minors of the matrix

$$\begin{pmatrix} 0 & y_0 & y_1 \\ y_1 & -y_2 & -y_3 \\ -y_3 & y_4 & y_5 \\ y_5 & -y_6 & 0 \end{pmatrix} \tag{7.8}$$

hence is an arithmetically Cohen–Macaulay variety of dimension $4 = 2 \cdot 3 - 2$.

The expected multiplicity of f is $n - 1 - \dim V(f)^* = 5 - 4 = 1$, which coincides with the effective multiplicity above.

Example 7.11 ($m = 4$) Look at the 3-leap catalecticant

$$C(4, 3) = \begin{pmatrix} x_0 & x_1 & x_2 & x_3 \\ x_3 & x_4 & x_5 & x_6 \\ x_6 & x_7 & x_8 & x_9 \\ x_9 & x_{10} & x_{11} & x_{12} \end{pmatrix}.$$

Here one has $H(f) = f^5 (\det \mathcal{H}_3)^2$, where $\mathcal{H}_3$ is the Hankel matrix

$$\mathcal{H}_3 = C(3, 1) = \begin{pmatrix} x_0 & x_3 & x_6 \\ x_3 & x_6 & x_9 \\ x_6 & x_9 & x_{12} \end{pmatrix}.$$

Note that the 2×4 matrix corresponding to $\mathcal{H}_3$ by the Gruson–Peskine principle (Section 7.3) is the submatrix of $C_{4,3}$ with first and fourth columns.

The dual variety $V(f)^* \subset (\mathbb{P}^{12})^*$ has dimension $6 = 2 \cdot 4 - 2$ and initial defining degree 2. Moreover, the subideal generated in degree 2 is perfect of codimension 3. The expected multiplicity of f is $n - 1 - \dim V(f)^* = 11 - 6 = 5$, which coincides with the effective multiplicity above.

Remark 7.12 The last two preceding examples fit the realm of hollow-filled $m \times m$ linear sections of the full generic matrix (Definition 4.9), thereby a priori $\dim V(f)^* \geq 2m - 2$ by Lemma 8.1. It would be pertinent to actually prove equality here for $(m - 1)$-leap catalecticants.

Finally, we consider one more case, where r has an intermediate value.

Example 7.13

$$C(4, 2) = \begin{pmatrix} x_0 & x_1 & x_2 & x_3 \\ x_2 & x_3 & x_4 & x_5 \\ x_4 & x_5 & x_6 & x_7 \\ x_6 & x_7 & x_8 & x_9 \end{pmatrix}.$$

The effective multiplicity of f is 2. The complementary factor (of degree 12) is rather unreachable for inspection, except that it belongs to the gradient ideal of f.

The dual $V(f)^*$ is an arithmetically Cohen–Macaulay subvariety of $(\mathbb{P}^9)^*$ of dimension $6 = 2 \cdot 4 - 2$, with 4-linear resolution, hence has degree $\binom{3+4-1}{3} = 20$. The expected multiplicity is $8 - 6 = 2$, which coincides with the effective multiplicity.

Remark 7.14 At this stage we are unable to make a guess about the structural properties of the dual variety $V(f)^*$, much less about when it is arithmetically Cohen–Macaulay. It would look like a good guess about its dimension is that it is $m - 1$ in the Hankel case (Conjecture 8.9) and $2m - 2$ otherwise.

7.4.2   The Homaloidal Problem

Consider the square generic r-leap catalecticant matrix $C(m, r)$ of (7.2).

Conjecture 7.15 (Arbitrary characteristic) The determinant of $C(m, r)$ is homaloidal if and only if $r = m - 1$ or $r = m$.

We discuss the terms of the above conjecture.

(a) The "only if" part of the conjecture.

  The evidence is essentially gathered by careful examination of small case-by-case computer verification. The core of this intervention is to make sure that, unless $r = m$ or $r = m - 1$, one does not have enough defining relations of bidegree $(1, s)(s \geq 1)$ in order that the corresponding Jacobian dual matrix have maximal rank (we know this at least if $r = 1$, the Hankel case [4, Theorem 3.3.5]).

(b) The "if" part of the conjecture.

  - For $r = m$, we are in the case of the fully generic matrix, in which case the result of Section 3.3.1 applies.
  - For $r = m - 1$, upon permutation of columns and of variables, the matrix $C(m, m - 1)$ becomes the linear section $\mathbf{L}(m, 1)$ introduced in Section 4.4.3 up to renaming the variables of the ground polynomial ring. Namely, permute the columns in the order $1 \rightsquigarrow 2, 2 \rightsquigarrow 3, \ldots, m - 1 \rightsquigarrow m$, then permute the variables $x_1, \ldots, x_m$ cyclically, i.e., $x_m \mapsto x_{m-1} \mapsto \ldots \mapsto x_1 \mapsto x_m$.

  Thus, the result follows from the conjectured Proposition 4.26, (ii).

(c) A related guess, perhaps poorly substantiated, is that the linear rank of the gradient ideal of $f := \det C(m, m - 1)$ is $m(m - 1) - (m - 3) = (m - 1)^2 + 2$. Thus, if f is going to be homaloidal it would better admit at least $m - 3$ additional relations of bidegree $(1, e)$, for some e of degree ≥ 2. A good guess is that enough of these in order to comply with the criterion of [7, Theorem 3.2.22] actually have $e = 2$ (i.e., are syzygetic relations [7, Section 8.4.1.2]).

Exercises

7.16 Consider the catalecticant

$$C(2, n, n - 3) = \begin{pmatrix} x_1 & x_2 & \ldots & x_{n-2} & x_{n-1} & x_n \\ x_{n-2} & x_{n-1} & \ldots & x_{2n-5} & x_{2n-4} & x_{2n-3} \end{pmatrix}.$$

1. Prove that the secant to the variety defined by the 2-minors of $C(2, n, n - 3)$ is a hypersurface defined by the determinant of the catalecticant

$$C(3, n-3) = \begin{pmatrix} x_1 & x_2 & x_3 \\ x_{n-2} & x_{n-1} & x_n \\ x_{2n-5} & x_{2n-4} & x_{2n-3} \end{pmatrix}.$$

(HINT: if needed resort to [10].)

2. Discuss the details of the cases $n = 4$ and $n = 5$.

The list of problems below will require hard computer assistance for most of it.

7.17 Consider the matrix $\mathcal{Y}$ of (7.8) with entries in the polynomial ring $R :=$ $k[y_0, \ldots, y_6]$.

1. Show that $I_2(\mathcal{Y})$ has height 3 and that $R/I_2(\mathcal{Y})$ has depth zero.
2. What is the shape of a primary decomposition of $I_2(\mathcal{Y})$?
3. Show that $I_2(\mathcal{Y})$ is set theoretically the singular locus of the dual variety defined by $I_3(\mathcal{Y})$.

7.18 Compute and discuss the details of Example 7.9.

7.19 (char$(k) = 0$) Let J denote the gradient ideal of $f = \det C(4, 3)$.

1. Show that J has 11 independent linear syzygies.
2. Show that J admits 4 syzygies with coefficients (entries) in J that are independent modulo the Koszul syzygies. Deduce that J admits four independent Rees relations of bidegree $(1, 2)$.

    (HINT: for the second part use [7, Lemma 8.4.5].)
3. Show that J has maximal analytic spread.

    (HINT: prove that the Hessian of f does not vanish – if the computer does not do it, take a conveniently sufficient sparse degeneration that allows a computer calculation.)

7.20 Compute the details of Example 7.13.

References

1. A. Conca, Straightening law and powers of determinantal ideals of Hankel matrices. Adv. Math. **138**, 263–292 (1998) 203
2. L. Gruson, C. Peskine, Courbes de L'Espace Projectif: Variétés de Sécantes 1–32, in *Enumerative Geometry and Classical Algebraic Geometry*, ed. by P. Le Barz, Y. Hervier. Progress in Mathematics, vol. 24 (Birkhäuser, Boston, 1982) 203, 205
3. J. Herzog, N.N. Trung, Gröbner bases and multiplicity of determinantal and Pfaffian ideals. Adv. Math. **96**, 1–37 (1992) 208
4. M. Mostafazadehfard, Hankel and sub-Hankel determinants – a detailed study of their polar ideals, PhD Thesis. Universidade Federal de Pernambuco, Recife (2014) 212
5. M. Mostafazadehfard, A. Simis, Homaloidal determinants. J. Algebra **450**, 59–101 (2016) 203
6. L.D. Nam, The determinantal ideals of extended Hankel matrices. J. Pure App. Algebra **215**, 1502–1515 (2011) 203, 204, 205

7. Z. Ramos, A. Simis, *Graded Algebras in Algebraic Geometry*. Expositions in Mathematics, vol. 70 (De Gruyter, Berlin, 2022) 212, 213
8. J.G. Semple, J.A. Tyrrell, Specialization of Cremona transformations. Mathematika **15**, 171–177 (1968) 202
9. A. Simis, *Commutative Algebra* (De Gruyter Textbook, Berlin, 2020) 206
10. A. Simis, B. Ulrich, On the ideal of an embedded join. J. Algebra **226**, 1–14 (2000) 213
11. B. Sturmfels, Gröbner bases and Stanley decompositions of determinantal rings. Math. Z. **205**, 137–144 (1990) 206

Chapter 8
The Dual Variety of a Linear Determinantal Hypersurface

Abstract This chapter, although of a marked geometric nature, is firmly grounded on previous chapters. It gives a good grip on the dimension of the dual variety to the determinantal hypersurfaces of many of the linear sections considered in previous chapters. It also gives a notion of the structure of the homogeneous defining ideal of the dual variety in its natural embedding – the latter by no means being a trivial issue as the dual variety is most of the times deficient. By and large, this discussion calls upon the structure of certain ladder ideals, the question of the generating degrees of the defining ideal standing up. As discussed in a previous chapter, an interesting question in general is whether the defining polynomial is a factor of its Hessian determinant with the *expected multiplicity* (according to Segre).

8.1 Hollow-Filled, Hollow, and Semi-Hollow Linear Sections

Throughout, unless stated otherwise, we assume that the defining polynomial f is reduced (i.e., the ideal (f) is radical).

The reader will refer to Section 4.2 for the terminology and basic knowledge in this section. Getting familiarized with the notation and main results of Section 2.1.4 is of great help.

8.1.1 The Generic Case

8.1.1.1 Hollow-Filled

Lemma 8.1 *Let $\mathbf{L}$ denote a hollow-filled linear section of the $m \times m$ generic matrix and let $f = \det \mathbf{L}$. Then $\dim V(f)^* \geq 2m - 2$. Moreover, if $\mathbf{L}$ is endowed with a sparse decomposition, then $\dim V(f)^* = 2m - 2$.*

Proof Corollary 4.12, gives the bound rank $H(f)(\mathrm{mod}\, f) \geq 2m$. Thus, Theorem 2.7 implies $\dim V(f)^* \geq 2m - 2$.

Z. Ramos, A. Simis, *Determinantal Ideals of Square Linear Matrices*,
https://doi.org/10.1007/978-3-031-55284-7_8

In addition, if $\mathbf{L}$ is endowed with a sparse decomposition, then by Proposition 2.14 one has rank $H(f)(\mathrm{mod}\, f) \leq 2m$. Hence, Theorem 2.7 again gives $\dim V(f)^* = 2m - 2$.   □

Unfortunately, we do not have any structural result for the dual variety in the case of arbitrary hollow-filled sections, not even a good grip on its expected dimension. In the particular case of one term linear section we prove some crispy results in Section 8.2.

8.1.1.2   Hollow

Consider the hollow linear section of the $m \times m$ generic matrix as in (4.7):

$$
\mathfrak{H}_m = \begin{pmatrix}
x_{1,1} & x_{1,2} & \cdots & x_{1,m-1} & x_{1,m} \\
x_{2,1} & 0 & \cdots & 0 & x_{2,m} \\
x_{3,1} & 0 & \cdots & x_{3,m-1} & 0 \\
\vdots & \vdots & \ddots & \vdots & \vdots \\
x_{m-1,1} & 0 & x_{m-1,3} & 0 & 0 \\
x_{m,1} & x_{m,2} & 0 & 0 & 0
\end{pmatrix}.
$$

Set $\tilde{\mathbf{X}} := (\mathfrak{H}_m)_{\mathcal{I}}$ in the notation of Section 4.2.1, i.e., the set of nonzero entries of $\mathfrak{H}_m$. Thus, $\mathfrak{H}_m$ is naturally a matrix over the polynomial ring $k[\tilde{\mathbf{X}}]$. Let $\tilde{\mathbf{Y}} = \{y_{i,j}\}$ stand for the dual variables of $\tilde{\mathbf{X}}$.

Write $\mathfrak{f} := \det \mathfrak{H}_m$, an irreducible polynomial (Proposition 4.10 (a)).

Theorem 8.2 *With the above notation, one has*:

(a) $\dim V(\mathfrak{f})^* = 2m - 2$.
(b) $V(\mathfrak{f})^* \subset \mathbb{P}^{3(m-1)\vee}$ *is a codimension* $m - 1$ *complete intersection*.

Proof

(a) This follows from Lemma 8.1 since $\mathfrak{H}_m$ is a linear section endowed with a sparse decomposition.
(b) Endow $\mathfrak{H}_m$ with the sparse decomposition where $\mathbf{X}_1 = \tilde{\mathbf{X}}$. By Theorem 2.11, the homogeneous defining ideal $\mathfrak{D}$ of $V(\mathfrak{f})^*$ in its embedding in $\mathbb{P}^{3(m-1)\vee}$ contains the ideal $\mathfrak{I}_2(\tilde{\mathbf{Y}})$ of generic 2×2 minors in the dual variables (Remark 2.12 (1)).

On the other hand, $\mathfrak{D}$ has codimension $m - 1$ by (a) and is prime since $\mathfrak{f}$ is irreducible. Therefore, it suffices to prove:

CLAIM 1 $\mathfrak{I}_2(\tilde{\mathbf{Y}})$ is a prime ideal of codimension $m - 1$.

Explicitly, one has

$$\mathfrak{I}_2(\tilde{\mathbf{Y}}) = (y_{1,1}y_{2,m} - y_{1,m}y_{2,1}, \dots, y_{1,1}y_{i,m+2-i} - y_{1,m+2-i}y_{i,1}, \dots, y_{1,1}y_{m,2} - y_{1,2}y_{m,1}).$$

To prove the claim, let $\prec$ stand for the reverse lexicographic term order on $k[\tilde{\mathbf{Y}}]$ induced by the following order of the variables

$$y_{1,1} > y_{1,2} > \cdots > y_{1,m} > y_{2,1} > y_{2,m} > y_{3,1} > y_{3,m-1} > \cdots > y_{m,1} > y_{m,2},$$

which is the natural order of the nonzero entries along the rows. Then,

$$(y_{1,m}y_{2,1}, \dots, y_{1,m+2-i}y_{i,1}, \dots, y_{1,2}y_{m,1}) \subset \text{in}_\prec(\mathfrak{I}_2(\tilde{\mathbf{Y}})).$$

But, ht $(y_{1,m}y_{2,1}, \dots, y_{1,m+2-i}y_{i,1}, \dots, y_{1,2}y_{m,1}) = m - 1$. Thus,

$$m - 1 \le \text{ht in}_\prec(\mathfrak{I}_2(\tilde{\mathbf{Y}})) = \text{ht}\,\mathfrak{I}_2(\tilde{\mathbf{Y}}) \le \mu(\mathfrak{I}_2(\tilde{\mathbf{Y}})) = m - 1.$$

Therefore, $\mathfrak{I}_2(\tilde{\mathbf{Y}})$ is a complete intersection of codimension $m - 1$.

By a similar token, $y_{1,1}$ is regular on $k[\mathbf{Y}]/\mathfrak{I}_2(\tilde{\mathbf{Y}})$. Now, since

$$k[\mathbf{Y}]\left[y_{1,1}^{-1}\right]/\mathfrak{I}_2(\tilde{\mathbf{Y}})k[\mathbf{Y}]\left[y_{1,1}^{-1}\right] \simeq k\left[y_{1,1}, \dots, y_{1,m}, y_{2,1}, \dots, y_{m,1}, y_{1,1}^{-1}\right]$$

is a domain, then so is $k[\mathbf{Y}]/\mathfrak{I}_2(\tilde{\mathbf{Y}})$.

This takes care of Claim 1 and concludes the proof of the item. $\square$

8.1.1.3   Semi-Hollow

Recall that we call semi-hollow a rather special case of a hollow-filled section. Namely, we are dealing now with the matrix $\mathbf{L}(m)[r]$ introduced in Section 4.5 with $0 \le r \le m - 2$.

Our main result on the structure of the dual variety $V(f)^*$ of $V(f)$ for $f = \det \mathbf{L}(m)[r]$ answers affirmatively a question posed by F. Russo as to whether the codimension of the dual variety of a homogeneous polynomial in its polar image can be arbitrarily large when its Hessian determinant vanishes. In addition it shows that this can happen in the case of nicely structured linear sections of determinantal varieties.

In the notation of Section 4.1.1, let $\mathcal{L}_r$ denote the ladder of $\mathbf{L}(m)[r]$ read in the dual variables:

$$
\begin{pmatrix}
y_{1,1} & \cdots & y_{1,m-r} & \cdots & y_{1,m-1} & y_{1,m} \\
\vdots & \vdots & \vdots & \vdots & \vdots & \vdots \\
y_{m-r,1} & \cdots & y_{m-r,m-r} & \cdots & y_{m-r,m-1} & y_{m-r,m} \\
y_{m-r+1,1} & \cdots & y_{m-r+1,m-r} & \cdots & y_{m-r+1,m-1} & \\
\vdots & \vdots & \vdots & \vdots & & \\
y_{m-1,1} & \cdots & y_{m-1,m-r} & \cdots & & \\
y_{m,1} & \cdots & y_{m,m-r} & & &
\end{pmatrix}
$$

Theorem 8.3 *Let $0 \le r \le m - 2$ and $f := \det \mathbf{L}(m)(r)$. Then:*

(i) *The dual variety $V(f)^*$ in its natural embedding is a ladder determinantal variety of codimension $(m - 1)^2 - \binom{r+1}{2}$ defined by 2-minors; in particular it is arithmetically Cohen–Macaulay and its codimension in the polar image of $V(f)$ is $(m - 1)^2 - r(r + 1)$.*

(ii) *$V(f)^*$ is arithmetically Gorenstein if and only if $r = m - 2$.*

Proof

(i) Once again, we are in the presence of a linear section endowed with a sparse decomposition, hence, by Lemma 8.1, $\dim V(f)^* = 2m - 2$.

Now, let $\mathfrak{D} \subset k[y_{i,j} \mid 2 \le i + j \le 2m - r]$ denote the homogeneous defining ideal of the dual variety $V(f)^*$ in its natural embedding. Since f is irreducible (Proposition 4.28) and $\dim V(f)^* = 2m - 2$, $\mathfrak{D}$ is a prime ideal of codimension $(m - 1)^2 - \binom{r+1}{2}$. On the other hand, by Theorem 2.11, $I_2(\mathcal{L}_r) \subset \mathfrak{D}$ and is a prime ideal of codimension $(m - 1)^2 - \binom{r+1}{2}$ (Theorem 4.1 (i)). Thus, $I_2(\mathcal{L}_r) = \mathfrak{D}$.

(ii) This follows from Theorem 4.1 (ii). $\square$

8.1.2   The Generic Symmetric Case

8.1.2.1   Symmetric Hollow-Filled Section

Lemma 8.4 *Let $\mathbf{L}$ denote a symmetric hollow-filled linear section of the $m \times m$ generic symmetric matrix and let $f = \det \mathbf{L} \neq 0$. Then $\dim V(f)^* \ge m - 1$. Moreover, if $\mathbf{L}$ is endowed with a sparse decomposition, then $\dim V(f)^* = m - 1$.*

Proof By Corollary 5.11, we have

$$
\operatorname{rank} H(f)(\bmod f) \ge m + 1.
$$

Thus, from formula (2.5) of Theorem 2.7 we have $\dim V(f)^* \ge m - 1$.

In addition, if $\mathbf{L}$ is endowed with a sparse decomposition, then by Proposition 2.14, we have also

$$\operatorname{rank} H(f)(\bmod f) \le 2m.$$

Hence, again from formula (2.5) of Theorem 2.7 we have dim $V(f)^* = m - 1$. $\quad\square$

As in the generic case, we do not know any structural result for the dual variety for arbitrary hollow-filled sections. Yet, in the particular case of one term linear section we obtain some optimum results in Section 8.3.

8.1.2.2 Symmetric Hollow Section

We refer to the definitions and notation of Section 5.3. Thus, let $\mathfrak{f}$ denote the determinant of the symmetric hollow linear section of the $m \times m$ generic symmetric matrix over a field k

$$\mathfrak{H} = \mathfrak{H}_m = \begin{pmatrix} x_{1,1} & x_{1,2} & \cdots & x_{1,m-1} & x_{1,m} \\ x_{1,2} & 0 & \cdots & 0 & x_{2,m} \\ x_{1,3} & 0 & \cdots & x_{3,m-1} & 0 \\ \vdots & \vdots & \ddots & \vdots & \vdots \\ x_{1,m-1} & 0 & x_{3,m-1} & 0 & 0 \\ x_{1,m} & x_{2,m} & 0 & 0 & 0 \end{pmatrix}. \tag{8.1}$$

Writing $t := \lfloor m/2 \rfloor + 1$, the nonzero entries of $\mathfrak{H}$ are

$$\{x_{1,1}, \ldots, x_{1,m};\, x_{2,m}, x_{3,m-1}, \ldots, x_{u,m-u+2}, \ldots, x_{t,m-t+2}\}.$$

Let $\mathbf{X} = \{x_{i,j} \mid 1 \le i \le j \le m\}$ be the set of entries of the $m \times m$ generic symmetric matrix and set $\mathbf{Y} = \{y_{i,j} \mid 1 \le i \le j \le m\}$ for their dual variables.

Recalling that $\mathfrak{f}$ is reduced (Proposition 5.9 (b)), one has:

Theorem 8.5 *With the above notation*:

(a) dim $V(\mathfrak{f})^* = m - 1$.
(b) $V(\mathfrak{f})^* \subset \mathbb{P}^{m+t-2*}$ *is a codimension* $\lfloor m/2 \rfloor$ *complete intersection*.
(c) $V(\mathfrak{f})^*$ *is non-deficient if and only if* $m = 3$.

Proof

(a) This follows from Lemma 8.4 as $\mathfrak{H}$ is endowed with a sparse decomposition.
(b) The argument is similar to the one in the proof of Theorem 8.2, but there will be some differences.

By Proposition 2.10, $\mathfrak{f}_{i,j} = \Delta_{i,j}$ up to nonzero positive scalar multipliers, where $\Delta_{i,j}$ is the signed (i, j)th cofactor of $\mathfrak{H}$. Thus, we can instead argue with the cofactors $\Delta_{i,j}$. Namely, we show that the homogeneous k-subalgebra

$$T := \frac{k[\Delta_{i,j} \mid (i,j) \in \mathcal{I} \cup \mathcal{J}]}{(\mathfrak{f}) \cap k[\Delta_{i,j} \mid (i,j) \in \mathcal{I} \cup \mathcal{J}]} \subset \frac{k[x_{i,j} \mid (i,j) \in \mathcal{I} \cup \mathcal{J}]}{(\mathfrak{f})}$$

is a codimension $\lfloor m/2 \rfloor$ complete intersection.

Reading the generic symmetric matrix in the dual variables $\mathbf{Y}$, let $\mathcal{L}_{m-3}$ denote the ladder upon the notation of Section 4.1.1. Letting $\widetilde{\mathbf{Y}}$ denote the non-null entries of the dual of $\mathfrak{H}$, set $\widetilde{\mathcal{L}_{m-3}}$ for the corresponding ladder. We focus on the ladder ideal $I_2(\widetilde{\mathcal{L}_{m-3}}) \subset k[\widetilde{\mathbf{Y}}]$. Then, let $D \subset I_2(\widetilde{\mathcal{L}_{m-3}})$ denote the subideal generated by the *irreducible* 2-minors. Note that such a generator of D is of the form

$$\det \begin{pmatrix} y_{1,1} & y_{1,u} \\ y_{1,m-u+2} & y_{m-u+2,u} \end{pmatrix}, \ 2 \le u \le t.$$

We will argue that the homogeneous defining ideal P of T coincides with D.

CLAIM 1 D is a prime ideal of codimension $t-1$.

To see this, let $\prec$ stand for the reverse lexicographic term order on $k[\widetilde{\mathbf{Y}}]$ induced by the following order of the variables

$$y_{1,1} > y_{1,2} > \cdots > y_{1,m} > y_{2,1} > y_{2,m} > y_{3,m-1} > \cdots > y_{t,m-t+2},$$

which is the natural order of the nonzero entries along the rows. Then,

$$(y_{1,m}y_{1,2}, \ldots, y_{1,m+2-i}y_{1,i}, \ldots, y_{1,m-t+2}y_{1,t}) \subset \mathrm{in}_{\prec}(D).$$

But, $\mathrm{ht}\,(y_{1,m}y_{1,2}, \ldots, y_{1,m+2-i}y_{1,i}, \ldots, y_{1,m-t+2}y_{1,t}) = t - 1$. Thus,

$$t - 1 \le \mathrm{ht}\,\mathrm{in}_{\prec}(D) = \mathrm{ht}\,D \le \mu(D) = t - 1.$$

Therefore, D is a complete intersection of codimension $t-1$.

By a similar token, $y_{1,1}$ is regular over $k[\widetilde{\mathbf{Y}}]/D$. Now, since

$$k[\widetilde{\mathbf{Y}}]\left[y_{1,1}^{-1}\right] / Dk[\widetilde{\mathbf{Y}}]\left[y_{1,1}^{-1}\right] \simeq k\left[y_{1,1}, \ldots, y_{1,m}, y_{1,1}^{-1}\right]$$

is a domain, then so is $k[\widetilde{\mathbf{Y}}]/D$.

This takes care of Claim 1.

CLAIM 2 The homogeneous defining ideal P of T contains D.

For this, using (5.19) and applying the adjugate formula (2.1) for $\mathfrak{H}_m$, yields the relations

$$x_{1,m-u+2}\Delta_{1,1} + x_{u,m-u+2}\Delta_{1,u} = 0 \quad \text{and}$$

$$x_{1,m-u+2}\Delta_{1,m-u+2} + x_{u,m-u+2}\Delta_{u,m-u+2} = \mathfrak{f}$$

for every $2 \le u \le t - 1$. Thus,

$$\begin{pmatrix} \Delta_{1,1} & \Delta_{1,u} \\ \Delta_{1,m-u+2} & \Delta_{u,m-i+2} \end{pmatrix} \begin{pmatrix} x_{1,m-u+2} \\ x_{u,m-u+2} \end{pmatrix} \equiv \mathbf{0} \ (\mathrm{mod}\ \mathfrak{f}) \quad (i = 2, \ldots, m).$$

In particular,

$$x_{1,m-u+2} \det \begin{pmatrix} \Delta_{1,1} & \Delta_{1,u} \\ \Delta_{1,m-u+2} & \Delta_{u,m-i+2} \end{pmatrix} \equiv 0 \ (\mathrm{mod}\ \mathfrak{f}) \quad (i = 2, \ldots, m).$$

But, by Proposition 5.9 (a) and (b), $x_{1,m-u+2}$ is regular module $\mathfrak{f}$. Thus, the 2-minor

$$\det \begin{pmatrix} y_{1,1} & y_{1,u} \\ y_{1,m-u+2} & y_{u,m-i+2} \end{pmatrix}$$

maps to zero modulo $\mathfrak{f}$ by the evaluation $y_{i,j} \mapsto \Delta_{i,j}$, as stated in the claim.

By (a), P is an ideal of codimension $m + t - 2 - V(\mathfrak{f})^* = m + t - 2 - m + 1 = t - 1$. Moreover, since $\mathfrak{f}$ is reduced we have that P is a radical ideal. Thus, $D \subset P$ is an inclusion of ideals of codimension $t - 1$ with D prime and P radical. Hence, $P = D$.

(c) From item (b), the dual variety has codimension one if and only if $m = 3$, hence is non-deficient exactly in this case. $\qquad\square$

8.1.2.3   Symmetric Semi-Hollow Section

In this part we refer to the symmetric semi-hollow linear section $\mathbf{L}(S)[r]$ introduced in Section 5.3.2. In contrast to the structural content of the later Theorem 8.19, here the dual variety of $\det \mathbf{L}(S)[r]$ will actually be a ladder determinantal variety, not a proper subvariety of one such.

Besides, it will produce additional examples where the codimension of the dual variety of a determinantal hypersurface f in its polar image can be arbitrarily large when the Hessian determinant $h(f)$ vanishes. The advantage of these new examples is that the ambient dimension can be smaller than in the similar examples with degenerations of the fully generic matrix.

Theorem 8.6 *Let* $0 \le r \le m - 2$ *and* $f := \det \mathbf{L}(S)[r]$. *Then:*

(a) *The dual variety* $V(f)^*$ *of* $V(f)$ *is a ladder determinantal variety of dimension $m - 1$ defined by 2-minors; in particular it is arithmetically Cohen–Macaulay and its codimension in the polar image of* $V(f)$ *is* $\binom{m}{2} - \mathfrak{o}(r)$.

(b) $V(f)^*$ *is arithmetically Gorenstein if and only if* $r = m - 2$.

Proof

(a) By Lemma 8.4 we have dim $V(f)^* = m - 1$.

By Proposition 2.10 as applied to the symmetric case and by a change of variables, we may assume that in the map (2.4) the partial derivatives have been replaced by the cofactors $\{\Delta_{i,j} \mid 2 \le i + j \le 2m - r, \ i \le j\}$. Then the kernel $P \subset k[\mathbf{Y}]$ of this new map still defines the dual variety $V(f)^*$ in its natural embedding.

Since dim $V(f)^* = m - 1$ then P has codimension $\binom{m}{2} - \mathfrak{o}(r)$. On the other hand, by Remark 2.12, P contains the ideal $I_2(\mathcal{L})$ of 2×2 minors of the ladder $\mathcal{L}$ as depicted below.

$$
\begin{pmatrix}
y_{1,1} & \cdots & y_{1,m-r} & \cdots & y_{1,m-1} & y_{1,m} \\
\vdots & \vdots & \vdots & \vdots & \vdots & \vdots \\
y_{1,m-r} & \cdots & y_{m-r,m-r} & \cdots & y_{m-r,m-1} & y_{m-r,m} \\
y_{1,m-r+1,} & \cdots & y_{m-r,m-r+1} & \cdots & y_{m-r+1,m-1} & \\
\vdots & \vdots & \vdots & \vdots & & \\
y_{1,m-1} & \cdots & y_{m-r,m-1} & \cdots & & \\
y_{1,m} & \cdots & y_{m-r,m} & & &
\end{pmatrix}
$$

Now, since $I_2(\mathcal{L})$ is a symmetric ladder determinantal ideal on a suitable symmetric generic matrix it is a Cohen–Macaulay prime ideal ([2, Theorem 1.13]). Moreover, its codimension is $\binom{m+1}{2} - \mathfrak{o}(r) - 1 - (m-1) = \binom{m}{2} - \mathfrak{o}(r)$ as follows from [2, Corollary 1.9], where $\mathfrak{o}(r) - 1$ is the codimension of the polar map (Theorem 5.17 (iii)).

Therefore, $I_2(\mathcal{L}) \subset P$ are prime ideals of the same codimension, and hence, $I_2(\mathcal{L}) = Q$.

(b) By the previous item, a homogeneous defining ideal of the dual variety is generated by the 2×2 minors of a ladder such as above. Observe that the smallest square matrix containing all the entries of the latter is an $m \times m$ matrix of the form $\mathbf{L}(\mathcal{S})(r)$. By [2, Theorem (b), p. 120], the ladder ideal is Gorenstein if and only if the inner corners of the ladder have indices (i, j) satisfying the equality $i + j = m + 1$. In the present case, the inner corners have indices satisfying the equation

$$
i+j = m-r+(m-1) = m-r+1+(m-2) = \cdots = m-1+(m-r) = 2m-r-1.
$$

Clearly this common value equals $m + 1$ if and only if $r = m - 2$. $\qquad\square$

8.1.3   The Generic Hankel Matrix

We emphasize that throughout the ground field k has characteristic zero (or sufficiently large as compared to the size m of the matrix).

8.1.3.1   Hollow

We now deal with the Hankel matrix from (6.10), where $\mathfrak{f} := \det \mathfrak{H}(\mathcal{H}_m)$ was shown not to be a factor of its Hessian.

Proposition 8.7 *Let $F := \mathfrak{f}_{\mathrm{red}} = x_{2,m}\tilde{q}$ in the notation there. Then $V(F)^* \simeq V(\tilde{q})$ up to variable renaming and conjugation.*

Proof Let $F_{i,j} := \frac{\partial F}{\partial x_{i,j}}$ and $\partial F = \{F_{i,j}\}$.

We claim that the homogeneous defining ideal of $V(F)^*$ is generated by the irreducible quadric

$$
\mathfrak{q} = \frac{1}{y_{1,1}^{m-2}} \det
\begin{pmatrix}
2y_{2,m} & y_{1,m} & \cdots & y_{1,3} & y_{1,2} \\
y_{1,m} & 0 & \cdots & 0 & 2y_{1,1} \\
\vdots & \vdots & \ddots & \vdots & \vdots \\
y_{1,3} & 0 & \cdots & 0 & 0 \\
y_{1,2} & 2y_{1,1} & \cdots & 0 & 0
\end{pmatrix}.
$$

On the one hand, a calculation shows that $\mathfrak{q} = \tilde{q}$ up to variable renaming and conjugation. On the other hand, as in the proof of Proposition 6.10, we have that the Hessian of F is not zero module F. Thus, by Theorem 2.7, $V(F)^*$ is a reduced hypersurface. In particular, to conclude the proof of the claim it is enough to show that $\mathfrak{q}$ belongs to the homogeneous defining ideal of $V(F)^*$.

By the equations obtained in the proof of item (ii) of Proposition 6.10, one has:

$$
F_{1,1} = x_{2,m}^2, \quad F_{1,u} = 2x_{1,m+2-u}\,x_{2,m} \ (2 \leq u \leq m), \quad \text{and} \quad F_{2,m} = q + x_{1,1}x_{2,m}.
$$

Evaluating the entries of $\mathfrak{q}$ at these expressions, yields

$$
\mathfrak{q}(\partial F) = \frac{1}{x_{2,m}^{2(m-2)}} \det
\begin{pmatrix}
2q + 2x_{1,1}x_{2,m} & 2x_{1,2}x_{2,m} & \cdots & 2x_{1,m-1}x_{2,m} & 2x_{1,m}x_{2,m} \\
2x_{1,2}x_{2,m} & 0 & \cdots & 0 & 2x_{2,m}^2 \\
\vdots & \vdots & \ddots & \vdots & \vdots \\
2x_{1,m-1}x_{2,m} & 0 & \cdots & 0 & 0 \\
2x_{1,m}x_{2,m} & 2x_{2,m}^2 & \cdots & 0 & 0
\end{pmatrix}
$$

$$
= \frac{2^m}{x_{2,m}^{2(m-2)}} \left[x_{2,m}^m \mathfrak{f} + x_{2,m}^{2(m-1)} \tilde{q} \right] = 2^m x_{2,m}^2 \tilde{q} = 2^m x_{2,m} F.
$$

Hence, $\mathfrak{q}$ belongs to the homogeneous defining ideal of $V(F)^*$ as desired. $\qquad\square$

8.1.3.2 Semi-Hollow Like

In this part we look at the case of the linear section $\mathcal{H}_m[r]$ of the generic Hankel matrix. Note that it does not exactly fit any of the previously variants of hollow sections since the linear forms in slots other than the three basic strands belong to these strands, thus explaining the above terminology.

In any case, we have the following result, which is the analogue of Lemma 8.4.

Proposition 8.8 *Let* $0 \leq r \leq m - 2 \, (m \geq 3)$, *and set* $f := \det \mathcal{H}_m[r]$. *Then* $\dim V(f)^* \geq m - 1$.

Proof By formula (2.5) of Theorem 2.7, it suffices to show that $H(f)$ has rank at least $m + 1$ modulo f (note, as a slight control, that $m + 1 \leq 2m - r - 1$ for $r \leq m - 2$).

Consider the Hankel matrix $\mathcal{H}_m(m - 2)$ obtained by further setting

$$x_{m+2} = \cdots = x_{2m-(r+1)} = 0.$$

The Hessian matrix of $\det \mathcal{H}_m(m-2)$ can be viewed as the $(m+1) \times (m+1)$ submatrix Θ of $H(f)$ of the first $m + 1$ rows and columns modulo $(x_{m+2}, \ldots, x_{2m-(r+1)})$. By [1, Theorem 4.4(iii)], $\det \Theta$ modulo $(x_{m+2}, \ldots, x_{2m-(r+1)})$ is a nonzero scalar multiple of $x_{m+1}^{(m+1)(m-2)}$. Thus, $\det \Theta$ does not vanish. In addition, $\det \Theta$ does not vanish module f because f does not have a pure term in x_{m+1}. Therefore, $H(f)$ has rank at least $m + 1$ modulo f. $\qquad\square$

Certain essential questions come up, which we choose to state as:

Conjecture 8.9 Let $0 \leq r \leq m - 2 \, (m \geq 3)$, and $f = \det \mathcal{H}_m[r]$. Then:

(i) $\dim V(f)^* = m - 1$. In particular, $V(f)^*$ is non-deficient if and only if $r = m - 2$, and, moreover the expected multiplicity of f as a factor of its Hessian determinant is $2m - r - 2 - (m - 1) - 1 = m - r - 2$.

(ii) f is a factor of its Hessian determinant with the expected multiplicity.

(iii) Let $\mathfrak{D}$ denote the homogeneous defining ideal of $V(f)^*$ in its Plücker embedding. Then:

    (a) The initial degree of $\mathfrak{D}$ is m and the subideal $(\mathfrak{D}_m)$ generated in the initial degree has codimension $m - r - 1$.

    (b) If $r = 0$ (generic), $\mathfrak{D}$ is equigenerated in degree m and has m-linear free resolution.

    (c) If $0 \neq r = m - 3$, then $(\mathfrak{D}_m)$ is a linearly presented codimension 2 perfect ideal.

(iv) $V(f)^*$ is arithmetically Cohen–Macaulay if and only if $r = m - 2$.

Remark 8.10 The conjectured statement in (i) accommodates the generic case as well as the sub-Hankel degeneration ($r = m - 2$) – for the latter, f is not a factor of its Hessian determinant (cf. [1, Theorem 4.4 (iii)]).

The value of the dimension in item (i) follows from Proposition 8.8 and the conjectured codimension in item (iii) (a). We emphasize the generating degrees in the present case of Hankel sections, which are often different from the case of many other linear sections for which the dual variety of the respective determinant tend to have a substantial amount of generators in degree 2.

As for (iv), one has an affirmative answer in one direction, as follows. A generic Hankel matrix whose entries are affected by nonzero scalars will be called a *Hankel matrix with multipliers*.

Proposition 8.11 *If* $r = m - 2$, *then* $V(f)^*$ *is the determinant of a sub-Hankel matrix with multipliers. In particular,* $V(f)^*$ *is arithmetically Cohen–Macaulay.*

Proof Recall that one is assuming characteristic zero throughout. Let $\bar{R} := R/f$ and $\bar{J} := J/f$ where J is the gradient ideal of f. By Section 6.4.1 and Euler's formula, the syzygy matrix of $\bar{J}$ contains an $(m + 1) \times (m + 1)$ submatrix with the following shape:

$$\varphi := \begin{pmatrix} x_1 & (m-1)x_1 & 2x_2 & 2x_3 & 2x_4 & \dots & 2x_m \\ x_2 & (m-2)x_2 & *x_3 & *x_4 & *x_5 & \dots & x_{m+1} \\ x_3 & (m-3)x_3 & *x_4 & *x_5 & *x_6 & \dots & 0 \\ \vdots & \vdots & \vdots & \vdots & \vdots & \dots & \vdots \\ x_{m-3} & 3x_{m-3} & *x_{m-2} & *x_{m-1} & *x_m & \dots & 0 \\ x_{m-2} & 2x_{m-2} & *x_{m-1} & *x_m & x_{m+1} & \dots & 0 \\ x_{m-1} & x_{m-1} & *x_m & x_{m+1} & 0 & \dots & 0 \\ x_m & 0 & x_{m+1} & 0 & 0 & \dots & 0 \\ x_{m+1} & -x_{m+1} & 0 & 0 & 0 & \dots & 0 \end{pmatrix} \quad (\mathrm{mod}\, f),$$

where $*$ are unspecified nonzero coefficients. Since φ is a submatrix of the syzygy matrix of $\bar{J}$, its rank is at most m. Expanding the determinant of φ by Laplace along the last row, one gets $x_{m+1} \det \widetilde{\varphi} \equiv 0\,(\mathrm{mod}\, f)$, where

$$\widetilde{\varphi} = \begin{pmatrix} mx_1 & 2x_2 & 2x_3 & 2x_4 & \dots & 2x_m \\ (m-1)x_2 & *x_3 & *x_4 & *x_5 & \dots & x_{m+1} \\ \vdots & \vdots & \vdots & \vdots & \dots & \vdots \\ 4x_{m-3} & *x_{m-2} & *x_{m-1} & *x_m & \dots & 0 \\ 3x_{m-2} & *x_{m-1} & *x_m & x_{m+1} & \dots & 0 \\ 2x_{m-1} & *x_m & x_{m+1} & 0 & \dots & 0 \\ x_m & x_{m+1} & 0 & 0 & \dots & 0 \end{pmatrix}.$$

Since $x_{m+1} \not\equiv 0\,(\mathrm{mod}\, f)$ and $\bar{R}$ is a domain, then $\det \widetilde{\varphi} \in (f)$. Note that $\det \widetilde{\varphi} \neq 0$. In particular, for reasons of degree, $\det \widetilde{\varphi}$ is a nonzero scalar multiple of f.

Let B be the unique $(m + 1) \times (m + 1)$ linear matrix with entries in $k[t_1, \dots, t_{m+1}] = k[\mathbf{t}]$ such that $(\mathbf{t})\varphi = (\mathbf{x})B$. In particular, B has the following shape:

$$\begin{pmatrix}
t_1 & (m-1)t_1 & 0 & 0 & 0 & \ldots & 0 \\
t_2 & (m-2)t_2 & 2t_1 & 0 & 0 & \ldots & 0 \\
t_3 & (m-3)t_3 & \bullet t_2 & 2t_1 & 0 & \ldots & 0 \\
t_4 & (m-4)t_4 & \bullet t_3 & \bullet t_2 & 2t_1 & \ldots & 0 \\
\vdots & \vdots & \vdots & \vdots & \vdots & \ldots & \vdots \\
t_m & 0 & \bullet t_{m-1} & \bullet t_{m-2} & \bullet t_{m-3} & \ldots & 2t_1 \\
t_{m+1} & -t_{m+1} & t_m & t_{m-1} & t_{m-2} & \ldots & t_2
\end{pmatrix},$$

where $\bullet$ are unspecified nonzero coefficients. We note that B^t is a submatrix of the Jacobian dual matrix of $\bar{J}$ in the sense of [5, Chapter 3].

Let $k[t_1, \ldots, t_{m+1}]/P$ denote the homogeneous coordinate ring of $V(f)^*$. By Proposition 8.8, the latter is non-deficient, i.e., P is a principal (prime) ideal.

Since the rank of the Jacobian dual matrix of $\bar{J}$ modulo P is at most m we have $\det B \in P$. Expanding $\det B$ by Laplace along the first row yields $\det B = t_1 \det \widetilde{B}$, where

$$\widetilde{B} = \begin{pmatrix}
-t_2 & 2t_1 & 0 & 0 & \ldots & 0 \\
-2t_3 & \bullet t_2 & 2t_1 & 0 & \ldots & 0 \\
-3t_4 & \bullet t_3 & \bullet t_2 & 2t_1 & \ldots & 0 \\
\vdots & \vdots & \vdots & \vdots & \ldots & \vdots \\
-(m-1)t_m & \bullet t_{m-1} & \bullet t_{m-2} & \bullet t_{m-3} & \ldots & 2t_1 \\
-mt_{m+1} & t_m & t_{m-1} & t_{m-2} & \ldots & t_2
\end{pmatrix},$$

a sort of Toeplitz matrix whose nonzero entries are indeterminates possibly affected by nonzero multipliers. Taking its mirror image (exchange rows equidistant from the extremes) yields the following sub-Hankel matrix with multipliers

$$\widetilde{H} := \begin{pmatrix}
-mt_{m+1} & t_m & t_{m-1} & t_{m-2} & \ldots & t_2 \\
-(m-1)t_m & \bullet t_{m-1} & \bullet t_{m-2} & \bullet t_{m-3} & \ldots & 2t_1 \\
-(m-2)t_{m-1} & \bullet t_{m-2} & \bullet t_{m-3} & \bullet t_{m-4} & \ldots & 0 \\
\vdots & \vdots & \vdots & \vdots & \ldots & \vdots \\
-2t_3 & \bullet t_2 & 2t_1 & 0 & \ldots & 0 \\
-t_2 & 2t_1 & 0 & 0 & \ldots & 0
\end{pmatrix}. \tag{8.2}$$

In particular, $t_1 \det \widetilde{H} \equiv 0 \,(\mathrm{mod}\, P)$. Since P is prime and $t_1 \notin P$, then $\det \widetilde{H} \in P$.

On the other hand, an argument analogous to the one in the proof Proposition 6.11 (ii), bringing in the untold nonzero scalars $\bullet$, shows that $\det \widetilde{H}$ is irreducible. Thus, $P = (\det \widetilde{H})$.   □

Remark 8.12 A generic sparse Hankel matrix with multipliers can sometimes be conjugate to the its counterpart without multipliers $\neq 1$ (see, e.g., Proposition 8.7). Not so, however, the ones coming up in the last proposition. We note that, in particular, the defining matrix of the ideal $(\mathfrak{D}_m)$ in Conjecture 8.9 (iii) (c) is a nonsquare sparse Hankel matrix with multipliers. One wonders if generic sparse Hankel matrices with multipliers admit a sufficiently stable theory.

8.2   One Term Linear Section of the Generic Matrix

Recall that the section title refers to the linear section $\mathbf{L}(m)_\lambda$ of Section 4.3, where λ designates a nonzero linear form not involving $x_{m,m}$. We will only deal with the case where $\lambda = x_{m-1,m-1}$ and, hence, remove the subscript altogether.

Some of the methods will closely resembles those of Section 8.1.1.3, but there will eventually be differences regarding the conclusions. For example, in marked contrast, here the dual variety will no longer be a ladder determinantal variety.

The core of the discussion will be the ladder $\mathcal{L}_c(m)$ on the $m \times m$ generic matrix, as discussed in Theorem 4.1:

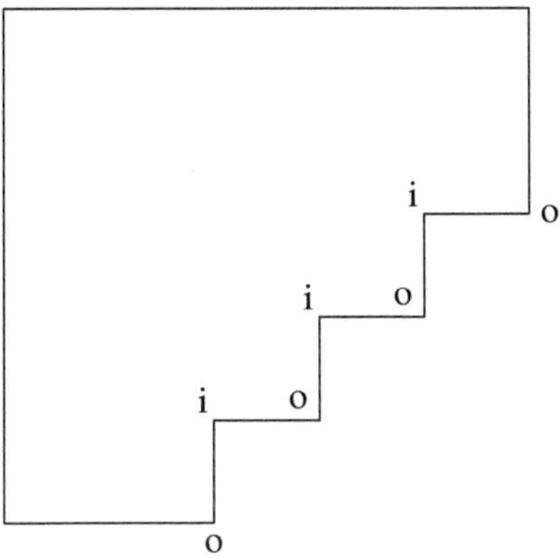

in the case where $c = 2$, with a focus upon its ideal of 2-minors $I_2(\mathcal{L}_2(m))$. In the sequel, by a harmless column permutation, we assume that the section is that of one term symmetric with respect to the main diagonal. Abiding by this, the ladder in question has the following shape:

$$
\begin{pmatrix}
y_{1,1} & y_{1,2} & \cdots & y_{1,m-2} & y_{1,m} & y_{1,m-1} \\
y_{2,1} & y_{2,2} & \cdots & y_{2,m-2} & y_{2,m} & y_{2,m-1} \\
\vdots & \vdots & \ddots & \vdots & \vdots & \vdots \\
y_{m-2,1} & y_{m-2,2} & \cdots & y_{m-2,m-2} & y_{m-2,m} & y_{m-2,m-1} \\
y_{m-1,1} & y_{m-1,2} & \cdots & y_{m-1,m-2} & y_{m-1,m} & \\
y_{m,1} & y_{m,2} & \cdots & y_{m,m-2} & &
\end{pmatrix}
$$

We write $\mathcal{L}_2(m) = \mathcal{L}_2$ since m is fixed in the discussion.

Lemma 8.13 *The homogeneous defining ideal of $V(f)^*$ contains the ideal $I_2(\mathcal{L}_2)$.*

Proof Endow $\mathbf{L}(m)$ with the decomposition $\mathbf{X} = \mathbf{X}_1 \cup \mathbf{X}_2$, where $\mathbf{X}_1 = \mathbf{X} \setminus \{x_{m-1,m-1}\}$ and $\mathbf{X}_2 = \{x_{m-1,m-1}\}$. We can apply Theorem 2.11, observing that assumption (b) there is automatic since f is irreducible. Then the result is read off Remark 2.12. $\square$

Next, we need the following result:

Lemma 8.14 *Let φ denote the $(m-1) \times m$ submatrix φ of $\mathbf{L}(m)$ that omits the first row. Then the ideal $I_{m-1}(\varphi)$ is of linear type.*

Proof The argument hinges on the specificity of the Hilbert–Burch kind of matrix. Namely, according to [3, Proposition 9.3, Remark 9.4 and Corollary 10.9], in this codimension and under the Cohen–Macaulay hypothesis, it suffices to show that $\mu(I_\wp) \leq \dim S_\wp$, for every prime ideal $\wp \subset S$. This property is equivalent to having that the codimension of the ideal $I_t(\varphi)$ of t-minors is at least rank $\varphi - t + 2 = m - t + 1$, for $1 \leq t \leq m - 1$ (see, e.g., [3, Lemma 8.2]). Letting Φ denote the fully generic predecessor of φ, it is well-known that $I_t(\Phi)$ has codimension $(m-t+1)(m-t) \geq m-t+1$. Since $I_t(\Phi)$ is a prime ideal, the form $x_{m-1,m-1}-x_{m,m}$ is a regular element thereof, hence $I_t(\Phi)$ specializes to $I_t(\varphi)$, whose codimension then has the required lower bound. $\square$

Remark 8.15 For $m \neq 4$, the lower bound $\dim V(f)^* \geq 2m - 2$ in the case of a one term section can be derived directly from the results on the hollow-filled section (Corollary 4.12). The discussion to follow includes the case $m = 4$, which is not hollow-filled (in the canonically defined decomposition).

Theorem 8.16 ($m \geq 3$) *Let $f = \det \mathbf{L}(m)$. Then $\dim V(f)^* = 2m - 2$. In particular, the expected multiplicity of f as a factor of $h(f)$ is $m(m-2)-1$.*

Proof We divide the argument in two parts:

1. $\dim V(f)^* \geq 2m - 2$.

Drawing upon the formula (2.5) of Theorem 2.7, it suffices to show that $H(f)$ has a submatrix of rank at least $2m$ modulo f.

Let φ be the matrix as in Lemma 8.14 where it follows that its m maximal minors are analytically independent elements, and being forms of the same degree, are algebraically independent over k. Therefore, their Jacobian matrix $\mathfrak{J}$ with respect to the entries of φ has rank m. Thus, let $\mathfrak{J}'$ denote an $m \times m$ submatrix of $\mathfrak{J}$ such that $\det \mathfrak{J}' \neq 0$ – this is in particular a submatrix of the entire Hessian matrix $H(f)$ of f. Now, $f \in (x_{1,1}, \ldots, x_{1,m})$ while $\det \mathfrak{J}' \notin (x_{1,1}, \ldots, x_{1,m})$. This means that $\det \mathfrak{J}' \neq 0$ even modulo f.

Write $H(f)$ in the block form

$$H(f) = \begin{pmatrix} \mathbf{0} & \mathfrak{J} \\ \mathfrak{J}^t & * \end{pmatrix},$$

where $\mathfrak{J}$ is written in the order of the variables starting with $\{x_{1,1}, \ldots, x_{1,m}\}$ and $\mathfrak{J}^t$ denotes the transpose of $\mathfrak{J}$. Since $\mathfrak{J}'$ above is a submatrix of $\mathfrak{J}$ of rank m modulo f, then $H(f)$ has rank at least $2m$ modulo f.

2. $\dim V(f)^* \leq 2m - 2$.

Letting $\mathcal{L}_2$ be the ladder in Lemma 8.13, we have seen (Section 4.1.1) that $I_2(\mathcal{L}_2)$ is a Cohen–Macaulay prime ideal. Moreover, by (4.1) its codimension in the ring $k[\mathcal{L}_2]$ is $(m-1)^2 - \binom{3}{2} = m(m-2) - 2$. Clearly, its extension to the full polynomial ring $B := k[y_{i,j} \mid 1 \leq i \leq j \leq m, (i,j) \neq (m,m)]$ is still prime of codimension $m(m-2) - 2$.

Thus, P contains a prime subideal of codimension $m(m-2) - 2$.

In addition, direct checking shows that the following two quadrics

$$g := y_{1,1}y_{m,m-1} - y_{1,m-1}y_{m,1}, \ h := y_{1,1}y_{m-1,m-1} - y_{m-1,1}y_{1,m-1} - y_{m,1}y_{1,m}$$

belong to P. We further claim that $\{g, h\}$ is a regular sequence modulo $I_2(\mathcal{L}_2)B$.

It suffices to prove the assertion locally at the powers of $y_{1,1}$ since $I_2(\mathcal{L}_2)B$ is prime and $y_{1,1} \notin I_2(\mathcal{L}_2)B$. Now, locally at $y_{1,1}$ and at the level of the ambient rings, f, g are like the variables $y_{m,m-1}$ and $y_{m-1,m-1}$, hence one has

$$B_{y_{1,1}}/(g, h) \simeq A_{y_{1,1}}.$$

Setting $I = I_2(\mathcal{L}_2)$ for lighter reading, it follows that

$$\left(B_{y_{1,1}}/IB_{y_{1,1}}\right)/(g, h)\left(B_{y_{1,1}}/IB_{y_{1,1}}\right) \simeq B_{y_{1,1}}/(g, h, I)$$

$$\simeq \left(B_{y_{1,1}}/(g, h)\right)/((g, h, I)/(g, h))$$

$$\simeq A_{y_{1,1}}/IA_{y_{1,1}}.$$

But clearly, $B_{y_{1,1}}/IB_{y_{1,1}} \simeq (A_{y_{1,1}}/IA_{y_{1,1}})[y_{m-1,m-1}, y_{m,m-1}]_{y_{1,1}}$, hence

$$\dim A_{y_{1,1}}/IA_{y_{1,1}} = \dim B_{y_{1,1}}/IB_{y_{1,1}} - 2.$$

This shows that

$$\dim \left(B_{y_{1,1}}/IB_{y_{1,1}}\right)/(g, h)\left(B_{y_{1,1}}/IB_{y_{1,1}}\right) = \dim B_{y_{1,1}}/IB_{y_{1,1}} - 2$$

and since $B_{y_{1,1}}/IB_{y_{1,1}}$ is Cohen–Macaulay, this proves the claim.

Summing up we have shown that P has codimension at least $m(m-2) - 2 + 2 = m(m-2)$. Therefore, $\dim V(f)^* = m^2 - 2 - \text{cod } V(f)^* \leq m^2 - 2 - m(m-2) = 2m - 2$, as was to be shown. $\square$

Corollary 8.17 ($m \geq 3$) *Let $f = \det L(m)$. Then f is a factor of its Hessian determinant $h(f)$ and has the expected multiplicity as such.*

Proof By the main result of Theorem 8.16 and once more by (2.5), one has

$$\text{rank } H(f) \ (\text{mod } f) = 2m.$$

But since $m \geq 3$ then $2m < m^2 - 1$, hence $h(f) \equiv 0 \ (\text{mod } f)$. Therefore, $h(f)$ is a multiple of f and is nonzero by Theorem 4.22 (i). On the other hand, the multiplicity of f as a factor of $h(f)$ is at least the expected multiplicity $m(m - 2) - 1$ (cf. [1, Section 2.1, p. 16]). Counting, the degree of the residual factor $h(f)/f^{m(m-2)-1}$ is 2. Again, since $\deg(f) = m \geq 3$ then the expected multiplicity obtained in Theorem 8.16 coincides with the multiplicity of f as a factor of $h(f)$. $\qquad\square$

Remark 8.18

(1) The factor $h(f)/f^{m(m-2)-1}$ coincides with the determinant of the 2×2 submatrix with rows $m - 1, m$ and columns $m - 1, m$.
(2) The precise structure of the dual variety is not very clear. Its homogeneous defining ideal is not Cohen–Macaulay and contains many minimal generators which are quadric trinomials and cubic forms.
(3) It would seem that Theorem 8.16 or Corollary 8.17, or perhaps both, had been claimed in [4] by geometric means, but for our misfortune we were not able to trace in the mentioned work the precise statements expressing these contents.

8.3   One Term Linear Section of the Generic Symmetric Matrix

We refer to the material and notation of Section 5.2.

Theorem 8.19 ($m \geq 3$) *Let* $\mathbf{L}(S)$ *denote a one term diagonal section of the* $m \times m$ *generic symmetric matrix* S *and let* $f = \det \mathbf{L}(S)$. *Then* $\dim V(f)^* = m - 1$.

Proof First, the lower bound $\dim V(f)^* \geq m - 1$.

If $m \neq 4$, as in the case of the generic matrix, $\mathbf{L}(S)$ is a hollow-filled linear section (Remark 8.15). In this case one can apply Corollary 5.11 to infer the required lower bound.

For the case where $m = 4$, one resorts to computer assistance in order to find a suitable 5×5 submatrix of $H(f)$ having rank 5 modulo f.

Therefore, it now suffices to show that $\dim V(f)^* \leq m - 1$.

As has been pointed out galore, let $P \subset k[\mathbf{Y}] := k[y_{i,j} \mid 1 \leq i \leq j \leq m, \ (i, j) \neq (m, m)]$ denote the homogeneous defining ideal of the dual variety $V(f)^*$ in its natural embedding, that is, one has an isomorphism of graded k-algebras induced by the assignment $y_{i,j} \mapsto \partial f / \partial x_{i,j}$:

$$k[\mathbf{Y}]/P \simeq T/(f) \cap T,$$

where $T = k[\partial f / \partial x_{i,j} \mid 1 \leq i \leq j \leq m, \ (i, j) \neq (m, m)]$.

Let us first effect a preliminary reduction, as follows: for any pair (i, j) of indices, set

$$c_{i,j} = \begin{cases} 1 \text{ if } i = j \\ 2 \text{ if } i \neq j \end{cases}.$$

Then Proposition 2.10 gives $\partial f/\partial x_{ij} = c_{ij}\Delta_{i,j}$ for $(i, j) \neq (m - 1, m - 1)$, and $\partial f/\partial x_{m-1,m-1} = \Delta_{m-1,m-1} + \Delta_{m,m}$.

Now apply the harmless change of variables in the polynomial ring $k[\mathbf{Y}]$ that fixes $y_{m-1,m-1}$ and maps $y_{ij} \mapsto c_{ij}^{-1} y_{ij}$, for $(i, j) \neq (m - 1, m - 1)$. Let $Q \subset k[\mathbf{Y}]$ denote P under this change. Clearly, Q still defines $V(f)^*$, this time around by elimination of the linear system of the following cofactors:

$$\{\Delta_{ij} \mid 1 \leq i \leq j \leq m, \ (i, j) \neq (m, m), (m - 1, m - 1)\} \cup \{\Delta_{m-1,m-1} + \Delta_{m,m}\}.$$

Consider the ladder $\mathcal{L}$:

$$\begin{pmatrix} y_{1,1} & y_{1,2} & \cdots & y_{1,m-2} & y_{1,m} & y_{1,m-1} \\ y_{1,2} & y_{2,2} & \cdots & y_{2,m-2} & y_{2,m} & y_{2,m-1} \\ \vdots & \vdots & \ddots & \vdots & \vdots & \vdots \\ y_{1,m-2} & y_{2,m-2} & \cdots & y_{m-2,m-2} & y_{m-2,m} & y_{m-2,m-1} \\ y_{1,m} & y_{2,m} & \cdots & y_{m-2,m} & y_{m-1,m} & \\ y_{1,m-1} & y_{2,m-1} & \cdots & y_{m-2,m-1} & & \end{pmatrix}$$

Note that, as done before in the generic case, the last two columns have been permuted, and in order to preserve symmetry, so have the last two rows. As in the generic case, rewriting this way makes it easier to get a hold of sufficiently many explicit minimal generators of Q coming from $I_2(\mathcal{L})$, recalling that $I_2(\mathcal{L}) \subset Q$ by Remark 2.12.

Now consider the quadric $h := y_{1,1} y_{m-1,m-1} - y_{1,m-1}^2 - y_{1,m}^2$. Since $I_2(\mathcal{L})$ is generated in degree 2 and none of its generators involves $y_{m-1,m-1}$, then h cannot be a k-linear combination of these generators. Thus, $h \notin I_2(\mathcal{L})$. But since $I_2(\mathcal{L})$ is a prime ideal it follows that Q has codimension at least $\binom{m}{2} - 1$. Therefore,

$$\dim V(f)^* = \binom{m + 1}{2} - 2 - \operatorname{codim} Q \leq \binom{m + 1}{2} - 2 - \binom{m}{2} + 1 = m - 1,$$

as was to be shown. $\square$

As to the precise structure of the ideal Q, Remark 8.18 (2) applies here as well.

We end this part by considering the question as to whether f is a factor of its Hessian determinant $h(f)$ with multiplicity ≥ 1. In distinction to this effective multiplicity, the expected multiplicity (according to Segre) is in this setup defined as $\binom{m+1}{2} - 2 - \dim V(f)^* - 1 = \binom{m}{2} - 2$, where $V(f)^*$ denotes the dual variety to the hypersurface $V(f)$ (Section 3.2.1).

Proposition 8.20 ($m \geq 3$)   *The determinant* $f = \det \mathbf{L}(S)$ *is a factor of its Hessian* $h(f)$ *with effective multiplicity equal to the expected multiplicity* $\binom{m}{2} - 2$.

Proof By the main result of Theorem 8.19 and (2.5), one has

$$\operatorname{rank} H(f) \pmod{f} = m - 1 + 2 = m + 1.$$

But since $m \geq 3$ then $m + 1 < \binom{m+1}{2} - 1$, as one readily verifies, and hence, $h(f) \equiv 0 \pmod{f}$. Therefore, $h(f)$ is a multiple of f and is nonzero by Theorem 5.6 (iii). At the other end, the multiplicity of f as a factor of $h(f)$ is at least the expected multiplicity $\binom{m}{2} - 2$ (cf. [1, Section 2.1, p. 16]). Counting degrees we find that the multiplicity of the residual factor $h(f)/f^{\binom{m}{2}-2}$ is 2. Since $\deg(f) = m \geq 3$, the statement is clear. $\qquad\square$

Exercises

8.21 Give a geometric argument predicting the result of Theorem 8.2 (b) to the effect that the dual variety is a complete intersection.

The following problems will require computer assistance in several parts.

8.22 Give the exact shape of the matrix $\tilde{H}$ of (8.2) in the range $3 \leq m \leq 5$. Compare with a calculation by computer assistance.

8.23 Prove the assertion (1) of Remark 8.18.

8.24 Give the structure of the defining ideal of the dual variety in Theorem 8.16 in terms of explicit generators, in the cases of $m = 3, 4$. Prove that the ideal is not Cohen–Macaulay.

8.25 Consider the following 4×4 sparse matrix:

$$L := \begin{pmatrix} x_{1,1} & 0 & x_{1,3} & x_{1,4} \\ x_{2,1} & 0 & 0 & x_{2,4} \\ x_{31} & 0 & x_{3,3} & 0 \\ 0 & x_{4,2} & 0 & 0 \end{pmatrix}$$

and let $f = \det L$. Show that f is reduced and the homogeneous defining ideal $\mathfrak{D}$ of the dual variety $V(f)^* \subset \mathbb{P}^{7*}$ contains the reduced quadrics

$$y_{1,1}y_{2,4} - y_{1,4}y_{2,1}, \ y_{1,1}y_{3,3} - y_{1,3}y_{3,1},$$

$$y_{4,2}y_{1,1}, \ y_{4,2}y_{1,3}, \ y_{4,2}y_{1,4}, \ y_{4,2}y_{2,1},$$

$$y_{4,2}y_{2,4}, \ y_{4,2}y_{3,1}, \ y_{4,2}y_{3,3}.$$

Can you prove that these generate $\mathfrak{D}$ without resorting to computer assistance?

References

1. C. Ciliberto, F. Russo, A. Simis, Homaloidal hypersurfaces and hypersurfaces with vanishing Hessian. Adv. Math. **218**, 1759–1805 (2008) 224, 230, 232
2. A. Conca, Ladder determinantal rings. J. Pure Appl. Algebra **98**, 119–134 (1995) 222
3. J. Herzog, A. Simis, W. Vasconcelos, Koszul homology and blowing-up rings, in *Proceedings of the Commutative Algebra*, Trento, ed. by S. Greco, G. Valla. Lecture Notes in Pure and Applied Mathematics, vol. 84 (Marcel-Dekker, New York City, 1983), pp. 79–169 228
4. J.M. Landsberg, L. Manivel, N. Ressayre, Hypersurfaces with degenerate duals and the geometric complexity theory program. Comment. Math. Helv. **88**, 469–484 (2013) 230
5. Z. Ramos, A. Simis, *Graded Algebras in Algebraic Geometry*. Expositions in Mathematics, vol. 70 (De Gruyter, Berlin, 2022) 226

Part III
Other Classes of Linear Sections

Chapter 9
Hilbert–Burch Linear Sections

Abstract This chapter takes an exception to the previous chapters in that the focus is on linear sections of $m \times (m-1)$ generic matrices instead of square ones. Because of the celebrated theorem of Hilbert–Burch associated to codimension 2 perfect ideals and such matrices, we call the latter by abuse Hilbert–Burch matrices. The first part deals with such matrices whose entries are general linear forms in a d-dimensional polynomial ring over a field with focus on the watershed case where $m \geq d$. In this situation the symbolic Rees algebra of the ideal of $(m-1)$-minors is thoroughly discussed, and the full nature of the cases $m = d$ and $m = d+1$ is displayed. A notable subsumed Cremona map based on certain inversion factors is given full description. In the sequel $m \times (m-1)$ Hankel linear sections are discussed with an emphasis on the related algebraic invariants. A large section is dedicated to Hilbert–Burch linear sections when $d = 3$, where the central theme is the discussion of the chaos invariant, a number based on the behavior of the ideals of lower minors. Other notions are shown to be related, such as fat points and reciprocal ideals of hyperplane arrangements. The chapter ends with a thorough discussion of linearly presented monomial ideals in three variables.

9.1 General Linear Forms

9.1.1 The Overall Picture

Fix an integer $m \geq 2$. Throughout this section, $\mathbf{L}$ denotes an $m \times (m-1)$ matrix over a polynomial ring $R = k[x_1, \ldots, x_d]$, with $m(m-1) \geq d$, whose entries are general linear forms.

We refer to Section 1.5.1 for the basic background. If no confusion arises, we often call $\mathbf{L}$ a general linear section.

Proposition 9.1 *Let $\mathbf{L}$ be an $m \times (m-1)$ general linear section with entries in a polynomial ring $R = k[x_1, \ldots, x_d]$ over a field k, with $d \geq 3$ and $m \geq 2$. Set $I = I_{m-1}(\mathbf{L})$. Then:*

© The Author(s), under exclusive license to Springer Nature Switzerland AG 2024   237
Z. Ramos, A. Simis, *Determinantal Ideals of Square Linear Matrices*,
https://doi.org/10.1007/978-3-031-55284-7_9

(a) *I has height 2 and $I_{m-2}(\mathbf{L})$ has height* $\min\{6, d\}$.
(b) (char(k)= 0) *R/I satisfies the condition (R_r) of Serre, with* $r = \min\{3, d-3\}$; *in particular, if $d \geq 4$, then R/I is normal and I is a prime ideal.*

Proof

(a) This follows from Theorem 1.15.
(b) The condition R_r follows from Proposition 1.17. Thus, if $d \geq 4$, then R/I satisfies R_1. At the other end, R/I is Cohen–Macaulay. It follows that, for $d \geq 4$, R/I is normal and, since I is homogeneous, R/I must be a domain. (If $d = 3$, then I is still a radical ideal.) $\square$

For the next result recall that $\mathcal{R}(I)$ denotes the Rees algebra of an ideal $I \subset R$ (Section 2.2.1). Let $\mathcal{J}$ stand for its presentation ideal over a polynomial ring $R[\mathbf{y}] = R[y_1, \ldots, y_m]$.

Proposition 9.2 *Let $\mathbf{L}$ and I be as in* Proposition 9.1. *Then:*

(a) *I is of linear type if and only if $m \leq d$.*
(b) *I satisfies the G_d condition.*
(c) *If $m > d$, then $\ell(I) = d$, $r(I) = d$, $\mathcal{J} = (I_1(\mathbf{x}\,B), I_d(B))$, and $\mathcal{R}(I)$ is Cohen–Macaulay.*

Proof

(a) Apply the result of Theorem 1.15 to this case. We claim that

$$\min\{d, (m - j + 1)(m - j)\} \geq m - j + 1, \quad \text{for } 1 \leq j \leq m - 1.$$

This is obvious if the minimum is attained by $(m-j+1)(m-j)$; if the minimum is d instead, then $m \leq d$ certainly implies $m - j + 1 \leq m$. This shows that I satisfies G_∞; hence, by Proposition 2.22, I is of linear type. The converse is evident since the linear type property implies the inequality $\mu(I) \leq \dim R$.
(b) For $m - d + 1 \leq j \leq m - 1$, we have $d \geq m - j + 1$ and $(m - j + 1)(m - j) \geq m - j + 1$. Thus,

$$\min\{d, (m - j + 1)(m - j)\} \geq m - j + 1$$

for $m - 1 \geq j \geq m - d + 1$. Hence, by Theorem 1.15 and Proposition 2.20, we have that I satisfies the G_d condition as claimed.
(c) This follows from item (b) and Theorem 2.24. $\square$

For a module M over a Noetherian ring R, we denote by $\mathrm{pd}_R(M)$ the projective (homological) dimension of M.

Corollary 9.3 *Let $\mathbf{L}$ and I be as in* Proposition 9.1. *Then,* $\mathrm{pd}_R(R/I^r) \leq d - 1$ *for every* $1 \leq r \leq d - 2$.

Proof By Theorem 9.2 (a), in the notation of Theorem 2.23, the minimal graded free resolution of I^r is given by $\mathcal{K}_r$. Hence,

$$\mathrm{pd}_R(I^r) = \min\{m-1, r\} \leq r \leq d - 2.$$

In particular,

$$\mathrm{pd}_R(R/I^r) = \mathrm{pd}_R(I^r) + 1 \leq d - 1. \qquad \square$$

Recall (Section 2.2.5) that $I^{(r)}$ denotes the symbolic power of order $r \geq 0$ of the ideal $I \subset R$.

Proposition 9.4 *Let* $\mathbf{L}$ *and* I *be as in* Proposition 9.1. *Given an integer* $r \geq 0$ *such that* $I^r \neq I^{(r)}$, *then* $I^{(r)}/I^r$ *is an* $(\mathbf{x})$-*primary* R-*module (in other words,* $I^{(r)}$ *is the* $(\mathbf{x})$-*saturation of* I^r).

Proof The assertion is equivalent to having $I^{(r)}{}_P = I^r{}_P$ for every prime $P \neq (\mathbf{x})$. Since $\mathrm{ht}\, P \leq d - 1$ for any such prime, the assertion is clear if $d = 3$. Hence, assume that $d \geq 4$. Thus, by Proposition 9.1, I is a prime ideal. Set

$$t_\infty := \min\{1 \leq t \leq m - 1 \mid I_t(\mathbf{L}) \subset P\}.$$

Since $\mathrm{ht}\, P \leq d - 1$, then, by Theorem 1.15, $\mathrm{ht}\, I_j(\mathbf{L}) = (m - j + 1)(m - j)$ for $t_\infty \leq j \leq m-1$. Inverting a $(t_\infty - 1)$-minor of $\mathbf{L}$ in R_P, we get $I_P = I_{m - t_\infty}(\widetilde{\mathbf{L}})$ for a suitable $(m - t_\infty + 1) \times (m - t_\infty)$ matrix $\widetilde{\mathbf{L}}$ over R_P. Moreover, $I_j(\mathbf{L})_P = I_{j - t_\infty + 1}(\widetilde{\mathbf{L}})$ for every $t_\infty \leq j \leq m - 1$. Hence,

$$\mathrm{ht}\, I_i(\widetilde{\mathbf{L}}) = \mathrm{ht}\, I_{i + t_\infty - 1}(\mathbf{L})_P$$

$$= \mathrm{ht}\, I_{i + t_\infty - 1}(\mathbf{L}) \qquad\qquad (\text{because } I_{i + t_\infty - 1}(\mathbf{L}) \subset P)$$

$$= (m - i - t_\infty + 1 + 1)(m - i - t_\infty + 1)$$

$$\geq m - t_\infty + 1 - i + 2, \ \text{for } 1 \leq i \leq (m - t_\infty + 1) - 2.$$

Then, $I_P^{(r)} = I_P^r$ follows from [1, Theorem 3.5 (b)] (and its predecessor [23, Theorem 3.4]). $\qquad \square$

Theorem 9.5 *Let* $\mathbf{L}$ *and* I *be as in* Proposition 9.1. *Then* $\mathrm{Ass}\,(R/I^r) = \mathrm{Ass}\,(R/I)$ *for* $1 \leq r \leq d - 2$.

Proof By Proposition 9.4, $\mathrm{Ass}\,(R/I^r) \subset \mathrm{Ass}\,(R/I) \cup \{(\mathbf{x})\}$. On the other hand, by Corollary 9.3, depth $R/I^r > 0$. Hence, $\mathrm{Ass}\,(R/I^r) = \mathrm{Ass}\,(R/I)$. $\qquad \square$

Theorem 9.6 *Let* $\mathbf{L}$ *and* I *be as in* Proposition 9.1, *with* $m \geq d \geq 3$. *Then:*

(a) *The rational map* $\mathfrak{F} : \mathbb{P}^{d-1} \dashrightarrow \mathbb{P}^{m-1}$ *defined by the* $(m - 1)$-*minors of* $\mathbf{L}$ *is birational onto its image.*

(b) *The map $\mathfrak{F}$ admits at least $\binom{m-1}{d-1}$ source inversion factors, each associated to a certain minimal representative of the inverse map; moreover, any one among them is an element of the symbolic power $I^{(m-1)}$ of degree $(m-1)(d-1)-1$.*

Proof

(a) By Theorem 2.27, it suffices to prove that the dimension of the k-subalgebra of R generated by the maximal minors of $\mathbf{L}$ has dimension d, i.e., that I has maximal analytic spread. But this follows from Proposition 9.2.

(b) Let B denote a matrix constructed as in (2.10) with respect to the given matrix $\mathbf{L}$. By Proposition 9.2 and Remark 2.25, any $(d-1)\times d$ submatrix of the transpose B^t has rank $d-1$ over the special fiber of I. Thus, by [19, Theorem 3.2.22], for any $(d-1)\times d$ submatrix of B^t, its d (ordered, signed) maximal minors are the coordinates of a representative of the inverse map; thus, there are $\binom{m-1}{d-1}$ such representatives.

By construction, the degree of any one of these representatives (i.e., of its coordinates as elements of the special fiber) is exactly $d-1$. Then each such representative gives rise to a source inversion factor that is an element of the symbolic power $I^{(d-1)}$ and has degree $(m-1)(d-1)-1$ (Theorem 2.30). $\square$

Remark 9.7

(1) For $d=3$, item (a) of the above proposition holds true solely under the assumptions that $\operatorname{ht} I_{m-1}(\mathbf{L}) = 2$ and $\operatorname{ht} I_1(\mathbf{L}) = 3$. This and more will be proved in Section 9.3.

(2) Since $\mathbf{L}$ is a matrix of general linear entries, one expects B as above to be sufficiently general. It would be interesting to give an argument to the effect that the elements of the symbolic power obtained in item (b) are independent minimal generators.

We single out the following outcome for convenience.

Corollary 9.8 *Let $\mathbf{L}$ and I be as in* Proposition 9.1, *with $m \geq d \geq 3$. Then $I^{(r)} = I^r$ for $1 \leq r \leq d-2$, and $D_j \in I^{(d-1)} \setminus I^{d-1}$, where D_j $(j=1,\ldots,\binom{m-1}{d-1})$ are the source inversion factors associated to a chosen set of minimal representatives of the inverse map to the birational map defined by the maximal minors of $\mathbf{L}$.*

Proof The first assertion follows immediately from Theorem 9.5, and the second assertion stems from Theorem 9.6. $\square$

We end the section with a simple normal torsion-freeness criterion.

Proposition 9.9 *Let $\mathbf{L}$ and I be as in* Proposition 9.1. *Then I is normally torsion-free if and only if $m < d$.*

Proof Suppose that $m < d$. If $d = 3$, the assertion is immediate because the ideal I is a complete intersection. If $d \geq 4$, then, by Proposition 9.1(b), I is a prime ideal. Moreover, for $1 \leq j \leq m-2$, we have

$$\text{ht } I_j(\mathbf{L}) = \min\{d, (m - j + 1)(m - j)\}.$$

But $(m - j + 1)(m - j) \geq m - j + 2$ and $d \geq m + 1 \geq m - j + 2$ for $1 \leq j \leq m - 2$. Then

$$\text{ht } I_j(\mathbf{L}) \geq m - j + 2, \quad 1 \leq j \leq m - 2.$$

Therefore, by [1, Theorem 3.5 (b)] (and its predecessor [23, Theorem 3.4]), the ideal I is normally torsion-free. The converse follows from Theorem 9.6 (b). $\square$

9.1.2   The Symbolic Rees Algebra

Thus far, the available features of the theory work for $m \geq d$. In this part we come to grips with a richer amount of information for the particular case $m = d$ or $m = d+1$. The overall goal is the structure of the corresponding symbolic Rees algebras, a long way to go. For obvious reasons, the case where $m = d$ is called the Cremona case, while the one where $m = d + 1$ is the implicitization case. Structure results on the symbolic algebra for $m \geq d+2$ are this far unknown to the authors (see Remark 9.22 below).

9.1.2.1   Cremona Case ($m = d$)

Proposition 9.10 (char(k) $= 0$) *Let* $R = k[x_1, \ldots, x_d]$ *be a standard graded polynomial ring over* k, *and let* $\mathbf{L}$ *be a* $d \times (d - 1)$ *matrix of general linear forms. Then* $I_{d-1}(\mathbf{L})$ *is the base ideal of a Cremona map of* $\mathbb{P}^{d-1}$, *and the associated source inversion factor is* $\frac{1}{d-1} \det \Theta$, *where* Θ *denotes the Jacobian matrix of the* $(d - 1)$-*minors of* $\mathbf{L}$.

Proof The first assertion follows from Proposition 9.1. The second assertion is a consequence of Proposition 2.29. $\square$

Here is the main theorem in the case $m = d$:

Theorem 9.11 *Let* $\mathbf{L}$ *denote a* $d \times (d - 1)$ *matrix of general linear forms over* $R = k[x_1, \ldots, x_d]$, *with* $d \geq 3$. *Set* $I := I_{d-1}(\mathbf{L}) \subset R$, *and let* $\mathcal{R}^{(I)}$ *denote its symbolic Rees algebra. Then:*

(a) $\mathcal{R}^{(I)}$ *is a Gorenstein normal domain.*
(b) (char(k) $= 0$) $\mathcal{R}^{(I)}$ *is generated by the* $(d - 1)$-*minors of* $\mathbf{L}$, *viewed in degree 1, and by the source inversion factor of the Cremona map defined by these minors, viewed in degree* $d - 1$. *Moreover, this inversion factor coincides with a nonzero scalar multiple of the Jacobian determinant of the very minors.*

Proof

(a) First, by the proof of [21, Corollary 2.4 (b)], one has an isomorphism $\mathcal{R}^{(I)} \simeq \mathcal{R}(I)[t^{-1}] = R[It, t^{-1}]$; hence $\mathcal{R}^{(I)}$ is finitely generated. Then, by [20, Corollary 3.4], it is a quasi-Gorenstein Krull domain since ht $I = 2$. Moreover, $\mathcal{R}(I)$ is Cohen–Macaulay since for $m = d$ the ideal I is of linear type and its symmetric algebra is a complete intersection. Therefore, $\mathcal{R}^{(I)}$ is Cohen–Macaulay as well and hence is a Gorenstein normal domain.

(b) To get the explicit generation, let $\eth_1, \ldots, \eth_d \in k[\mathbf{y}]$ be forms of the same degree, with gcd $= 1$, defining the inverse map, and let $D \in R$ denote the corresponding source inversion factor. Write $J = (\eth_1, \ldots, \eth_d) \subset k[\mathbf{y}]$. By definition, one has

$$D = \eth_i (\Delta_1, \ldots, \Delta_d)/x_i, \ 1 \le i \le d,$$

where $\mathbf{\Delta} := \{\Delta_1, \ldots, \Delta_d\}$ are the (signed) minors generating I. Identifying the two Rees algebras $\mathcal{R}_R(I) = R[It] \subset R[t]$ and $\mathcal{R}_{k[\mathbf{y}]}(J) = k[\mathbf{y}][Ju] \subset k[\mathbf{y}][u]$ by a k-isomorphism that maps $y_i \mapsto \Delta_i t$ and $x_i \mapsto \eth_i u$, D is identified with $\eth_1/x_1$ in the common field of fractions. Drawing upon Proposition 9.4 (here we need char$(k) = 0$), the symbolic Rees algebra is generated by It and Dt^{d-1} as a consequence of [26, Corollary 7.4.3 (b)] (note that the notation for the two ideals is reversed in the latter).

The additional statement follows from Proposition 9.10 (again in characteristic 0). $\square$

As an application of the results so far in the case $m = d$, we settle in affirmative the following:

Conjecture ([7, Section 2]) (char$(k) = 0$) If I is the ideal of 2-minors of a generic (that is, random) 2×3 matrix of linear forms in three variables, then the annihilator of $I^{(r)}/I^r$ is $F_1(I)^e$, where e is the greatest integer $\le r/2$.

There is a misprint in the above statement since by definition the Fitting ideal $F_1(I)^e$ is the ideal of 2-minors of the matrix, which is the ideal I itself. The correct Fitting should be $F_2(I)^e$, the ideal of 1-minors of the matrix. But since the entries are general linear forms, this ideal is the irrelevant ideal $\mathfrak{m} := (x, y, z) \subset R := k[x, y, z]$ with k a field.

Let φ denote the given matrix. A consequence of Theorem 9.11 (b) above is that $I^{(2)} = (I^2, D)$, where $D \in R$ is the inversion factor of the Cremona map defined by the 2-minors of φ, and, moreover, for every $r \ge 1$, the following equalities hold:

$$I^{(r)} = \begin{cases} (I^{(2)})^{\frac{r}{2}} & \text{if } r \text{ is even} \\ I \, (I^{(2)})^{\frac{r-1}{2}} & \text{if } r \text{ is odd.} \end{cases}$$

By definition of the inversion factor, one has $D\mathfrak{m} \in I^2$. It follows that the annihilator of $I^{(2)}/I^2$ is $\mathfrak{m}$. We also know that $\deg(D) = 3$ since the inverse map to the Cremona map defined by the 2-minors is also defined by forms of degree 2, and so $\deg(D) = 2.2 - 1 = 3$.

Consider separately the even and the odd cases.

r EVEN. One has

$$\mathfrak{m}^{r/2} I^{(r)} = \mathfrak{m}^{r/2} (I^{(2)})^{r/2} = (\mathfrak{m} I^{(2)})^{r/2} \subset (I^2)^{r/2} = I^r.$$

Conversely, let $f \in \mathfrak{m}$ be a form such that $f I^{(r)} \subset I^r$. Since the annihilator of $I^{(2)}/I^2$ is the entire maximal ideal $\mathfrak{m}$, it suffices to show that $\deg(f) \geq r/2$. Since $I^{(r)} = (I^{(2)})^{r/2}$ and $I^{(2)} = (I^2, D)$, in particular, we get $f D^{r/2} \in I^r$. Reading degrees on both sides, one has that $\deg(f) + 3r/2 \geq 2r$. Therefore, $\deg(f) \geq r/2$; hence $f \in \mathfrak{m}^{r/2}$ as required.

r ODD. One has

$$\mathfrak{m}^{(r-1)/2} I^{(r)} = \mathfrak{m}^{(r-1)/2} I (I^{(2)})^{(r-1)/2} = I\, (\mathfrak{m} I^{(2)})^{(r-1)/2} \subset I\, (I^2)^{(r-1)/2}$$

$$= I\, I^{r-1} = I^r.$$

The hypothesis is that $f D^{(r-1)/2} I \subset I^r$. In particular, taking a minor Δ among the generators of I, we find $f\, D^{(r-1)/2} \Delta \subset I^r$. Again, reading degrees, we get the inequality $\deg(f) + 3((r-1)/2) + 2 \geq 2r$, from which follows that $\deg(f) \geq (r-1)/2$.

We conclude as before. Since in the odd case, $(r-1)/2 = \lfloor r/2 \rfloor$, we are done.

Remark 9.12 The above result was originally proved in [17] and has recently been generalized in [14, Section 5].

9.1.3   The Implicitization Case ($m = d + 1$)

We will now assume that $m = d + 1$.

9.1.3.1   Homological Preliminaries

We basically maintain the notation of previous parts. As above, $R = k[x_1, \ldots, x_d]$ and $\mathbf{L}$ denotes a $(d + 1) \times d$ matrix of general linear forms. Letting $I \subset R$ denote the perfect codimension 2 ideal generated by the maximal minors of $\mathbf{L}$, we have seen that I satisfies the condition G_d. By Theorem 2.23, $\mathcal{S}_{d-1}(I) \simeq I^{d-1}$ and a free resolution of I^{d-1} is

$$\mathcal{K}_{d-1} : 0 \to F_{d-1} \to F_{d-2} \to \ldots \to F_1 \to F_0 \to 0,$$

where

$$F_i := \bigwedge^i R^d \otimes_R S_{(d-1)-i}(R^{d+1})$$

and $\partial_i : F_i \to F_{i-1}$ is given by

$$\partial_i(e_1 \wedge \ldots \wedge e_i \otimes g) := \sum_{l=1}^{i} e_1 \wedge \ldots \wedge \widehat{e_l} \wedge \ldots \wedge e_i \otimes \varphi(e_l)g,$$

with $\{e_1, \ldots, e_d\}$ denoting a basis of R^d and $\varphi : R^d \to R^{d+1}$ standing for the map defined by the $(d+1) \times d$ presentation matrix $\mathbf{L} = (\ell_{ij})$ of the ideal I.

Consider the R-dual map to $\partial_{d-1} : F_{d-1} \to F_{d-2}$. Since I^{d-1} is generated in (standard) degree $d(d-1)$, after identification and taking into account the degrees shift, the dual map is of the form

$$\eta := \partial_{d-1}^* : R^N((d+1)(d-1)-1) \to R^d((d+1)(d-1)), \tag{9.1}$$

where $N = (d+1)\binom{d}{2}$.

Let $M = \operatorname{coker} \eta$. Shifting by $-((d+1)(d-1))$, we get a homogeneous presentation

$$R^N(-1) \xrightarrow{\eta} R^d \to M(-(d+1)(d-1)) \to 0. \tag{9.2}$$

Theorem 9.13 *With the above notation, there is a homogeneous isomorphism*

$$M(-(d+1)(d-1)) \simeq R^d/(\mathbf{x})R^d = k^d.$$

Proof Picking up from the above preliminaries, let us make explicit the dual map to $\partial_{d-1} : F_{d-1} \to F_{d-2}$. Note that

$$F_{d-1} = \bigwedge^{d-1} R^d \otimes_R S_0(R^{d+1}) \simeq R^d, \quad F_{d-2} = \bigwedge^{d-2} R^d \otimes_R S_1(R^{d+1}) \simeq R^{\binom{d}{d-2}} \otimes_R R^{d+1}.$$

Applying these identifications, the basis vector $e_1 \wedge \cdots \wedge \widehat{e_k} \wedge \cdots \wedge e_d$ gets identified with e_k, and we write $a_{1,\ldots,\hat{k},\ldots,\hat{i},\ldots,d}$ for a basis vector of $R^{\binom{d}{d-2}}$ corresponding to $e_1 \wedge \cdots \wedge \widehat{e_j} \wedge \cdots \wedge \widehat{e_l} \wedge \cdots \wedge e_d$. Furthermore, let $\{b_1, \ldots, b_{d+1}\}$ stand for a basis of R^{d+1}. With this notation, for $k = 1, \ldots, d$, the map is quite simply

$$e_k \mapsto \sum_{l=1}^{d-1} a_{1,\ldots,\hat{k},\ldots,\hat{i},\ldots,d} \otimes \varphi(e_l) = \sum_{l=1}^{d-1} a_{1,\ldots,\hat{k},\ldots,\hat{i},\ldots,d} \otimes \sum_{i=1}^{d+1} \ell_{il} b_i$$

$$= \sum_{i=1}^{d+1} \sum_{l=1}^{d-1} \ell_{il}\, a_{1,\ldots,\hat{k},\ldots,\hat{i},\ldots,d} \otimes b_i,$$

where $\mathbf{L} = (\ell_{ij})$ is as above.

Going back to η (9.1), this gives that the matrix of η in (9.2) has the following block shape:

$$\left(M_{d-1,d}|\ldots|M_{1,d}|\ldots|M_{j-1,j}|\ldots|M_{1,j}|\ldots|M_{1,2}\right), \tag{9.3}$$

where, for $1 \leq i \leq j \leq d$, M_{ij} is the following $d \times (d+1)$ matrix up to signs:

$$M_{i,j} = \begin{pmatrix} 0 & 0 & \ldots & 0 \\ \vdots & \vdots & \ldots & \vdots \\ \ell_{1,i} & \ell_{2,i} & \ldots & \ell_{(d+1),i} \\ \vdots & \vdots & \ldots & \vdots \\ \ell_{1,j} & \ell_{2,j} & \ldots & \ell_{(d+1),j} \\ \vdots & \vdots & \ldots & \vdots \\ 0 & 0 & \ldots & 0 \end{pmatrix} \begin{matrix} \\ \\ \leftarrow (d+1-j)\text{th row} \\ \\ \leftarrow (d+1-i)\text{th row.} \\ \\ \end{matrix}$$

Next let $\widetilde{M_{i,j}}$ denote the submatrix of $M_{i,j}$ consisting of the first d columns, and consider the following block submatrix of (9.3):

$$(\widetilde{M_{d-1,d}}|\ldots|\widetilde{M_{1,d}}\,|\,\widetilde{M_{d-2,d-1}}),$$

consisting of d square blocks of order d each; in particular, this matrix has d^2 columns.

CLAIM. The R-submodule of R^d generated by the columns of the above matrix coincides with $(\mathbf{x})R^d$.

For this, since the columns have standard degree 1, it is sufficient to show that the columns are k-linearly independent as elements of the k-vector space $((\mathbf{x})R)_1$.

Suppose that a nontrivial k-linear combination of these columns vanishes, with coefficients $\alpha_1, \ldots, \alpha_{d^2} \in k$. Grouping the coefficients corresponding to the variables $x_1, \ldots, x_d$, one gets a $d \times d$ linear system $\mathcal{L}$ with coefficients in k such that $\{x_1, \ldots, x_d\}$ is a nonzero solution. But then every row of the system gives a k-linear relation of these variables. Clearly this is only possible if all the coefficients of this system vanish. Write this condition as a new square linear system, this time around of order d^2 with solution $\{\alpha_1, \ldots, \alpha_{d^2}\}$ and appropriate coefficients in k. Since the latter coefficients are nothing but the coefficients of all linear forms $\ell_{i,j}$, they can be expressed as partial derivatives of these forms, so the corresponding $d^2 \times d^2$ matrix has the following shape (up to signs):

$$\Theta = \begin{pmatrix} \Theta_{d-1} & \Theta_{d-2} & \Theta_{d-3} & \dots & \Theta_2 & \Theta_1 & \mathbf{0} \\ \Theta_d & \mathbf{0} & \mathbf{0} & \dots & \mathbf{0} & \mathbf{0} & \Theta_{d-2} \\ \mathbf{0} & \Theta_d & \mathbf{0} & \dots & \mathbf{0} & \mathbf{0} & \Theta_{d-1} \\ \mathbf{0} & \mathbf{0} & \Theta_d & \dots & \mathbf{0} & \mathbf{0} & \mathbf{0} \\ \vdots & \vdots & \vdots & \dots & \vdots & \vdots & \vdots \\ \mathbf{0} & \mathbf{0} & \mathbf{0} & \dots & \Theta_d & \mathbf{0} & \mathbf{0} \\ \mathbf{0} & \mathbf{0} & \mathbf{0} & \dots & \mathbf{0} & \Theta_d & \mathbf{0} \end{pmatrix},$$

where Θ_i is the transpose of the Jacobian matrix of $\{\ell_{i,1}, \dots, \ell_{i,d}\}$ and $\mathbf{0}$ denotes the null matrix of order d. Note that $\det \Theta_i \neq 0$ since $\{\ell_{i,1}, \dots, \ell_{i,d}\}$ is a k-linearly independent set (Proposition 3.2 (ii)).

Now, the system $\mathfrak{L}$ has only the trivial solution if and only if $\det \Theta \neq 0$. One can see that, after appropriate elementary row operations, the above determinant is non-vanishing if and only if the following matrix has nonzero determinant:

$$\begin{pmatrix} I & 0 & 0 & \dots & 0 & 0 & \Theta_{d-2} \\ 0 & I & 0 & \dots & 0 & 0 & \Theta_{d-1} \\ 0 & 0 & I & \dots & 0 & 0 & 0 \\ \vdots & \vdots & \vdots & \dots & \vdots & \vdots & \vdots \\ 0 & 0 & 0 & \dots & I & 0 & 0 \\ 0 & 0 & 0 & \dots & 0 & I & 0 \\ 0 & 0 & 0 & \dots & 0 & 0 & \Omega \end{pmatrix},$$

where I stands for the $d \times d$ identity matrix and $\Omega = \Theta_{d-1}\Theta_{d-2} - \Theta_{d-2}\Theta_{d-1}$. Thus, $\det \Theta \neq 0$ if and only if $\det \Omega \neq 0$. Now, the entries of the matrices Θ_{d-1} and Θ_{d-2} are among the coefficients of the entries of the matrix $\mathbf{L} = (\ell_{ij})$; hence these are sufficiently general matrices over k.

We argue that $\det \Omega \neq 0$.

In order to prove that such a determinant is nonzero for general matrices, since this is an open condition in the parameter space of the coefficients, it suffices to prove it for particular matrices. Consider the following matrices:

$$A_{0,2} = \begin{pmatrix} 0 & 1 \\ 1 & 0 \end{pmatrix} \quad \text{and} \quad B_{0,2} = \begin{pmatrix} a & 0 \\ 0 & b \end{pmatrix}$$

$$A_{0,3} = \begin{pmatrix} 0 & 1 & 0 \\ 1 & 0 & 1 \\ 0 & 0 & 1 \end{pmatrix} \quad \text{and} \quad B_{0,3} = \begin{pmatrix} a & 0 & 0 \\ 0 & b & 0 \\ 0 & 1 & 1 \end{pmatrix},$$

for unspecified $a, b \in k$. Note that

$$\det(A_{0,2}B_{0,2} - B_{0,2}A_{0,2}) = (a-b)^2 \quad \text{and} \quad \det(A_{0,3}B_{0,3} - B_{0,3}A_{0,3}) = -a^2 + ab + a - b.$$

Since k is infinite, we can choose $a, b \in k$ such that both $D_2 := \det(A_{0,2}B_{0,2} - B_{0,2}A_{0,2})$ and $D_3 := \det(A_{0,3}B_{0,3} - B_{0,3}A_{0,3})$ are nonzero. Now, if d is even, say, $d = 2s$, consider the following $d \times d$ matrices:

- A is the block diagonal matrix with s blocks equal to $A_{0,2}$.
- B is the block diagonal matrix with s blocks equal to $B_{0,2}$.

If d is odd, say, $d = 2s + 1$, set

- A is the block diagonal matrix such that the first block is $A_{0,3}$ and the other $s - 1$ blocks are equal to $A_{0,2}$.
- B is the block diagonal matrix such that the first block is $B_{0,3}$ and the other $s - 1$ blocks are equal to $B_{0,2}$.

Thus,

$$\det(AB - BA) = \begin{cases} D_2^s \neq 0, & \text{if } d = 2s \\ D_3 D_2^{s-1} \neq 0, & \text{if } d = 2s + 1. \end{cases}$$

This takes care of the claim.

Now, to conclude, we have shown that the image of the map η in (9.2) is the R-submodule $(\mathbf{x})R^d$. Therefore, $M(-(d+1)(d-1)) \simeq R^d/(\mathbf{x})R^d$ as required. $\square$

Example 9.14 The above discussion has many common points with [26, Section 8.2], which treats the case of linearly presented perfect ideals in dimension $d = 3$. However, the above proof draws on the hypothesis that $\mathbf{L}$ is a matrix of general linear forms – and, in fact, it may be false for arbitrary linearly presented ideals. We are indebted to A. Tchernev for having provided us with the following counter-example to Theorem 9.13 in the context of arbitrary linear sections:

$$\varphi = \begin{pmatrix} x_1 & x_2 & x_3 \\ x_2 & x_3 & 0 \\ x_3 & 0 & x_1 \\ 0 & x_1 & x_2 \end{pmatrix}. \tag{9.4}$$

Here the vector space dimension of the linear forms in $\mathrm{Im}(\eta)$ is 8, where η denotes the correspondingly defined matrix as in the theorem. We note that by changing six out of the nine nonzero entries of the Tchernev matrix into general 1-forms, the resulting matrix gives the maximal value 9 for the vector space dimension of the linear forms in $\mathrm{Im}(\eta)$.

Proposition 9.15 *Let* $\mathbf{L}$ *denote a* $(d + 1) \times d$ *matrix of general linear forms over* $R = k[x_1, \ldots, x_d]$, *with* $d \geq 3$. *Set* $I = I_d(\mathbf{L}) \subset R$. *Then*

$$I^{(d-1)}/I^{d-1} \simeq k^d(-(d(d-1)-1)),$$

as graded R-*modules.*

Proof By Theorem 9.13, one has a (shifted) homogeneous isomorphism

$$M(-d) \simeq k^d(d(d-1)-1).$$

On the other hand, by definition there is a homogeneous isomorphism

$$M \simeq \mathrm{Ext}_R^d(R/I^{d-1}, R).$$

Therefore, it obtains

$$
\begin{aligned}
I^{(d-1)}/I^{d-1} &\simeq H_{(\mathbf{x})}^0(R/I^{d-1}) \quad \text{(since } I^{(d-1)}/I^{d-1} \text{ has finite length)} \\
&\simeq \mathrm{Hom}_R(\mathrm{Ext}_R^d(R/I^{d-1}, R(-d)), E(k)) \quad \text{(by graded local duality)} \\
&\simeq \mathrm{Hom}_R(M(-d), E(k)) \simeq \mathrm{Hom}_R(k^d(d(d-1)-1), E(k)) \\
&\simeq \mathrm{Hom}_R(k, E(k))^{\oplus d}(-(d(d-1)-1)) \simeq k^d(-(d(d-1)-1)),
\end{aligned}
$$

where the last isomorphism is given in [2, Lemma 3.2.7 (b)].   □

Example 9.16 Proposition 9.15 fails for arbitrary perfect ideals of codimension 2 admitting linear presentation. For $d = 3$, Example 9.14 is a counter-example. Cooking up some of these examples requires some extra care to make sure that I is a radical ideal; otherwise the whole known repository of symbolic power theory crumbles down. Therefore, slightly perturbing a linear $(d+1) \times d$ matrix whose d-minors generate a radical ideal may lead us astray.

Further details of this example will be worked out in the exercises.

9.1.4   A Notable Subsumed Cremona Map: The Inversion Factors

In the case where $m = d$, the Cremona map defined by the maximal minors of $\mathbf{L}$ played a prominent role. In the case where $m = d + 1$, we bring out yet another Cremona map, this time around defined by a set of source inversion factors of the implicitization birational map $\mathbb{P}^{d-1} \dashrightarrow \mathbb{P}^d$.

We start afresh by letting $\mathbf{L}$ stand for an $m \times (m-1)$ matrix of general linear forms over a polynomial ring $k[\mathbf{x}] = k[x_1, \ldots, x_d]$, where $m = d + 1$. Let us bring up the matrix B of (2.10) which, in this case, is a $d \times d$ matrix of linear forms in $k[\mathbf{y}] = k[y_1, \ldots, y_d, y_{d+1}]$. Note that, similarly, there is a relation for the transpose B^t such as

$$\mathbf{y}\,\mathbf{L}' = \mathbf{x}\,B^t, \tag{9.5}$$

for a suitable $m \times (m-1)$ matrix $\mathbf{L}'$. Since its entries are obtained from $\mathbf{L}$ by the rearrangement of the coefficients of the entries of the latter, $\mathbf{L}'$ too is a matrix whose entries are general linear forms in $\mathbf{x}$. Thus, by Theorem 9.6 (a), its maximal minors define a birational map onto the image.

Let $\boldsymbol{\Delta} = \{\Delta, \ldots, \Delta_{d+1}\}$ (respectively, $\boldsymbol{\Delta}' = \{\Delta'_1, \ldots, \Delta'_{d+1}\}$) denote the set of maximal minors of $\mathbf{L}$ (respectively, $\mathbf{L}'$). At the other end, let $\{D_1(\mathbf{x}), \ldots, D_d(\mathbf{x})\} \subset R$ (respectively, $\{D'_1(\mathbf{x}), \ldots, D'_d(\mathbf{x})\} \subset R$) stand for the corresponding set of source inversion factors associated to a set of minimal representatives of the inverse map to the map defined by $\boldsymbol{\Delta}$ (respectively, $\boldsymbol{\Delta}'$), obtained in accordance to Theorem 9.6.

For reasons that will become clear along the proof of the next theorem, we change the habitat of $\mathbf{L}'$ and its related constructs to a duplicate $R' := k[\mathbf{x}'] = k[x'_1, \ldots, x'_d]$ of $k[\mathbf{x}]$. Thus, Δ'_i and D'_i are forms in R'.

As usual, the maximal minors are taken with the correct signs. If no confusion arises, we consider $\mathbf{x}$ both as the set of source variables and as the vector of these variables.

The following is the main result of this subsection.

Theorem 9.17 ($\mathrm{char}(k) = 0$) *With the above notation, set* $I := (\Delta_1, \ldots, \Delta_{d+1}) \subset R = k[\mathbf{x}] = k[x_1, \ldots, x_d]$. *Then:*

(i) *The set* $\{D_1(\mathbf{x}), \ldots, D_d(\mathbf{x})\} \subset R$ *(respectively,* $\{D'_1(\mathbf{x}'), \ldots, D'_d(\mathbf{x}')\} \subset k[\mathbf{x}']$*) generates an ideal of codimension 2.*

(ii) *The map* $\mathfrak{D} = (D_1(\mathbf{x}) : \cdots : D_d(\mathbf{x})) : \mathbb{P}^{d-1} \dashrightarrow \mathbb{P}^{d-1}$ *is a Cremona map with inverse map* $(D'_1(\mathbf{x}') : \cdots : D'_d(\mathbf{x}'))$.

(iii) *The R-module* $I^{(d-1)}/I^{d-1}$ *is minimally generated by the residue classes of* $D_1(\mathbf{x}), \ldots, D_d(\mathbf{x})$.

(iv) *There exists an element* $E \in I^{(d(d-1)-1)}$ *such that the source inversion factor of* $\mathfrak{D}$ *coincides with* $E^{(d-1)}$.

   Moreover, E coincides with the Jacobian determinant of $D_1(\mathbf{x}), \ldots, D_d(\mathbf{x})$ *whenever the latter is irreducible.*

(v) *The minimal graded free resolution of the ideal* $(D_1(\mathbf{x}), \ldots, D_d(\mathbf{x})) \subset R$ *is*

$$0 \to R(-d^2) \xrightarrow{\mathbf{x}^t} R(-(d^2-1))^d \xrightarrow{\Psi} R(-(d(d-1)-1))^d \to R, \qquad (9.6)$$

where Ψ denotes the Jacobian dual matrix B^t of the signed d-minors of $\mathbf{L}$ evaluated orderly at these signed minors, while $\mathbf{x}^t$ stands for the transpose of the vector $\mathbf{x}$.

Proof

(i) We only discuss the ideal $(D_1(\mathbf{x}), \ldots, D_d(\mathbf{x}))$ since the line of argument is analogous for $(D'_1(\mathbf{x}'), \ldots, D'_d(\mathbf{x}'))$.

   One has $(D_1(\mathbf{x}), \ldots, D_d(\mathbf{x})) \subset I$ by Corollary 9.8; hence the codimension of $(D_1(\mathbf{x}), \ldots, D_d(\mathbf{x}))$ is at most 2. Thus, it suffices to show that it is at least 2. For this, set $\boldsymbol{\Delta} = \{\Delta, \ldots, \Delta_{d+1}\}$ as above and $\mathfrak{B} := \det B$, where B is as in (2.10).

CLAIM 1. $\left(\frac{\partial \mathfrak{B}}{\partial y_1}(\mathbf{\Delta}), \ldots, \frac{\partial \mathfrak{B}}{\partial y_{d+1}}(\mathbf{\Delta})\right) \subset (D_1(\mathbf{x}), \ldots, D_d(\mathbf{x}))$.

For this, write $\mathbf{L} = (\ell_{i,j})$. Then

$$B = \begin{pmatrix} \sum_{r=1}^{d+1} \frac{\partial \ell_{r,1}}{\partial x_1} y_r & \sum_{r=1}^{d+1} \frac{\partial \ell_{r,2}}{\partial x_1} y_r & \cdots & \sum_{r=1}^{d+1} \frac{\partial \ell_{r,d}}{\partial x_1} y_r \\ \sum_{r=1}^{d+1} \frac{\partial \ell_{r,1}}{\partial x_2} y_r & \sum_{r=1}^{d+1} \frac{\partial \ell_{r,2}}{\partial x_2} y_r & \cdots & \sum_{r=1}^{d+1} \frac{\partial \ell_{r,d}}{\partial x_2} y_r \\ \vdots & \vdots & \ddots & \vdots \\ \sum_{r=1}^{d+1} \frac{\partial \ell_{r,1}}{\partial x_d} y_r & \sum_{r=1}^{d+1} \frac{\partial \ell_{r,2}}{\partial x_d} y_r & \cdots & \sum_{r=1}^{d+1} \frac{\partial \ell_{r,d}}{\partial x_d} y_r \end{pmatrix}.$$

Expanding $\mathfrak{B} = \det B$ yields

$$\mathfrak{B} = \sum_{\sigma} \operatorname{sgn}(\sigma) \left(\sum_{r=1}^{d+1} \frac{\partial \ell_{r,\sigma(1)}}{\partial x_1} y_r\right) \cdots \left(\sum_{r=1}^{d+1} \frac{\partial \ell_{r,\sigma(d)}}{\partial x_d} y_r\right),$$

while its derivative, for $1 \leq n \leq d+1$, is

$$\frac{\partial \mathfrak{B}}{\partial y_n} = \sum_{\sigma} \operatorname{sgn}(\sigma) \left[\sum_{1 \leq s \leq d} \left(\sum_{r=1}^{d+1} \frac{\partial \ell_{r,\sigma(1)}}{\partial x_1} y_r\right) \cdots \left(\frac{\partial \ell_{n,\sigma(s)}}{\partial x_s}\right) \cdots \left(\sum_{r=1}^{d+1} \frac{\partial \ell_{r,\sigma(d)}}{\partial x_d} y_r\right)\right]$$

$$= \sum_{1 \leq s \leq d} \left[\sum_{\sigma} \operatorname{sgn}(\sigma) \left(\sum_{r=1}^{d+1} \frac{\partial \ell_{r,\sigma(1)}}{\partial x_1} y_r\right) \cdots \left(\frac{\partial \ell_{n,\sigma(s)}}{\partial x_s}\right) \cdots \left(\sum_{r=1}^{d+1} \frac{\partial \ell_{r,\sigma(d)}}{\partial x_d} y_r\right)\right].$$

Note that for any given $1 \leq s \leq d$, the expression inside the square brackets in the last line above is the determinant of the matrix

$$B_s := \begin{pmatrix} \sum_{r=1}^{d+1} \frac{\partial \ell_{r,1}}{\partial x_1} y_r & \sum_{r=1}^{d+1} \frac{\partial \ell_{r,2}}{\partial x_1} y_r & \cdots & \sum_{r=1}^{d+1} \frac{\partial \ell_{r,d}}{\partial x_1} y_r \\ \vdots & \vdots & \ddots & \vdots \\ \frac{\partial \ell_{n,1}}{\partial x_s} & \frac{\partial \ell_{n,2}}{\partial x_s} & \cdots & \frac{\partial \ell_{n,d}}{\partial x_s} \\ \vdots & \vdots & \ddots & \vdots \\ \sum_{r=1}^{d+1} \frac{\partial \ell_{r,1}}{\partial x_d} y_r & \sum_{r=1}^{d+1} \frac{\partial \ell_{r,2}}{\partial x_d} y_r & \cdots & \sum_{r=1}^{d+1} \frac{\partial \ell_{r,d}}{\partial x_d} y_r \end{pmatrix}.$$

Computing this determinant once again, this time around by a Laplace expansion along the sth row of B_s gives $\det B_s = \sum_{t=1}^{d} \frac{\partial \ell_{n,t}}{\partial x_s} \sigma_t^{[s]}$, where $\sigma_t^{[s]}$ denotes the $(d-1)$-minor of B_s omitting the sth row and the tth column. Coming from the other end, for given s, $(\sigma_1^{[s]} : \cdots : \sigma_d^{[s]})$ is a representative of the inverse map to the map defined by $\mathbf{\Delta}$ [19, Theorem 3.2.22, Supplement].

By the definition of $D_s(\mathbf{x})$, we can assume that it is the source inversion factor corresponding to this representative; hence $\sigma_t^{[s]}(\boldsymbol{\Delta}) = x_t D_s(\mathbf{x})$, for $1 \leq s, t \leq d$. Assembling the information, we get

$$\frac{\partial \mathfrak{B}}{\partial y_n}(\boldsymbol{\Delta}) = \det B_1(\boldsymbol{\Delta}) + \cdots + \det B_d(\boldsymbol{\Delta})$$

$$= \sum_{t=1}^{d} \frac{\partial \ell_{n,t}}{\partial x_1} \sigma_t^{[1]} \boldsymbol{\Delta} + \cdots + \sum_{t=1}^{d} \frac{\partial \ell_{n,t}}{\partial x_d} \sigma_t^{[d]} \boldsymbol{\Delta}$$

$$= \left(\sum_{t=1}^{d} \frac{\partial \ell_{n,t}}{\partial x_1} x_t \right) D_1(\mathbf{x}) + \cdots + \left(\sum_{t=1}^{d} \frac{\partial \ell_{n,t}}{\partial x_d} x_t \right) D_d(\mathbf{x}),$$

which proves the claim.

We next prove the following claim:

CLAIM 2. The ideal $\left(\frac{\partial \mathfrak{B}}{\partial y_1}(\boldsymbol{\Delta}), \ldots, \frac{\partial \mathfrak{B}}{\partial y_{d+1}}(\boldsymbol{\Delta}) \right)$ has codimension 2.

We proceed in two steps. First we argue that the ideal $I_d(\Theta)$ of maximal minors of Θ has codimension 2. Indeed, since $\boldsymbol{\Delta}$ are minors of a matrix of general linear forms, they are sufficiently general d-forms, and so are any of their derivatives (the entries of Θ). Therefore, the assertion follows from Theorem 1.15.

Next, we show that $\left(\frac{\partial \mathfrak{B}}{\partial y_1}(\boldsymbol{\Delta}), \ldots, \frac{\partial \mathfrak{B}}{\partial y_{d+1}}(\boldsymbol{\Delta}) \right) = I_d(\Theta)$.

In fact, one has a presentation $k[\boldsymbol{\Delta}] \simeq k[\mathbf{y}]/(\mathfrak{B})$, where $\mathfrak{B} := \det B$. Since $\mathfrak{B}(\boldsymbol{\Delta}) = 0$, the chain rule of derivatives gives the short complex

$$R \xrightarrow{\partial} R^{d+1} \xrightarrow{\Theta} R^d, \tag{9.7}$$

where ∂ is the transpose of

$$\left[\frac{\partial \mathfrak{B}}{\partial y_1}(\boldsymbol{\Delta}) \, \ldots \, \frac{\partial \mathfrak{B}}{\partial y_{d+1}}(\boldsymbol{\Delta}) \right].$$

Now, in particular, the rank of Θ is d – this fact alone does not require general forms, being a consequence of having $\dim k[\boldsymbol{\Delta}] = d$ and $\mathrm{char}(k) = 0$ (via Proposition 3.2 (ii)). Therefore, $\ker(\Theta)$ is generated by the single (column) vector whose jth coordinate is the d-minor of Θ omitting the jth column of Θ further divided by the gcd of all the d-minors. But having a proper common divisor is a closed condition in the parameter space of the coefficients. Thus, $\ker(\Theta)$ is generated in degree $(d-1)d$ (the degree of a d-minor of Θ). On the other hand, a simple calculation shows that the coordinates of ∂ are also of degree $d(d-1)$. Since $\mathrm{Im}(\partial) \subset \ker(\Theta)$ is easily seen to be generated in degree $d(d-1)$ as well, the required equality follows.

(ii) Let $\boldsymbol{\Delta}' = \{\Delta'_1, \ldots, \Delta'_{d+1}\} \subset R'$ stand for the d-minors of the matrix $\mathbf{L}'$ as explained above. Since the entries of $\mathbf{L}'$ are general linear forms, just like $\mathbf{L}$, its d-minors define a birational map onto the image, with B as its Jacobian dual matrix. Thus, for any $j \in \{1, \ldots, d\}$, the (i, j)-cofactors $\{B_{j,1}, \ldots, B_{j,d}\}$ of B, taken modulo $\det B = \mathfrak{B}$, define an inverse to the map defined by $\mathbf{\Delta}'$. By [19, Lemma 4.2.1] this yields the following structural congruencies:

$$\Delta_i'(B_{j,1}, \ldots, B_{j,d}) \equiv E_j y_i \mod (\det B), \tag{9.8}$$

where E_j denotes the corresponding target inversion factor.

Claim. $I_1(\mathbf{x}\, B(\mathbf{\Delta}'))$ is contained in the presentation ideal of the Rees algebra of the ideal $(D_1, \ldots, D_d) \subset k[\mathbf{x}]$, defined over the ring $k[\mathbf{x}, \mathbf{x}']$.

To see this it suffices to prove that the entries of the matrix $\mathbf{x}\, B$ vanish by evaluating $y_n \mapsto \Delta_n'(D_1(\mathbf{x}), \ldots, D_d(\mathbf{x}))$, $n = 1, \ldots, d + 1$, or, equivalently, by evaluating $y_n \mapsto \Delta_n'(x_d D_1(\mathbf{x}), \ldots, x_d D_d(\mathbf{x}))$, $1 \leq n \leq d + 1$. Going back to the signed d-minors $\mathbf{\Delta} = \{\Delta_1, \ldots, \Delta_{d+1}\}$ of $\mathbf{L}$, one has the relations

$$x_d D_i = B_{i,d}^t(\Delta_1, \ldots, \Delta_{d+1}) = B_{d,i}(\Delta_1, \ldots, \Delta_{d+1}), \tag{9.9}$$

since D_i is an inversion factor for $\mathbf{\Delta}$, where $B_{i,j}$ is the cofactor of B^t corresponding to the entry indexed by (i, j) and $B_{i,d}^t(\Delta_1, \ldots, \Delta_{d+1})$ is the result of evaluating this cofactor at $\mathbf{\Delta}$. Since $B_{i,j} = B_{j,i}^t$, setting $\mathbf{D} = \{D_1(\mathbf{x}), \ldots, D_d(\mathbf{x})\}$ yields

$$\mathbf{x}\, B(\mathbf{\Delta}(x_d \mathbf{D})) = \mathbf{x} \begin{pmatrix} \sum_{i=1}^{d+1} \frac{\partial \ell_{i,1}}{\partial x_1} \Delta_i'(x_d \mathbf{D}) & \cdots & \sum_{i=1}^{d+1} \frac{\partial \ell_{i,d}}{\partial x_1} \Delta_i'(x_d \mathbf{D}) \\ \vdots & \ddots & \vdots \\ \sum_{i=1}^{d+1} \frac{\partial \ell_{i,1}}{\partial x_3} \Delta_i'(x_d \mathbf{D}) & \cdots & \sum_{i=1}^{d+1} \frac{\partial \ell_{i,d}}{\partial x_d} \Delta_i'(x_d \mathbf{D}) \end{pmatrix}$$

$$= \left(\sum_{i=1}^{d+1} \ell_{i,1} \Delta_i'(x_d \mathbf{D}), \ldots, \sum_{i=1}^{d+1} \ell_{i,d} \Delta_i'(x_d \mathbf{D}) \right)$$

$$= E_d(\mathbf{\Delta}) \left(\sum \ell_{i,1} \Delta_i, \ldots, \sum \ell_{i,d} \Delta_i \right) = (0, \ldots, 0),$$

where the third equality from top follows from (9.8) and (9.9) and the last equality uses that $\mathbf{L}$ is a syzygy matrix of $\mathbf{\Delta}$. This proves the claim.

As a consequence, the matrix $B^t(\mathbf{\Delta}')$ is a submatrix of the full Jacobian dual matrix of $\mathbf{D}$. On the other hand, we have $\det(B^t(\Delta')) = (\det B^t)(\Delta') = 0$ since $\det B$ is a polynomial relation of $\mathbf{\Delta}'$. Therefore, $B^t(\mathbf{\Delta}')$ has rank $\leq d - 1$. But since $\mathbf{\Delta}'$ defines a birational map, not all $(d - 1)$-minors vanish modulo $\det B$. Thus, $B^t(\mathbf{\Delta}')$ has rank $d - 1$. For even more reason, the rank of the Jacobian dual matrix of $\mathbf{D}$ is $\geq d - 1$ (hence $= d - 1$, its maximal possible value). Using again the criterion of [19, Theorem 3.2.22], we derive that $\mathbf{D}$ defines a Cremona map.

Now, we prove the additional statement of this item. Let s denote the minimal number of generators of the defining ideal of the Rees algebra of $\mathbf{D}$ of bidegree $(1, *)$, with $*$ representing any integer value ≥ 1. Then the full Jacobian dual matrix of $\mathbf{D}$ is an $s \times d$ matrix over $k[\mathbf{y}]$ of rank $d - 1$, which, as we have just shown, contains the $d \times d$ submatrix $B^t(\mathbf{\Delta}')$. By [19, Theorem 3.2.22, Supplement] we know that the inverse map to the Cremona map defined by $\mathbf{D}$ is defined by the $(d-1)$-minors of any $(d-1) \times d$ submatrix of rank $d - 1$ of the Jacobian dual matrix of $\mathbf{D}$, further divided by their gcd. Since $B^t(\mathbf{\Delta}')$ has rank $d-1$, one can take, say, the submatrix of $B^t(\mathbf{\Delta}')$ formed with the first $d-1$ rows of $B^t(\mathbf{\Delta}')$. Write $\partial_i(\mathbf{x}')$ for the $(d-1)$th minor omitting the ith column. Then we get $\partial_i(\mathbf{x}') = B^t_{d,i}(\mathbf{\Delta}') = B_{d,i}(\mathbf{\Delta}') = x_d D'_i(\mathbf{x}')$, where $D'_i(\mathbf{x}')$ as before denotes the corresponding source inversion factor of the birational map defined by $\mathbf{\Delta}'$. It follows that $(D'_1(\mathbf{x}') : \cdots : D'_n(\mathbf{x}'))$ defines the inverse map to $\mathfrak{D}$.

(iii) By Theorem 9.6, one has $(I^{d-1}, D_1(\mathbf{x}), \ldots, D_d(\mathbf{x})) \subset I^{(d-1)}$. On the other hand, by Proposition 9.15, $I^{(d-1)}/I^{d-1}$ is minimally generated by d elements of degree $d(d - 1) - 1$. To conclude that the residues of $D_1(\mathbf{x}), \ldots, D_d(\mathbf{x})$ on $I^{(d-1)}/I^{d-1}$ form a set of minimal generators of the latter, it suffices to show that they are k-linearly independent. By part (i) they are even k-algebraically independent.

(iv) By (i) and (ii), $\{D'_1 = D'_1(\mathbf{x}'), \ldots, D'_d = D'_d(\mathbf{x}')\}$ generate an ideal of codimension 2 defining the inverse map to $\mathfrak{D}$. Write

$$h_i = h_i(z_1, \ldots, z_d) := z_i D'_i \ (= B_{i,i}(\mathbf{\Delta}')), \ \text{for } 1 \leq i \leq d,$$

and evaluate h_i at $x_i \mathbf{D} = (x_i D_1(\mathbf{x}), \ldots, x_i D_d(\mathbf{x}))$ via $z_j \mapsto x_i D_j(\mathbf{x})$:

$$h_i(x_i \mathbf{D}) = x_i D_i D'_i(x_i \mathbf{D})$$
$$= x_i^{d(d-1)} D_i D'_i(\mathbf{D})$$
$$= x_i^{d(d-1)+1} D_i G,$$

where $G := x_i^{-1} D'_i(D_1, \ldots, D_d)$ is the source inversion factor of the Cremona map defined by $\mathbf{D}$.

On the other hand, one has

$$h_i(x_i \mathbf{D}) = B_{i,i}(\Delta'_1(x_i \mathbf{D}), \ldots, \Delta'_d(x_i \mathbf{D}))$$
$$= B_{i,i}(E_i(\mathbf{\Delta})\Delta_1, \ldots, E_i(\mathbf{\Delta})\Delta_{d+1})$$
$$= E_i(\mathbf{\Delta})^{d-1} B_{i,i}(\Delta_1, \ldots, \Delta_{d+1})$$
$$= E_i(\mathbf{\Delta})^{d-1} x_i D_i,$$

where $\{E_i,\ i = 1, \ldots, d\}$ is a set of independent target inversion factors of the birational map defined by $\boldsymbol{\Delta}'$, as in (9.8). This implies the relation

$$x_i^{d(d-1)} G = E_i(\boldsymbol{\Delta})^{d-1}. \tag{9.10}$$

Extracting $(d-1)$th roots yields

$$x_i^d G^{1/d-1} = E_i(\boldsymbol{\Delta}). \tag{9.11}$$

Since E_i has degree $d(d-1)-1$, then $(x_1^d, \ldots, x_d^d) G^{1/d-1} \subset I^{d(d-1)-1}$, from which follows $E := G^{1/d-1} \in I^{(d(d-1)-1)}$.

The supplementary statement follows from Theorem 2.29 by assuming the irreducibility of $\det \Theta(\mathbf{D})$. This is because as it divides a power of E then it will divide E itself, and since $\deg(\det \Theta(\mathbf{D})) = \deg(E)$, they coincide up to a nonzero scalar.

(v) Set $D_i = D_i(\mathbf{x})$ for short. We first check that (9.6) is indeed a complex. For this, using that $\{D_1, \ldots, D_d\}$ is a set of inversion factors of the birational map defined by $\boldsymbol{\Delta}$, the adjugate matrix of $\Psi = B^t(\boldsymbol{\Delta})$ is

$$\mathrm{adj}(\Psi) = \begin{pmatrix} x_1 D_1 & x_1 D_2 & \ldots & x_1 D_d \\ x_2 D_1 & x_2 D_2 & \ldots & x_2 D_d \\ \vdots & \vdots & \ldots & \vdots \\ x_d D_1 & x_d D_2 & \ldots & x_d D_d \end{pmatrix}. \tag{9.12}$$

Since Ψ has rank $d-1$, the adjugate relation gives

$$\mathrm{adj}(\Psi)\, \Psi = \mathbf{0} \tag{9.13}$$

and

$$\Psi\, \mathrm{adj}(\Psi) = \mathbf{0}. \tag{9.14}$$

From (9.12), (9.13) implies that Ψ is a matrix of syzygies of $\mathbf{D}$, while (9.14) gives that $\mathbf{X}^t$ is a second syzygy thereof. This shows that one has indeed a complex. To finish we check the required Fitting codimension by the Buchsbaum–Eisenbud acyclicity criterion. The verification at the tail of the complex is immediate, while at the middle the codimension of

$$I_{d-1}(\Psi) = I_1(\mathrm{adj}(\Psi)) = (\mathbf{x})(D_1, \ldots, D_d)$$

is 2 because (i) showed that the ideal $(D_1, \ldots, D_d)$ has codimension 2. $\qquad \square$

Remark 9.18

(i) Lacking the hypothesis of general linear entries, most of Theorem 9.17 crumbles down. The complex (9.7) may fail to be exact, and the source inversion factors may have a proper common divisor (see Example 9.16).

(ii) The authors believe that there must be an elegant proof of Theorem 9.17 (i), e.g., by finding a suitable parametric open set for Θ and showing it is nonempty by exhibiting one single member.

9.1.4.1   The Structure of the Symbolic Rees Algebra

With the notation of the previous part, we file the degree numerology so far:

- $\deg(D_i') = \deg(D_i) = d(d-1) - 1$, for $1 \leq i \leq d$.
- $\deg(G) = (d(d-1) - 1)d(d-1) - d(d-1) = (d-1)d(d(d-1) - 2)$
  This follows from (9.10).
- $\deg(E) = \deg(G)/(d-1) = d(d(d-1) - 2)$.

The following result is a first approximation to the expected defining relations of the symbolic Rees algebra. We will use capital letters for several presentation variables to appear. Thus, to keep the previous presentation variables of the inverse map on the same foot, we change them to capital letters as well.

Proposition 9.19 *Let* $\mathbf{L}$ *denote a* $(d+1) \times d$ *general linear matrix over* $R = k[x_1, \ldots, x_d]$, *with* $d \geq 3$. *Set* $I := I_{d-1}(\mathbf{L}) \subset R$, *and let* $\mathcal{R}^{(I)}$ *denote its symbolic Rees algebra. Let* $D_1, \ldots, D_d \in I^{(d-1)}$ *and* $E \in I^{(d(d-1)-1)}$ *be as above. Let* $\mathbf{Y} = \{Y_1, \ldots, Y_{d+1}\}, \mathbf{Z} = \{Z_1, \ldots, Z_d\}, W$ *be sets of mutually independent indeterminates over* R. *Consider the surjective homomorphism of* R-*algebras:*

$$\pi : R[\mathbf{Y}, \mathbf{Z}, W] \twoheadrightarrow R[It, D_1 t^{d-1}, \ldots, D_d t^{d-1}, E t^{d(d-1)-1}]$$

such that $Y_j \mapsto \Delta_j t$, $Z_r \mapsto D_r t^{d-1}$ *and* $W \mapsto E t^{d(d-1)-1}$. *Then, in the notation of Theorem* 9.17, *one has:*

(i) $\ker(\pi)$ *contains the following polynomials:*

(1) *The entries of the matrix* $\mathbf{x} B = \mathbf{Y} \mathbf{L}$ (d *such polynomials*).

(2) *The entries of the matrix* $\mathbf{Z} B^t$ (d *such polynomials*).

(3) *The entries of the matrix* $\mathbf{x}^t \mathbf{Z} - \mathrm{adj}(B^t)$ (d^2 *such polynomials*).

(4) *The* d *polynomials* $\{x_1 W^{d-1} - D_1'(\mathbf{Z}), \ldots, x_d W^{d-1} - D_d'(\mathbf{Z})\}$, *where* $D_1', \ldots, D_d'$ *are forms defining the inverse of* $D_1, \ldots, D_d$.

(5) *The* $d+1$ *polynomials* $\{Y_1 W - \Delta_1'(\mathbf{Z}), \ldots, Y_{d+1} W - \Delta_{d+1}'(\mathbf{Z})\}$, *where* Δ_j' *is as in Theorem* 9.17.

(ii) *The* d *polynomials of item* (4) *can be replaced by* d *polynomials of the form* $x_i W - Q_i(\mathbf{Y}, \mathbf{Z})$, *where* $Q_i(\mathbf{Y}, \mathbf{Z})$ *is a polynomial in* $k[\mathbf{Y}, \mathbf{Z}]$ *of the shape*

$$Q_i(\mathbf{Y}, \mathbf{Z}) = \sum_{\substack{\{j_1, \ldots, j_{d-2}\} \subset \{1, \ldots, d+1\} \\ t_1 + \ldots + t_{d-2} = d-2}} Y_{j_1}^{t_1} Y_{j_2}^{t_2} \cdots Y_{j_{d-2}}^{t_{d-2}} P_{t_1, \ldots, t_{d-2}}(\mathbf{Z}).$$

In particular, $(\mathbf{x})E \subset I^{d-2}\,(I^{(d-1)})^{d-1}.$

Proof (ii) The first four blocks were discussed before, namely:

(1) These are equations defining the Rees algebra $R[It]$ of I on the polynomial ring $k[\mathbf{x}, \mathbf{Y}]$. Since $R[It]$ is a subalgebra of $R[It, D_1 t^{d-1}, \ldots, D_d t^{d-1}, E t^{d(d-1)-1}]$, then the equations obviously vanish under π.
(2) Note that the matrix B evaluated by $Y_j \mapsto \Delta_j t$ is a syzygy matrix of $\{D_1, \ldots, D_d\}$ by Theorem 9.17 (iv). Since Z_i maps to $D_i t^{d-1}$, the vanishing of $I_1(\mathbf{Z}B^t)$ is clear as well by the same token.
(3) The argument relies on the details of Theorem 9.17 (v), especially (9.12)
(4) These equations under π just express the fact that $G = E^{d-1}$ is a source inversion factor of the Cremona map defined by $\{D_1, \ldots, D_d\}$.

   Thus, we only have to deal with block (5).
(5) To discuss these equations, recall the relation obtained in (9.8):

$$\Delta_j'(B_{d,1}(\mathbf{\Delta}), \ldots, B_{d,d}(\mathbf{\Delta})) = E_d(\mathbf{\Delta})\Delta_j. \tag{9.15}$$

On the other hand, we have

$$\begin{aligned}
\Delta_j'(B_{d,1}(\mathbf{\Delta}), \ldots, B_{d,d}(\mathbf{\Delta})) &= \Delta_j'(B_{d,1}^t(\mathbf{\Delta}), \ldots, B_{d,d}^t(\mathbf{\Delta})) \\
&= \Delta_j'(x_d D_1, \ldots, x_d D_d) \\
&= x_d^d \Delta_j'(D_1, \ldots, D_d).
\end{aligned}$$

Therefore,

$$E_d(\mathbf{\Delta})\Delta_j = x_d^d \Delta_j'(\mathbf{D}).$$

Collecting the two resulting expressions yields

$$x_d^d \Delta_j E = x_d^d \Delta_j'(\mathbf{D}),$$

and hence

$$\Delta_j E = \Delta_j'(\mathbf{D}), \tag{9.16}$$

thus yielding the asserted relations.

(ii) Recall that $\Delta_1' = \Delta_1'(\mathbf{Z}), \ldots, \Delta_{d+1}' = \Delta_{d+1}'(\mathbf{Z})$ denote the d-minors of the matrix $\mathbf{L}'$ where $\mathbf{Z}$ replaces $(\mathbf{x})'$. We claim that for any collection of nonnegative integers $t_1, \ldots, t_s$, with $s \leq d + 1$, and for every subset $\{j_1, \ldots, j_s\} \subset \{1, \ldots, d + 1\}$, the polynomials

$$Y_{j_1}^{t_1} \cdots Y_{j_s}^{t_s} W^{t_1 + \ldots + t_s} - \Delta'_{j_1}(\mathbf{Z})^{t_1} \cdots \Delta'_{j_s}(\mathbf{Z})^{t_s} \in \ker(\pi)$$

belong to the ideal generated by the polynomials from block (5) in item (i).

We proceed by induction on s.

The result is clear for $s = 1$ because $Y_j W - \Delta'_j \in \ker(\pi)$ by item (i) and is a factor of $Y_j^t W^t - \Delta'^{\,t}_j$, for any t.

Thus, assume that $s > 1$ and that, without loss of generality, $t_1 \neq 0$ (the result is trivially satisfied if all t's are null). Write

$$(Y_{j_1}^{t_1} W^{t_1} - \Delta'^{t_1}_{j_1})Y_{j_2}^{t_2} \cdots Y_{j_s}^{t_s} W^{t_2 + \ldots + t_s} = Y_{j_1}^{t_1} Y_{j_2}^{t_2} \cdots Y_{j_s}^{t_s} W^{t_1 + \ldots + t_s} - \Delta'^{t_1}_{j_1} Y_{j_2}^{t_2} \cdots Y_{j_s}^{t_s} W^{t_2 + \ldots + t_s}$$

$$= Y_{j_1}^{t_1} Y_{j_2}^{t_2} \cdots Y_{j_s}^{t_s} W^{t_1 + \ldots + t_s} - \Delta'^{t_1}_{j_1} \cdots \Delta'^{t_s}_{j_s} + \Delta'^{t_1}_{j_1} \cdots \Delta'^{t_s}_{j_s} - \Delta'^{t_1}_{j_1} Y_{j_2}^{t_2} \cdots Y_{j_s}^{t_s} W^{t_2 + \ldots + t_s}$$

$$= \left(Y_{j_1}^{t_1} Y_{j_2}^{t_2} \cdots Y_{j_s}^{t_s} W^{t_1 + \ldots + t_s} - \Delta'^{t_1}_{j_1} \cdots \Delta'^{t_s}_{j_s}\right) - \Delta'^{t_1}_{j_1} \left(Y_{j_2}^{t_2} \cdots Y_{j_s}^{t_s} W^{t_2 + \ldots + t_s} - \Delta'^{t_2}_{j_2} \cdots \Delta'^{t_s}_{j_s}\right).$$

Applying the inductive hypothesis to the two ends of this strand of equalities shows that the polynomial

$$Y_{j_1}^{t_1} Y_{j_2}^{t_2} \cdots Y_{j_s}^{t_s} W^{t_1 + \ldots + t_s} - \Delta'_{j_1}(\mathbf{Z})^{t_1} \cdots \Delta'_{j_s}(\mathbf{Z})^{t_s}$$

also belongs to $\ker(\pi)$. In particular, taking $s = d - 2$ and $t_1, \ldots, t_{d-2}$ any partition of $d - 2$, the polynomial

$$Y_{j_1}^{t_1} Y_{j_2}^{t_2} \cdots Y_{j_{d-2}}^{t_{d-2}} W^{d-2} - \Delta'_{j_1}(\mathbf{Z})^{t_1} \cdots \Delta'_{j_{d-2}}(\mathbf{Z})^{t_{d-2}} \tag{9.17}$$

belongs to $\ker(\pi)$.

On the other hand, as seen in Theorem 9.17 (ii), the forms $\{D'_1 \ldots, D'_d\}$ defining the inverse of the Cremona map defined by $\{D_1, \ldots, D_d\}$ also constitute a complete set of source inversion factors of the birational map defined by the d-minors $\Delta'_1, \ldots, \Delta'_d$ of $\mathbf{L}'$. Therefore, Theorem 9.6 gives

$$(D'_1, \ldots, D'_d) \subset (\Delta'_1, \ldots, \Delta'_{d+1})^{(d-1)},$$

and for even more reason

$$(D'_1, \ldots, D'_d) \subset (\Delta'_1, \ldots, \Delta'_{d+1})^{(d-2)} = (\Delta'_1, \ldots, \Delta'_{d+1})^{d-2}. \tag{9.18}$$

Fixing $i \in \{1, \ldots, d\}$, we can write

$$D'_i(\mathbf{Z}) = \sum_{t_1 + \ldots + t_{d-2} = d-2} P_{t_1, \ldots, t_{d-2}}(\mathbf{Z}) \Delta'_{j_1}(\mathbf{Z})^{t_1} \cdots \Delta'_{j_{d-2}}(\mathbf{Z})^{t_{d-2}}. \tag{9.19}$$

Thus, one gets that the polynomial

$$x_i W^{d-1} - D_i'(\mathbf{Z}) - \sum_{t_1+\ldots+t_{d-2}=d-2} P_{t_1,\ldots,t_{d-2}}(\mathbf{Z}) \left(Y_{j_1}^{t_1} Y_{j_2}^{t_2} \cdots Y_{j_{d-2}}^{t_{d-2}} W^{d-2} - \Delta_{j_1}'(\mathbf{Z})^{t_1} \cdots \Delta_{j_{d-2}}'(\mathbf{Z})^{t_{d-2}} \right)$$

$$= W^{d-2} \left(x_i W - \sum_{t_1+\ldots+t_{d-2}=d-2} P_{t_1,\ldots,t_{d-2}}(\mathbf{Z}) Y_{j_1}^{t_1} Y_{j_2}^{t_2} \cdots Y_{j_{d-2}}^{t_{d-2}} \right),$$

for arbitrary subsets $\{j_1, \ldots, j_{d-2}\} \subset \{1, \ldots, d+1\}$, belongs to $\ker(\pi)$. Since $\ker(\pi)$ is a prime ideal and $W \notin \ker(\pi)$, we conclude that

$$x_i W - \sum_{t_1+\ldots+t_{d-2}=d-2} P_{t_1,\ldots,t_{d-2}}(\mathbf{Z}) Y_{j_1}^{t_1} Y_{j_2}^{t_2} \cdots Y_{j_{d-2}}^{t_{d-2}} \in \ker(\pi),$$

as was to be shown.

The supplementary statement is clear. $\square$

The proofs of the next corollary and the subsequent theorem both depend on some hard Gröbner basis calculation, for which we will systematically refer to [17]. It would be most desirable to find an alternative argument.

Corollary 9.20 *With the notation of the previous lemma, the kernel of the presentation map*

$$\pi : R[\mathbf{Y}, \mathbf{Z}, W] \twoheadrightarrow R[It, D_1 t^{d-1}, \ldots, D_d t^{d-1}, E t^{d(d-1)-1}]$$

$$Y_j \mapsto \Delta_j t, \; Z_r \mapsto D_r t^{d-1}, \; W \mapsto E t^{d(d-1)-1}$$

is the ideal generated by the polynomials

$$I_1(\mathbf{x}B), \; I_1(\mathbf{Z}B^t), \; I_1(\mathbf{x}^t \mathbf{Z} - \mathrm{adj}(B^t)), \; Y_j W - \Delta_j'(\mathbf{Z}), \; x_i W - Q_i(\mathbf{Y}, \mathbf{Z}),$$

where $1 \le j \le d+1$ and $1 \le i \le d$.

Proof Let $\mathcal{P} \subset R[\mathbf{Y}, \mathbf{Z}, W]$ denote the ideal generated by those many equations in the statement. By Proposition 9.19, we have $\mathcal{P} \subset \ker(\pi)$.

CLAIM. $\mathrm{codim}\,\mathcal{P} = \mathrm{codim}\,\pi = 2d+1$.

Indeed, the algebra $R[It, D_1 t^{d-1}, \ldots, D_d t^{d-1}, E t^{d(d-1)-1}] \subset R[t]$ has the same dimension as the Rees algebra $R[It]$, which is $d+1$; this shows that $\ker(\pi)$ has codimension $2d+1$. As for $\mathcal{P}$, the crucial point is to show that W is a nonzerodivisor on $R[Y, Z, W]/\mathcal{P}$ – we refer to [17, Proposition 2.23 (a)] for a proof based on a hard Gröbner basis argument. Then, localizing at the powers of W, we have that $\mathcal{P}$ and $\mathcal{P}[W^{-1}] \subset k[\mathbf{x}, \mathbf{Y}, \mathbf{Z}, W, W^{-1}]$ have the same codimension. But in the latter the generators

$$\{Y_j - W^{-1}\Delta_j'(\mathbf{Z}), \; x_i - W^{-1}Q_i(\mathbf{Y}, \mathbf{Z}) \mid 1 \le j \le d+1, 1 \le i \le d\}$$

form a regular sequence of length $d + 1 + d = 2d + 1$.

This proves the claim.

Therefore, to show that $\mathcal{P} = \ker(\pi)$, it suffices to prove that $\mathcal{P}$ is a prime ideal, i.e., that the localization of $\mathcal{P}$ at the powers of W is prime. But, by the nature of the generators of $\mathcal{P}$, one gets an isomorphism of k-algebras

$$k[\mathbf{x}, \mathbf{Y}, \mathbf{Z}, W, W^{-1}]/\mathcal{P}[W^{-1}] \simeq k[\mathbf{Z}, W, W^{-1}]/\widetilde{\mathcal{P}[W^{-1}]} \tag{9.20}$$

by mapping $x_i \mapsto W^{-1}Q_i(\mathbf{Y}, \mathbf{Z})$ and subsequently $Y_j \mapsto W^{-1}\Delta'_j(\mathbf{Z})$. Since $k[\mathbf{Z}, W, W^{-1}]$ has dimension $d + 1$, we must conclude that $\widetilde{\mathcal{P}[W^{-1}]} = 0$. (As a slight control, one has that, e.g., $I_1(\mathbf{x}B)$ maps to $I_1(D'_1(\mathbf{Z}) \cdots D'_d(\mathbf{Z})) \cdot B(\Delta'_1(\mathbf{Z}), \ldots, \Delta'_{d+1}(\mathbf{Z}))$, which vanishes as seen in the proof of Theorem 9.17 (ii)). Therefore, $\mathcal{P}[W^{-1}]$ is a prime ideal, and hence so is $\mathcal{P}$. $\qquad\square$

The main upshot of this part is the following result.

Theorem 9.21 (char(k) = 0) *Let $I = I_d(\mathbf{L}) \subset R = k[x_1, \ldots, x_d]$ denote the ideal of maximal minors of a $(d + 1) \times d$ matrix $\mathbf{L}$ of general linear forms, and let $\mathcal{R}_R^{(I)}$ stand for its symbolic Rees algebra. Then, with the previous notation, one has*

$$\mathcal{R}_R^{(I)} = R[It, D_1 t^{d-1}, \ldots, D_d t^{d-1}, E t^{d(d-1)-1}] \subset R[t].$$

Proof We apply Proposition 2.26 to $C := R[It, D_1 t^{d-1}, \ldots, D_d t^{d-1}, E t^{d(d-1)-1}]$. Condition (i) of this proposition is verified (Proposition 9.4), so we have to deal with condition (ii), that is, we need to check that $(\mathbf{x})C$ has grade at least 2.

CLAIM. The grade of $(\mathbf{x})C$ is the same as the grade of its extension to the localization A_w at the powers of the image w of W.

To see this it is enough to show that w avoids some associated prime $\wp$ of $C/(\mathbf{x})C$ such that grade $(\mathbf{x})C$ = grade $\wp$. Actually, more is true, namely, W is a nonzerodivisor on $C/(\mathbf{x})C$. For a proof via the Gröbner basis theory, we refer once more to [17].

We are now left with computing the grade of $(\mathbf{x})C_w$. By (9.20), $C_w \simeq k[\mathbf{Z}, W, W^{-1}]$, which is a Cohen–Macaulay graded ring. Therefore, the grade of $(\mathbf{x})C_w$ coincides with its codimension. Now, the image of $(\mathbf{x})C_w$ by the same isomorphism is the ideal

$$\left(Q_1(W^{-1}\Delta'_j(\mathbf{Z}), \mathbf{Z}), \ldots, Q_n(W^{-1}\Delta'_j(\mathbf{Z}), \mathbf{Z})\right) \subset k[\mathbf{Z}, W, W^{-1}].$$

By homogeneity we can pull out W^{-1}. Then (9.19) shows that this ideal is generated by the defining forms $\{D'_1, \ldots, D'_n\}$ of the inverse map to the Cremona map defined by $D_1, \ldots, D_n$. By Theorem 9.17(i), they generate an ideal of codimension 2. $\qquad\square$

Remark 9.22

(1) In the above proofs it was crucial knowing that the variable W was not a zerodivisor on both C and $C/\mathbf{x}C$. A proof other than using the Gröbner basis theory would be very welcome.

(2) In the proof of Theorem 9.21 it would suffice to show that C satisfies the
Serre property (S_2). One may in fact wonder if the symbolic Rees algebra is
Cohen–Macaulay, in which case it would be a Gorenstein normal domain by
[20, Corollary 3.4]. We have verified this in the case where $d = 3$ by writing
an explicit regular sequence of length $d + 1 = 4$, whose terms can actually be
taken to be linear forms involving only the $\mathbf{x}$, $\mathbf{Z}$, and W – this exploits the fact
that in this dimension one can change to a grading where the $\{\mathbf{x}, \mathbf{Y}, \mathbf{Z}\}$-part is
standard and the variable W has weight 2. In this grading the Hilbert series is
$(1 + 7t + 13t^2 + 7t^3 + t^4)/(1 - t)^4$. For $d \geq 4$, some of such facilitating features
are not available. On the other hand, even for $d = 3$, the property that W is a
nonzerodivisor on $C/(\mathbf{x})C$ is really on the edge as the ideal $(\mathbf{x}, \mathbf{Y}, \mathbf{Z})C$ is an
associated prime ideal of $C/\mathbf{x}C$.

(3) In the case $m \geq d + 2$, it may happen that elements of $I^{(d-1)}$ have standard
degree less than $(m - 1)(d - 1) - 1$. The simplest such situation occurs with
$d = 3$ and $m = 7$, in which case $I^{(2)}$ admits three minimal generators of degree
10. This implies that, in this range, the inclusion $(\mathbf{x})I^{(r)} \subset I^r$ for every $r \geq 0$
fails. This is an indication that, for general values of m and n, it may be difficult
to guess bounds for the value of the saturation exponent.

(4) Computational evidence shows that in the smallest possible numerology beyond
the two cases dealt with, i.e., $d = 3, m = 5$, the behavior of the symbolic
powers is quite erratic: There are genuine generators in $I^{(r)}$ throughout the
range $2 \leq r \leq 5$. The subsequent symbolic powers have an unpredictable
behavior with genuine generators creeping up on irregular intervals and new
symbolic generators even in $I^{(23)}$. Thus, one may actually wonder whether for
$m \gg d$ the symbolic Rees algebra $\mathcal{R}^{(I)}$ of I fails to be finitely generated.

9.2 Hilbert–Burch Matrices of Catalecticant and Hankel Nature

9.2.1 Generic Hilbert–Burch Catalecticant Matrices

An r-leap catalecticant (or, shortly, r-catalecticant) has been introduced in Definition 7.1. In this subsection we focus on generic r-leap catalecticant of size
$m \times (m - 1)$, namely,

$$
C(m, m-1, r) = \begin{pmatrix}
x_1 & x_2 & x_3 & \cdots & x_{m-1} \\
x_{r+1} & x_{r+2} & x_{r+3} & \cdots & x_{m+r-1} \\
x_{2r+1} & x_{2r+2} & x_{2r+3} & \cdots & x_{m+2r-1} \\
\vdots & \vdots & \vdots & \ddots & \vdots \\
x_{(m-1)r+1} & x_{(m-1)r+2} & x_{(m-1)r+3} & \cdots & x_{(m-1)r+(m-1)}
\end{pmatrix}
$$

in the notation of Section 7.1. Thus, the entries span a k-vector space generating the polynomial ring $R = k[x_1, \ldots, x_d]$, with $d = (m-1)(r+1)$, where $1 \le r \le m-1$. The extreme values $r = 1$ and $r = m - 1$ yield, respectively, the ordinary generic Hankel matrix and the fully generic matrix.

Since we are going to draw quite a bit on the properties of 1-genericity, we assume the ground field to be algebraically closed, even if this is often irrelevant in many passages.

As pointed out in Proposition 7.3, r-catalecticant matrices are 1-generic. Using this, we prove the following theorem:

Theorem 9.23 *Let C denote the $m \times (m-1)$ generic r-leap catalecticant matrix as above over the polynomial ring $R = k[x_1, \ldots, x_d]$, with $1 \le r \le m-1$. Then:*

(a) $\mathrm{ht}\,(I_t(C)) \ge m - t + 2$ *for* $1 \le t \le m - 2$ *and* $\mathrm{ht}\,(I_{m-1}(C)) = 2$.
    Setting $I := I_{m-1}(C)$, *one has:*
(b) *R/I is a Cohen–Macaulay normal domain.*
(c) *I is an ideal of linear type; in particular, the Rees algebra $\mathcal{R}_R(I)$ is Cohen–Macaulay.*
(d) *I is normally torsion-free.*

Proof

(a) The result is clear for $t = 1$; hence assume that $2 \le t \le m - 1$. For t in this interval, consider the submatrix $[C]_t$ of C consisting of its first t columns. By Proposition 7.3, $[C]_t$ is 1-generic, and by Corollary 7.4, its ideal $I_t([C]_t)$ of t-minors (maximal minors) is prime and satisfies $\mathrm{ht}\,(I_t([C]_t)) = m - t + 1$.

   To conclude, it suffices to show that the inclusion $I_t([C]_t) \subset I_t(C)$ is proper for $t \le m - 2$. For this, let Δ stand for the lower rightmost t-minor of C. Since Δ has a term involving effectively the last variable, while no t-minor of $[C]_t$ (with $t \le m - 2$) has such a term, and since all minors in consideration live in the same degree, we clearly have $\Delta \notin I_t([C]_t)$.

(b) Cohen–Macaulayness is well-known. Since C is 1-generic, then every prime $Q \subset R/I$ such that $(R/I)_Q$ is not regular must contain the ideal $I_{m-2}(C)/I$ [6, Corollary 3.3]. Then $\mathrm{ht}\,(Q) \ge \mathrm{ht}\,(I_{m-2}(C)/I) \ge 4 - 2 = 2$, by (a). Therefore, R/I satisfies the Serre condition (R_1); hence R/I is a normal domain.

(c) The bounds in (a) imply that I satisfies the condition G_∞. Therefore, the main assertion follows from Proposition 2.22 (i), and the supplementary assertion follows from item (ii) of the same proposition.

(d) This follows from [1, Theorem 3.5(b)] (and its predecessor [23, Theorem 3.4]). For a possibly different pattern, see [17]. $\qquad\qquad\square$

9.2.2 Deformations of Sparse Hilbert–Burch Hankel Matrices

In this part we consider linear sections of the $m \times (m - 1)$ generic Hankel matrix which are actually deformations of the semi-hollow like Hankel matrices $\mathcal{H}(r)$ considered in Section 6.3, only here for size $m \times (m - 1)$.

For convenience, we fix an integer $m \leq d \leq 2(m-1)$, and let $R = k[x_1, \ldots, x_d]$ stand for the ground polynomial ring of the occurring variables. Thus, in the notation of Section 6.3, we think of r as being $2(m - 1) - d$.

Our concern is then the Hankel linear section

$$
\left(
\begin{array}{ccccccccc}
x_1 & \cdots & x_{d-m+1} & x_{d-m+2} & x_{d-m+3} & x_{d-m+4} & \cdots & x_{m-2} & x_{m-1} \\
x_2 & \cdots & x_{d-m+2} & x_{d-m+3} & x_{d-m+4} & x_{d-m+5} & \cdots & x_{m-1} & x_m \\
\vdots & & \vdots & \vdots & \vdots & \vdots & & \vdots & \vdots \\
x_{d-m+2} & \cdots & x_{2d-2m+2)} & x_{2d-2m+3} & x_{2d-2m+4} & x_{2d-2m+5} & \cdots & x_{d-1} & x_d \\
\hline
x_{d-m+3} & \cdots & x_{2d-2m+3} & x_{2d-2m+4} & x_{2d-2m+5} & x_{2d-2m+6} & \cdots & x_d & \lambda_1 \\
x_{d-m+4} & \cdots & x_{2d-2m+4} & x_{2d-2m+5} & x_{2d-2m+6} & x_{2d-2m+7} & \cdots & \lambda_1 & \lambda_2 \\
\vdots & & \vdots & \vdots & \vdots & \vdots & & \vdots & \vdots \\
x_{m-1} & \cdots & x_{d-1} & x_d & \lambda_1 & \lambda_2 & \cdots & \lambda_{2(m-1)-d-2} & \lambda_{2(m-1)-d-1} \\
x_m & \cdots & x_d & \lambda_1 & \lambda_2 & \lambda_3 & \cdots & \lambda_{2(m-1)-d-1} & \lambda_{2(m-1)-d}
\end{array}
\right),
$$

where $\lambda_1, \ldots, \lambda_{2(m-1)-d}$ are linear forms in R defining a deformation of the semi-hollow like Hankel section $\mathcal{H}(2(m - 1) - d)$. Assuming that m and d are fixed in the discussion, we denote the above matrix by $\mathcal{H}(\lambda)$.

This model has the following properties:

Theorem 9.24 *Let* $4 \leq m + 1 \leq d \leq 2(m - 1)$ *and set* $I := I_{m-1}(\mathcal{H}(\lambda))$. *Then:*

(a) $\operatorname{ht}(I_t(\mathcal{H}(\lambda))) \geq m - t + 2$ *for* $1 \leq t \leq m - 2$ *and* I *is a height 2 prime ideal.*
(b) R/I *is normal if* $d \geq m + 2$.
(c) I *is an ideal of linear type.*
(d) I *is normally torsion-free.*

Proof

(a) The case $t = 1$ is immediate. Next we let $[\mathcal{H}(\lambda)]_t$ denote the submatrix of $\mathcal{H}(\lambda)$ consisting of the first t columns, for values of t in the range $2 \leq t \leq m - 2$. For $t \leq d - m + 1$, the matrix $[\mathcal{H}(\lambda)]_t$ is a generic $m \times t$ Hankel matrix. Therefore, it is 1-generic, and hence its t-minors (maximal minors) generate a prime ideal of codimension $m - t + 1$. For $t > d - m + 1$, one has

$$
\frac{k[x_1, \ldots, x_d]}{I_t([\mathcal{H}(\lambda)]_t)} \simeq \frac{k[x_1, \ldots, x_d, x_{d+1}, \ldots, x_{d+s}]}{(x_{d+1} - \lambda_1, \ldots, x_{d+s} - \lambda_s, I_t([\mathcal{H}]_t))},
$$

where $s = t - (d - m + 1)$ and $[\mathcal{H}]_t$ denotes the generic $m \times t$ Hankel matrix. Since $s \leq t - 2$ (because $m + 1 \leq d$), then we can use [5, Theorem 1] to deduce again that $I_t([\mathcal{H}(\lambda)]_t)$ is a prime ideal of codimension $m - t + 1$. The second assertion of (a) can be deduced similarly.

Now, to complete the argument, it suffices to show that the inclusion $I_t([\mathcal{H}(\lambda)]_t) \subset I_t(\mathcal{H}(\lambda))$ is proper for $2 \leq t \leq m - 2$, since the first of these ideals is prime. We will consider two cases.

Case 1. $t \leq 2(m - 1) - d$.

In this case,

$$\Delta = \det \begin{pmatrix} x_{d-t+1} & x_{d-t+2} & \cdots & x_d \\ x_{d-t+2} & x_{d-t+3} & \cdots & \lambda_1 \\ \vdots & \vdots & \ddots & \vdots \\ x_d & \lambda_1 & \cdots & \lambda_{t-1} \end{pmatrix}$$

is a t-minor of $\mathcal{H}(\lambda)$ whose x_d-degree is t. On the other hand, since x_d does not occur in the first column of $\mathcal{H}(\lambda)_t$, then every t minor of $\mathcal{H}(\lambda)_t$ has x_d-degree at most $t - 1$. Hence, $\Delta \notin I_t([\mathcal{H}(\lambda)]_t)$.

Case 2. $t > 2(m - 1) - n$.

Consider the lower rightmost t-minor of $\mathcal{H}(\lambda)$

$$\Delta = \det \begin{pmatrix} x_{2(m-1)-2t+1} & \cdots & x_{2(m-1)-t} \\ \vdots & & \vdots \\ x_{2(m-1)-t} & \cdots & \lambda_{2(m-1)-d} \end{pmatrix}.$$

This is a t-minor of $\mathcal{H}(\lambda)$ whose $x_{2(m-1)-t}$-degree is t. Since $t \leq m - 2$, then $2(m-1)-t \geq m$. Hence, $x_{2(m-1)-t}$ does not occur in the first column of $\mathcal{H}(\lambda)_t$. Thus, every t minor of $\mathcal{H}(\lambda)_t$ has $x_{2(m-1)-t}$-degree at most $t - 1$. As before, $\Delta \notin I_t([\mathcal{H}(\lambda)]_t)$.

(b) Consider the transpose M of $\mathcal{H}(\lambda)$, a matrix of the same type with the λ's symmetrically rearranged around the main diagonal. Let M' denote the transpose of the $(m - 1) \times m$ generic Hankel matrix. Thinking of $M \subset M'$ as the respective k-subspaces spanned by the entries, one has $\mathrm{codim}_{M'} M = 2(m - 1) - d$. Our hypothesis implies that $\mathrm{codim}_{M'} M \leq m - 4$. Now, M' is 1-generic; hence the singular locus of $\mathrm{Proj}(R/I)$ is contained in the union of $\mathrm{Proj}(R/I_{m-2}(\mathcal{H}(\lambda)))$ and a certain set of codimension at least $m - 2 - \mathrm{codim}_{M'} M$ in $\mathrm{Proj}(R/I)$ – according to the discussion immediately following the statement in [6, Theorem 2.1 (3)] and its proof in [6, Proposition 3.1 and ff.]. The first of these two has codimension ≥ 2 in $\mathrm{Proj}(R/I)$ by the estimates of item (a). As for the second, its codimension in $\mathrm{Proj}(R/I)$ is now at least $m - 2 - \mathrm{codim}_{M'} M \geq m - 2 - (m - 4) = 2$. This shows that R/I satisfies Serre's property (R_1) and hence is normal.

(c) and (d) are proved exactly in the same way as in Theorem 9.23 $\qquad\square$

For $d = m + 1$, the normality of the ring R/I may depend on the linear forms $\lambda_1, \ldots, \lambda_d$ as shown in the following proposition:

Proposition 9.25 *If $d = m + 1$ and $\lambda_1 = \ldots = \lambda_{2(m-1)-d} = 0$, then R/I is not normal.*

Proof Consider the height 3 prime $P = (x_{m-1}, x_m, x_{m+1})$. Clearly, $I \subset P$ by direct inspection of the shape of the matrix. We claim that R_P/I_P is not regular. For this, we will prove that I_P is not generated by a subset of a regular system of parameters of R_P. Now, note that the upper left $(d - 3)$-minor of $\mathcal{H}(\lambda)$ has the form $x_{m-2}^{m-2} + q$, where $q \in P$ and hence does not belong to I. Therefore, up to elementary row and column operations over R_P, we see that $I_P = (\Delta_{m-1}, \Delta_m)$, where Δ_i denotes the $(m - 1)$-minor of $\mathcal{H}(\lambda)$ obtained by omitting the ith row. It suffices to show that $\Delta_{m-1} \in P^2$. But

$$\Delta_{m-1} = (-1)^m x_m \det \begin{pmatrix} x_2 & x_3 & \cdots & x_{m-1} \\ x_3 & x_4 & \cdots & x_m \\ x_4 & x_5 & \cdots & x_{m+1} \\ x_5 & x_6 & \cdots & 0 \\ \vdots & \vdots & \vdots & \vdots \\ x_{m-1} & x_m & \cdots & 0 \end{pmatrix} + (-1)^{m+1} x_{m+1} \det \begin{pmatrix} x_1 & x_3 & \cdots & x_{m-1} \\ x_2 & x_4 & \cdots & x_m \\ x_3 & x_4 & \cdots & x_{m+1} \\ x_4 & x_5 & \cdots & 0 \\ \vdots & \vdots & \vdots & \vdots \\ x_{m-1} & x_m & \cdots & 0 \end{pmatrix}.$$

Expanding these determinants along their last column clearly shows the claim.   □

9.3   Hilbert–Burch Linear Sections in Dimension 3

In this part we consider $n \times (n - 1)$ matrices whose entries are linear forms in the standard graded polynomial ring $R = k[x, y, z]$, but not necessarily general. For convenience, we trade the letter n for m in the notation throughout.

9.3.1   The Chaos Invariant

Definition 9.26 Let φ denote an $n \times (n - 1)(n \geq 3)$ matrix such that its entries are linear forms in the polynomial ring $R := k[x, y, z]$. Assume that ht $I_1(\varphi) = 3$ and ht $I_{n-1}(\varphi) = 2$. Then there is a uniquely defined integer $1 \leq u \leq n - 2$ such that

$$\text{ht } I_t(\varphi) = \begin{cases} 3, & 1 \leq t \leq u \\ 2, & u + 1 \leq t \leq n - 1. \end{cases} \tag{9.21}$$

We call $u = u(\varphi)$ the *chaos invariant* of φ.

One can introduce a local version of this invariant in the following way: We know that $I_{n-1}(\varphi)$ is a codimension 2 perfect ideal; hence its associated primes have codimension 2. Furthermore, for any $1 \leq t \leq n-2$, one has $I_{n-1}(\varphi) \subset I_t(\varphi)$. Let p stand for any associated prime of $I_{n-1}(\varphi)$. Then there exists a unique integer $1 \leq u_p \leq n-2$ such that $p \in \mathrm{Min}(I_{u_p+1}(\varphi)) \setminus \mathrm{Min}(I_{u_p}(\varphi))$. We will call $u_p = u_p(\varphi)$ the *chaos invariant of φ at p*.

Clearly, one has

$$u(\varphi) = \min\{u_p(\varphi) \mid p \in \mathrm{Ass}(R/I_{n-1}(\varphi))\}. \tag{9.22}$$

Note that the concept depends only on the Fitting ideals of φ, hence it is invariant by conjugation and change of variables (Section 1.1), so one often refers to it as an invariant of the corresponding codimension 2 perfect ideal $I_{n-1}(\varphi)$.

In the sequel we will systematically set $I := I_{n-1}(\varphi)$ and omit φ in both $u(\varphi)$ and $u_p(\varphi)$ if no confusion arises. The basic result about the chaos invariant goes as follows.

Lemma 9.27 *Let φ denote an $n \times (n-1)$ matrix whose entries are linear forms in the polynomial ring $R := k[x, y, z]$, such that $I_1(\varphi) = (x, y, z)$ and $\mathrm{ht}\, I = 2$. Let p stand for an associated prime of I. Then:*

(a) *The rank of φ over R/p is u_p.*
(b) $\mu(I_p) = n - u_p$.
(c) $I_p \subset p_p^{\,n-u_p-1}$.

Proof

(a) By definition of u_p, p is not a minimal prime of $I_{u_p}(\varphi)$, in particular, $I_{u_p}(\varphi) \not\subset p$. Therefore, there exists an $(u_p \times u_p)$-minor of φ not contained in p, i.e., the rank of φ over R/p is at least u_p. On the other hand, still by definition, $I_{u_p+1}(\varphi) \subset p$. Therefore, $I_t(\varphi) \subset p$ for every $t \geq u_p + 1$. Thus, the rank of φ over R/p is at most u_p.
(b) One has

$$\mu(I_p) = \mu((I/pI)_p) \quad \text{(Nakayama)}$$

$$= n - \mathrm{rk}_{R_p/p_p}(\varphi) = n - \mathrm{rk}_{R/p}(\varphi)$$

$$= n - u_p \quad \text{(by (a)).}$$

(c) This is an immediate consequence of knowing that $I_{u_p}(\varphi) \not\subset p$. For then, locally at p, some $(u_p \times u_p)$-minor of φ is invertible, so that up to conjugation φ has the form

$$\left(\begin{array}{c|c} \mathbb{I} & 0 \\ \hline 0 & \psi \end{array} \right), \tag{9.23}$$

where $\mathbb{I}$ is the identity matrix of order u_p and ψ is an $(n - u_p) \times (n - u_p - 1)$ matrix with entries in p_p. Therefore, any generator of $I_p(\varphi)$ belongs to $p_p{}^{n-u_p-1}$. $\qquad\square$

Remark 9.28 Although $I_p(\varphi)$ above has the same number of minimal generators as $p_p{}^{n-u_p-1}$, it may be the case that they are not all strictly of order $n - u_p - 1$ – see Theorem 9.39 (b) below for the case where $u_p = 1$.

Our next concern has to do with the value of the analytic spread $\ell(I) = \dim \mathcal{F}(I)$. Note that since I is equigenerated in degree $n - 1$, then $\dim \mathcal{F}(I) = \dim k[[I]_{n-1}]$. Clearly, $2 \leq \ell(I) \leq \dim R = 3$. Since I is in general not generically a complete intersection, the Cowsik–Nori theorem is not applicable to deduce that $\ell(I) \neq 2$, and neither is the previous work of Ulrich–Vasconcelos [25, Corollary 4.3] and Morey–Ulrich [15, Theorem 1.3] because requiring that I satisfy G_3 is tantamount to requiring that I be generically a complete intersection.

Instead, by a quirk using a more encompassing version of a birationality criterion, we can show that $\ell(I) = 3$ and more:

Theorem 9.29 *Let $R = k[x, y, z]$ be a polynomial ring over an algebraically closed field k, and let $I \subset R$ denote a height 2 perfect ideal presented by an $n \times (n - 1)$ matrix φ whose entries are linear forms in R. Assume that* $\mathrm{ht}\, I_1(\varphi) = 3$. *Then*:

(a) *The rational map $\mathfrak{F}$ defined by the maximal minors of φ is a birational map of* $\mathbb{P}^2$ *onto its image. In particular, the ideal I has maximal analytic spread.*
(b) *The inverse map to $\mathfrak{F}$ in* (a) *is defined by forms of degree* 2.
(c) $\mathrm{depth}(R/I^2) = 0$.

Proof Let $u := u(\varphi)$ as above denote the chaos invariant of φ, and let p denote a minimal prime of $I_{u+1}(\varphi)$. By Lemma 9.27 (a), there is a u-minor of φ not contained in $I_u(\varphi)$. Therefore, up to conjugation and change of variables, we may assume that $p := (y, z)$ and that φ has the following shape:

$$\varphi = \left(\begin{array}{ccccc|c} x + a_1 & & & & & \\ & x + a_2 & & & & \\ & & \ddots & & & \\ & & & x + a_u & & \\ \hline & & & & & \\ & & & & & \end{array} \right), \tag{9.24}$$

where $a_1, a_2, \ldots, a_u$ and the blank entries are linear forms in y and z solely.

(a) In order to prove the statement, we use the criterion of [19, Theorem 3.2.22] in terms of a certain Jacobian dual matrix. Namely, consider new variables $\mathbf{t} = \{t_1, \ldots, t_n\}$ over R and let B denote the uniquely defined matrix with entries in

$k[\mathbf{t}] = k[t_1, \ldots, t_n]$ satisfying the equality

$$(t_1 \cdots t_n)\, \varphi = (x\ y\ z)\, B. \tag{9.25}$$

Clearly, B turns out to be a $3 \times (n-1)$ matrix with linear entries. For convenience, we look at its transpose:

$$
B^t =
\begin{pmatrix}
t_1 & \ell_{1,2} & \ell_{1,3} \\
\vdots & \vdots & \vdots \\
t_u & \ell_{u,2} & \ell_{u,3} \\
0 & \ell_{u+1,2} & \ell_{u+1,3} \\
\vdots & \vdots & \vdots \\
0 & \ell_{n-1,2} & \ell_{n-1,3}
\end{pmatrix},
$$

which is a submatrix of the so-called Jacobian dual matrix of I as introduced in [19, Chapter 3]. Note that the inequality $u \le n-2$ implies the existence of at least one zero along the first column of B^t.

On the other hand $\ell_{u+1,2} \neq 0$. This is because otherwise φ would admit an entire column with entries depending only on z. Since the maximal minors of φ belong to the ideal generated by the entries of that column, this would contradict the standing assumption that $\mathrm{ht}\, I_{n-1}(\varphi) = 2$.

By a similar token, the maximal minors are k-linearly independent and hence admit no polynomial relation of degree 1 in $k[\mathbf{t}]$. But the full ideal P of polynomial relations of the maximal minors defines the image of the rational map defined by these minors and hence is prime. This finally gives

$$
\det \begin{pmatrix} t_u & \ell_{u,2} \\ 0 & \ell_{u+1,2} \end{pmatrix} \not\equiv 0 \ \mod P,
$$

saying that the matrix B^t has rank at least 2 over $R[\mathbf{t}]/P$. Therefore, so does the more encompassing Jacobian dual matrix as considered above. Moreover, by [19, Corollary 3.2.10], this rank is at most $\dim R - 1 = 2$. Therefore, the Jacobian dual matrix has rank exactly 2. This is condition (b) in the aforementioned criterion.

(b) We draw upon [19, Theorem 3.2.22, Supplement (ii)], by which a representative of the inverse map can be taken to be the 2×2 minors of an arbitrary 2×3 submatrix of rank 2 of the Jacobian dual matrix. In particular, one can choose such a matrix to be the horizontal slice

$$
\begin{pmatrix} t_u & \ell_{u,2} & \ell_{u,3} \\ 0 & \ell_{u+1,2} & \ell_{u+1,3} \end{pmatrix}
$$

of B^t.

(c) Since the inverse map to $\mathfrak{F}$ is defined by forms of degree 2 as in (b), then Theorem 2.30 says that the corresponding source inversion factor is an element of $I^{(2)} \setminus I^2$. Clearly, then $\operatorname{depth}(R/I^2) = 0$ since $I^{(2)}$ is the unmixed part of the second power. $\qquad\square$

Proposition 9.30 (k Algebraically Closed) *Let φ denote an $n \times (n-1)$ matrix such that its entries are linear forms in the polynomial ring $R := k[x, y, z]$, with* $\operatorname{ht} I_1(\varphi) = 3$ *and* $\operatorname{ht} I_{n-1}(\varphi) = 2$.

Setting $u = u(\varphi)$, one has:

(a) *If $n \geq 2(u+1)$, all the ideals $I_t(\varphi)$ for $u+1 \leq t \leq n-(u+1)$ have one single and the same minimal prime.*

Let q denote the uniquely universal minimal prime in (a). *Then:*

(b) *For every associated prime of I other than q, one has $u_p \geq n-(u+1)$ and $\mu(I_p) \leq u+1$.*

Proof

(a) Under the assumption that $n \geq 2(u+1)$, consider the chain of prime ideals

$$I \subset I_{n-(u+1)}(\varphi) \subset I_{n-(u+2)}(\varphi) \subset \cdots \subset \underbrace{I_{u+1}(\varphi)}_{2} \subset \underbrace{I_u(\varphi)}_{3},$$

where the tagged numbers denote codimensions arising from the definition of $u = u(\varphi)$. Fix a minimal prime q of $I_{u+1}(\varphi)$. Then q is both a minimal prime of $I_{n-(u+1)}(\varphi)$ and of I. By the definition of u_q, we have $u = u_q$. Then, by Lemma 9.27 (a), the rank of φ over R/q is $u_q = u$. Therefore, up to conjugation and change of variables, φ has the form of (9.24), where $q = (y, z)$, and $a_1, \ldots, a_u$ and the blank entries are linear forms involving only y and z.

Let $J \subset I_{n-(u+1)}(\varphi)$ denote the ideal generated by the $(n-(u+1))$-minors of the rightmost $n-(u+1)$ columns. Since the generators of J depend only on the variables y and z and $\operatorname{ht} J = 2$ (because $I \subset J \subset I_{n-(u+1)}(\varphi)$), then $q = (y, z)$ is the unique minimal prime of J. Hence, q is the unique minimal prime of $I_{n-(u+1)}(\varphi)$ as well.

By tracing up the above chain of ideals $I_t(\varphi)$, it is clear that q is also the unique minimal prime of each of them for $u+1 \leq t \leq n-(u+1)$.

(b) By assumption and item (a), p is a minimal prime of I but not of $I_{n-(u+1)}(\varphi)$. Therefore, $u_p \geq n-(u+1)$ by definition. Then, by Lemma 9.27 (b) one has $\mu(I_p) \leq u+1$. $\qquad\square$

Remark 9.31 The extreme case where $u(\varphi) = 1$ has been treated before in [13] – in this case, all the ideals of minors, except $I_1(\varphi)$ and I, have a unique and the same minimal prime. A full consideration of the results obtained in that paper will be taken up in a subsequent part.

9.3.2 Reduction Number and the Rees Algebra

Recall the Rees algebra $\mathcal{R}_R(I)$ of an ideal I of a ring R as introduced in Section 2.2.1. We will draw upon the following result of Cortadellas–Zarzuela [3], as slightly reformulated in [19, Theorem 6.2.29]:

Proposition 9.32 *Let R denote a Cohen–Macaulay local ring, and let $I \subset R$ stand for an ideal of height ≥ 2 satisfying the following properties:*

(1) *I is unmixed, and R/I and R/I^2 have different depths.*
(2) *$\ell(I) = \mathrm{ht}\,(I) + 1$.*
(3) *$r(I_p) \leq 1$ for every associated prime p of R/I of minimal height.*
(4) *$r(I) \leq 2$.*

Then $\mathrm{depth}\,\mathcal{R}_R(I) = \min\{\mathrm{depth}\,R/I, \mathrm{depth}\,R/I^2 + 1\} + \mathrm{ht}\,(I) + 1$.

Theorem 9.33 *Let $I \subset R = k[x, y, z]$ denote a codimension 2 perfect homogeneous ideal satisfying the following conditions:*

(a) *I is linearly presented.*
(b) *$r(I_p) \leq 1$ for every associated prime p of R/I.*

Consider the following conditions:

 (i) *$r(I) \leq 2$.*
 (ii) *The Rees algebra $\mathcal{R}_R(I)$ is Cohen–Macaulay.*
(iii) *The special fiber $\mathcal{F}(I)$ is Cohen–Macaulay.*

Then (i) $\Rightarrow$ (ii) *and* (iii) $\Rightarrow$ (i). *In particular, these conditions are equivalent under the assumption of* Conjecture 2.18.

Proof Note that since $k[x, y, z]$ is standard graded, assumption (a) implies that I is equigenerated.

(i) $\Rightarrow$ (ii) In order to apply Proposition 9.32, thus concluding that depth $\mathcal{R}_R(I) = 4 = \dim R + 1$, we note that conditions (1) and (3) follow from the standing hypotheses (specially, (a)) by drawing upon Theorem 9.29, while condition (2) is contained in hypothesis (b).

(iii) $\Rightarrow$ (i) We apply Corollary 2.16 by showing that its condition (iii) holds. To see this, we first show that this condition holds locally at every associated prime p of R/I. But locally at such p the ideal I_p is p_p-primary in R_p. By the assumption in (b), [12, Theorem 2.1] implies that the Hilbert–Samuel function $HS_{I_p}(t)$ of I_p is polynomial for any $t \geq 1$.

Let $e(M)$ denote the multiplicity (degree) of a graded module M over R. Letting $t \geq 1$, the associativity formula for multiplicities yields

$$e(R/I^t) = \sum_{p \in \mathrm{Min}(R/I)} \lambda(R_p/I_p^t)e(R/p) = \sum_{p \in \mathrm{Min}(R/I)} \lambda(R_p/I_p^t)$$

$$= \sum_{p \in \mathrm{Min}(R/I)} HS_{I_p}(t).$$

It follows that $e(R/I^t)$ has the values of a polynomial for every $t \geq 1$.

But the graded Hilbert function of $\mathcal{F}(I)$ has the form

$$H_{\mathcal{F}(I)}(t) = \mu(I^t) = \binom{\mu(I)\,t + 1}{2} - e(R/I^t),$$

for every $t \geq 1$, where $\mu(\_)$ denotes the minimal number of generators. Therefore, it has the values of a polynomial in every degree ≥ 1 and hence coincides with its Hilbert polynomial throughout. $\qquad\square$

9.3.3 A Notable 1-Generic Matrix

The matrix B obtained in (9.25) is useful not only as a major vehicle to test for birationality and to obtain a representative of the inverse map as described in [19, Chapter 3] but often to guess generators of the homogeneous defining ideal of the image of the map as well.

For convenience, we write it explicitly:

$$B = \begin{pmatrix} t_1 & t_2 & \ldots & t_u & 0 & \ldots & 0 \\ \ell_{1,2} & \ell_{2,2} & \ldots & \ell_{u,2} & \ell_{u+1,2} & \ldots & \ell_{n-1,2} \\ \ell_{1,3} & \ell_{2,3} & \ldots & \ell_{u,3} & \ell_{u+1,3} & \ldots & \ell_{n-1,3} \end{pmatrix}, \tag{9.26}$$

where $\ell_{i,2}$ and $\ell_{i,3}$ are certain linear forms in $k[\mathbf{t}] = k[t_1, \ldots, t_n]$.

The following lemma says how B behaves under conjugation and change of variables.

Lemma 9.34 *Let φ be an $n \times (n-1)$ matrix whose entries are linear forms in $R = k[x, y, z]$. Let B be the unique matrix with linear entries in the polynomial ring $k[t_1, \ldots, t_n]$ such that*

$$(t_1 \; \cdots \; t_n)\,\varphi = (x\ y\ z)\,B.$$

Given $P \in \mathrm{GL}_3(k)$ and $Q \in \mathrm{GL}_{n-1}(k)$, let $\tilde{B} = P B Q$ stand for the resulting transform. Then,

$$(t_1 \; \cdots \; t_n)\,\tilde{\varphi} = (x\ y\ z)\,\tilde{B},$$

where $\tilde{\varphi} = \varphi(\mathbf{x}\,P)\,Q$, $\mathbf{x} = (x\ y\ z)$, and $\varphi(\mathbf{x}\,P)$ denotes the matrix φ as read in the new variables in $\mathbf{x}\,P$.

Proof Set $\mathbf{x}' = (x'\ y'\ z') := \mathbf{x}\,P$ and $\mathbf{t} = (t_1\ \cdots\ t_n)$ for short. Then,

$$\mathbf{t}\,\tilde{\varphi} = \mathbf{t}\,\varphi(\mathbf{x}\,P)\,Q = \mathbf{t}\,\varphi(\mathbf{x}')\,Q = \mathbf{x}'\,B\,Q = \mathbf{x}\,P\,B\,Q = \mathbf{x}\,\tilde{B},$$

as claimed.   □

Set

$$B' := \begin{pmatrix} \ell_{u+1,2} & \cdots & \ell_{n-1,2} \\ \ell_{u+1,3} & \cdots & \ell_{n-1,3} \end{pmatrix}. \tag{9.27}$$

We have already remarked in the proof of Theorem 9.29 (b) that no entry of this matrix vanishes. We prove more:

Proposition 9.35 *The matrix B' is 1-generic.*

Proof As a rare occasion, we verify the stated property directly from its very definition.

Thus, we are to show that B' acquires no zero entry under the action of $\mathrm{GL}_2(k) \times \mathrm{GL}_{n-1-u}(k)$. Let us argue by contradiction, assuming that there exist $P' \in \mathrm{GL}_2(k)$ and $Q' \in \mathrm{GL}_{n-1-u}(k)$ such that $P'B'C'$ has a null entry.

Set

$$P := \begin{pmatrix} 1 & \\ & P' \end{pmatrix} \in \mathrm{GL}_3(k) \qquad \text{and} \qquad Q = \begin{pmatrix} \mathbb{I}_u & \\ & Q' \end{pmatrix} \in \mathrm{GL}_{n-1}(k),$$

where $\mathbb{I}_u$ denotes the identity matrix of order u.

Then the transform of B by the latter matrices has the shape

$$P\,B\,Q = \left(\begin{array}{ccc|c} t_1 & \cdots & t_u & \mathbf{0} \\ {*} & \cdots & {*} & P'B'Q' \end{array} \right), \tag{9.28}$$

where, by assumption, the rightmost block has a column with at least two zero entries.

Changing to coordinates x', y', and z' given by $(x'\ y'\ z') = (x\ y\ z)\,P$, Lemma 9.34 yields

$$(t_1\ \cdots\ t_n)\,\tilde{\varphi} = (x\ y\ z)\,P\,B\,Q,$$

where $\tilde{\varphi} = \varphi(\mathbf{x}\,P)\,Q$.

Since there exists a column of $P\,B\,Q$ with two zeros, then $\tilde{\varphi}$ has a column depending solely on one of the variables y' or z'. This implies that $I_{n-1}(\tilde{\varphi})$ is contained in the principal ideal generated by one of these variables – an absurd as $\operatorname{codim} I_{n-1}(\tilde{\varphi}) = \operatorname{codim} I_{n-1}(\varphi) = 2$.   □

The next is the main structural result of the 1-generic matrix B', retrieving the meaning of the chaos invariant.

Corollary 9.36 *Let φ stand for the matrix in (9.24), and let B' be the associated matrix as in (9.27). Then $I_2(B')$ is isomorphic to the homogeneous coordinate ring of a rational normal scroll of dimension $u + 1$ in $\mathbb{P}^{n-1}$, where u denotes the chaos invariant of φ. In particular, it is a Cohen–Macaulay normal domain.*

Proof This is a consequence of [11, Proposition 9.12] or rather from its subsequent generalization due to F. Schreyer, where it is shown that any 1-generic $2 \times r$ matrix, with $r \leq n - 1$, is conjugate to a matrix of the shape

$$
\begin{pmatrix}
x_0 \; x_1 \; \ldots \; x_{a_1-1} & y_0 \; y_1 \; \ldots \; y_{a_2-1} & \cdots & z_0 \; z_1 \; \ldots \; z_{a_s-1} \\
x_1 \; x_2 \; \ldots \; \; x_{a_1} & y_1 \; y_2 \; \cdots \; \; \; y_{a_2} & \cdots & z_1 \; z_2 \; \ldots \; \; z_{a_s}
\end{pmatrix}
$$

formed by s mutually independent Hankel blocks, with $r = a_1 + \cdots + a_s$. The 2-minors generate the homogeneous defining ideal of a rational normal scroll in projective space $\mathbb{P}^{r+s-1}$; hence it is a prime ideal and the corresponding homogeneous coordinate ring is normal.

We also note that the rational normal scroll has dimension s. Applying to B', with $r = n - u - 1$, one has that $I_2(B')$ has the expected codimension $n - u - 1 - 2 + 1 = n - u - 2$ in $k[\mathbf{t}] = k[t_1, \ldots, t_n]$; hence $\dim k[\mathbf{t}]/I_2(B') = u + 2$, and therefore $s = u + 1$ in our case. Then $k[\mathbf{t}]/I_2(B')$ is isomorphic to the homogeneous coordinate ring of a rational normal scroll of dimension $u + 1$ in $\mathbb{P}^{n-1}$. $\qquad\square$

A neat case goes in the next part.

9.3.4 Chaos Invariant One

The following results include some of the main results of [13] from a different angle, perhaps simpler too.

We start with a preliminary assertion about matrices with binary linear entries which may have interest on itself. It is possibly the best result about the Jacobian dual method in the binary case.

Lemma 9.37 (*k Algebraically Closed*) *Let M denote an $m \times m$ matrix with linear entries in the polynomial ring $k[y, z]$. Bringing in new variables $t_1, \ldots, t_m$ over k, let C denote the uniquely defined $2 \times m$ matrix whose entries are linear forms in the polynomial ring $k[t_1, \ldots, t_m]$ and such that*

$$
(t_1 \cdots t_m)\, M = (y \; z)\, C.
$$

If $\det M \neq 0$ and $I_{m-1}(M)$ is (y, z)-primary, then $I_2(C)$ has the expected codimension $m - 2 + 1 = m - 1$.

Proof First, since k is algebraically closed, up to conjugation and change of variables, M has the shape

$$\widetilde{M} := \left(\begin{array}{c|c} y & \alpha \\ \hline 0 & M' \end{array}\right),$$

where $\det M' \neq 0$ and the entries of $\alpha = (\alpha_2 z \cdots \alpha_d z)$ depend only on z, $\alpha_i \in k$.

This fact is most probably well-known being essentially a result about matrices in one single variable. For the sake of completeness, we give the following argument: Write $M = yA + zB$, for certain $m \times m$ scalar matrices A and B. Since $\det M$ is a nonzero binary form over an algebraically closed field, it has a zero $(a, b) \in k^2$, where, say, $a \neq 0$. Substituting gives $Av = -(b/a)Bv$, for suitable nonzero vector $v \in k^m$. Letting $\{v_1 = v, v_2, \ldots, v_m\}$ denote a vector basis of k^m, there exists an invertible $m \times m$ scalar matrix T, with v as its first column, and such that the first column of AT is the first column of BT times $-b/a$. In this way we get that the entries of the first column of TMT^{-1} are scalar multiples of $y + (b/a)z$. Additional elementary row operations on TMT^{-1} will make all entries of its first column vanish except one. By a coordinate change, we may assume that the latter is y. Suitable column operations involving the first column will turn into scalar multiples of z the remaining entries along the first row, as was to be shown.

Let us now switch to the proof of the main statement. We induct on m.

Clearly, M' is an $(m - 1) \times (m - 1)$ matrix satisfying the same hypothesis as M, namely, $\det M' \neq 0$ and $I_{m-2}(M)$ is (y, z)-primary. Writing, similarly,

$$(t_2 \cdots t_m)\, M' = (y\ z)\, C',$$

where C' is a uniquely defined $2 \times (m - 1)$ matrix of linear forms in $k[t_2, \ldots, t_m]$, by the inductive hypothesis the ideal $I_2(C')$ has codimension $m - 2$. Say,

$$C' := \begin{pmatrix} \lambda_{1,2} \cdots \lambda_{1,m} \\ \lambda_{2,2} \cdots \lambda_{2,m} \end{pmatrix},$$

with $\lambda_{ij} \in k[t_2, \ldots, t_m]$.

Then, taking into account the above shape of $\widetilde{M}$, C turns out to be up to conjugation and change of variables

$$\widetilde{C} := \begin{pmatrix} t_1 & \lambda_{1,2} & \cdots & \lambda_{1,m} \\ 0 & \lambda_{2,2} + \alpha_2 t_1 & \cdots & \lambda_{2,m} + \alpha_m t_1 \end{pmatrix}. \tag{9.29}$$

Therefore, it suffices to show that $I_2(\widetilde{C})$ has codimension $m - 1$.

For this, first note that the linear forms $\lambda_{2,2} + \alpha_2 t_1, \ldots, \lambda_{2n} + \alpha_m t_1$ are k-linearly independent. Indeed, otherwise up to conjugation and change of variables M would have two columns involving only one of the two variables y and z and that would imply that $\mathrm{cod}\, I_{m-1}(M) = 1$.

Now, if $P \supset I_2(\widetilde{C})$ is any prime ideal, then either it contains the above linear forms, in which case its codimension is at least $m - 1$, or else it contains t_1, and hence also $I_2(C')$. Thus, P contains $(I_2(C'), t_1)$ which has codimension $m-2+1 = m-1$ by the inductive hypothesis and the fact that t_1 is a nonzerodivisor thereof. $\square$

Proposition 9.38 *Let* $R = k[x, y, z]$ *be a polynomial ring over an algebraically closed field* k, *and let* $I \subset R$ *denote a height 2 perfect ideal presented by an* $n \times (n-1)$ *matrix* φ *whose entries are linear forms in* R. *Assume that* $\operatorname{ht} I_1(\varphi) = 3$ *and that* $u(\varphi) = 1$. *Then the ideal* $I_2(B)$ *has codimension* $n - 1$, *where* B *denotes the matrix* (9.26) *in this case.*

Proof Recall that

$$B = \begin{pmatrix} t_1 & 0 & \cdots & 0 \\ \ell_{2,1} & \ell_{2,2} & \cdots & \ell_{2,n-1} \\ \ell_{3,1} & \ell_{3,2} & \cdots & \ell_{3,n-1} \end{pmatrix}, \tag{9.30}$$

where $\ell_{2,i}$ and $\ell_{3,i}$ are certain linear forms in $k[\mathbf{t}] = k[t_1, \ldots, t_n]$. From the shape of the matrix φ as in (9.24), one can write

$$\varphi = \left(\frac{L}{M} \right),$$

where M is an $(n-1) \times (n-1)$ matrix with linear entries in $k[y, z]$. Let D denote the unique matrix with entries in $k[t_2, \ldots, t_n]$ satisfying the equality

$$(t_2 \cdots t_n) M = (y\ z) D.$$

In addition, $\det M \neq 0$ as it is one of the minimal generators of our main ideal $I = I_{n-1}(\varphi)$. On the other hand, $I_{n-2}(M)$ is (y, z)-primary since it contains the ideal I and hence has codimension 2. Therefore, one can apply Lemma 9.37 with $m = n - 1$ to conclude that $I_2(D)$ has codimension $n - 2$; hence $(I_2(D), t_1)$ has codimension $n - 1$. Thus, letting C denote the submatrix of B of the last two rows, it follows that the ideal $(I_2(C), t_1) = (I_2(D), t_1)$ has codimension $n - 1$.

Now, $I_2(B) = (I_2(C), t_1 I_1(B'))$, where

$$B' := \begin{pmatrix} \ell_{2,2} & \cdots & \ell_{2,n-1} \\ \ell_{2,3} & \cdots & \ell_{3,n-1} \end{pmatrix}. \tag{9.31}$$

Passing to radicals gives $\operatorname{ht} I_2(B) = \min\{\operatorname{ht}(I_2(C), t_1), \operatorname{ht}(I_2(C), I_1(B'))\}$. Therefore, it suffices to prove that the ideal $(I_2(C), I_1(B')) = I_1(B')$ has codimension at least $n - 1$. For this, by Proposition 9.35, B' is 1-generic; hence its entries span a linear vector space of dimension at least $n - 2 + 2 - 1 = n - 1$ (Theorem 1.22 (a)). Therefore $I_1(B')$ has codimension at least $n - 1$. $\square$

The following collects the main results on chaos invariant one.

Theorem 9.39 *Let* $R = k[x, y, z]$ *be a polynomial ring over an algebraically closed field* k, *and let* $I \subset R$ *denote a height* 2 *perfect ideal presented by an* $n \times (n-1)$ *matrix* φ *whose entries are linear forms in* R. *Assume that* ht $I_1(\varphi) = 3$ *and that* $u(\varphi) = 1$. *One has:*

(a) *Let* q *denote the unique minimal prime ideal of all the ideals* $I_t(\varphi)$, *for* $t \neq 1, n-1$, *as obtained in* Proposition 9.30 (a). *Then, for every associated prime* p *of* I *other than* q, I_p *is a complete intersection.*

(b) *Let* q *be as in* (a). *Then* I_q *admits a set of* $n-1$ *generators*

$$\{g_1, \ldots, g_r, h_{r+1}, \ldots, h_{n-1}\}, \ for \ some \ 1 \le r \le n-1,$$

with $g_1, \ldots, g_r \in q_q{}^{n-2} \setminus q_q{}^{n-1}$ *and* $h_{r+1}, \ldots, h_{n-1} \in q_q{}^{n-1} \setminus q_q{}^{n}$.
  In addition, $I_q = q_q{}^{n-2}$ *if and only if* $r = n-1$.

(c) *The special fiber* $\mathcal{F}(I)$ *is the homogeneous coordinate ring either of a rational normal scroll or of a cone over the Veronese embedding of* $\mathbb{P}^1$ *in* $\mathbb{P}^{n-2}$. *In particular,* $\mathcal{F}(I)$ *is Cohen–Macaulay and has minimal degree and* $r(I) \le 1$.

(d) *The Rees algebra of* I *is Cohen–Macaulay.*

(e) *The Rees algebra of* I *is of fiber type and normal.*

Proof

(a) This follows immediately from Proposition 9.30 (b).

(b) As before, we may assume that $q = (y, z)$ and that φ is in the canonical form as in (9.24):

$$\varphi = \left(\begin{array}{c|c} x + a & \\ \hline & H \end{array} \right),$$

where a and the blank entries are 1-forms involving only y and z and so are the entries of the $(n-1) \times (n-2)$ matrix H. Since $x + a$ is invertible locally at q, the ideal I_q is generated by the images of the maximal minors of φ fixing the first row. The latter have the form

$$(x + a)\Delta_i + D_i, \tag{9.32}$$

for $1 \le i \le n-1$, where Δ_i is the $(n-2)$-minor of H excluding the ith row and $D_i \in q^{n-1} \setminus q^n$. Since the $(n-1)$-minor of φ excluding the first row is nonzero and its Laplace expansion along the leftmost column is a linear combination of the Δ_i's, it follows that there is at least one non-vanishing Δ_i. On the other hand, since $x + a$ is invertible in R_q, then $(x + a)\Delta_i + D_i \in q_q{}^{n-2} \setminus q_q{}^{n-1}$ if and only if $\Delta_i \neq 0$. This proves the required statement.

To get the additional statement, it suffices to argue that if $r = n-1$, then I_q is generated by $n-1$ k-linearly independent elements of order $n-2$; hence the modules R_q/I_q and $R_q/q_q{}^{n-2}$ have the same length.

(c) Since $\mathcal{F}(I)$ has codimension $n - 3$, it follows from Proposition 9.35 and Corollary 9.36 that the homogeneous defining ideal of $\mathcal{F}(I)$ is generated by the 2-minors of a 1-generic $2 \times (n - 2)$ matrix of linear forms. By [11, Proposition 9.4 and Proposition 9.12], this matrix is conjugate to the defining matrix of a rational normal scroll or a cone over the standard rational normal curve in $\mathbb{P}^{n-2}$. The latter are varieties of minimal degree; hence $\mathcal{F}(I)$ is Cohen–Macaulay. The value $r(I) \leq 1$ follows from the formula of the degree of $\mathcal{F}(I)$ coming from its Hilbert series.

(d) By Theorem 9.33 and (c), we only have to prove that $r(I_p) \leq 1$ for every associated prime of I. If $p \neq q$, this follows from (a) above. We now deal with the case of the prime q. Still from (a) we know quite generally that the length $\lambda(R_p/I_p^t)$ is a polynomial function for any $t \geq 1$. On the other hand, by (c) we also know that the Hilbert function of $H_{\mathcal{F}(I)}(t)$ is polynomial for every $t \geq 1$. Since

$$\lambda(R_q/I_q^t) = H_{\mathcal{F}(I)}(t) - \binom{nt + 1}{2} + \sum_{p \in \mathrm{Ass}(R/I) \setminus \{q\}} \lambda(R_p/I_p^t),$$

it follows that $\lambda(R_q/I_q^t)$ is also polynomial for every $t \geq 1$. Applying [12, Theorem 2.1], we obtain that $r(I_q) \leq 1$.

(e) Let B denote the matrix of (9.30). Setting $\mathbf{x} = (x\, y\, z)$, due to the nature of B as in (9.25), the ideal $I_1(\mathbf{x}\, B) = I_1((t_1 \cdots t_n)\, \varphi)$ gets identified with the presentation ideal of the symmetric algebra of I.

CLAIM. The ideal $\mathcal{I} := (I_1(\mathbf{x}\, B), I_2(B'))$ coincides with the presentation ideal $\mathcal{J}$ of the Rees algebra of I.

For this, it suffices to show that $\mathcal{I}$ is a prime ideal of codimension $n - 1$ and $\mathcal{I} \subset \mathcal{J}$. First, the inclusion $I_1(\mathbf{x}\, B) \subset \mathcal{J}$ is an obvious general fact. Next, by an easy application of Cramer rule, one has $I_3(B) \subset \mathcal{J}$ as well. But $I_3(B) = t_1 I_2(B')$ and $\mathcal{J}$ is a prime ideal not containing t_1. Therefore, $I_2(B') \subset \mathcal{J}$ as well.

Now recall from Corollary 9.36 that $I_2(B')$ is the defining ideal of a rational normal scroll. On the other hand, one can write

$$\mathcal{I} = (t_1 x + \ell_{2,1} y + \ell_{3,1} z, I_2(B'')),$$

where

$$B'' := \begin{pmatrix} z & \ell_{2,2} & \dots & \ell_{2,n-1} \\ -y & \ell_{2,3} & \dots & \ell_{3,n-1} \end{pmatrix}.$$

Therefore, $I_2(B'')$ is also the defining ideal of a rational normal scroll. In particular, it is a Cohen–Macaulay prime ideal of codimension $n - 2$. Now, $I_2(B'')$ being a prime generated in degree 2 in the variables $y, z, \mathbf{t}$ forces $t_1 x + \ell_{2,1} y + \ell_{3,1} z$ to be a nonzerodivisor on it. It follows that $\mathcal{I}$ is Cohen–Macaulay of codimension $n - 1$. It remains to show that it is a prime ideal. We do this by showing that the ring $R[\mathbf{t}]/\mathcal{I}$ is normal (and hence, the Rees algebra of I will be normal).

It suffices to show the condition (R_1) of Serre's. Note that the Jacobian matrix Θ of $\mathcal{I}$ with respect to the variables $x, y, z, t_1, \ldots, t_n$ has the following shape:

$$\Theta = \begin{pmatrix} B & \mathbf{0} \\ \varphi & \Theta' \end{pmatrix},$$

where Θ' denotes the Jacobian matrix of $I_2(B')$. We have to show that the Jacobian ideal $(I_{n-1}(\Theta), \mathcal{I})$ has codimension at least $n - 1 + 2 = n + 1$. Inspection leads to the inclusion

$$(I_{n-3}(\Theta')\, I_2(B), I_2(B'), I) \subset (I_{n-1}(\Theta), \mathcal{I}).$$

Now, the ideal $(I_{n-3}(\Theta'), I_2(B'))$ has codimension at least $n + 1$ since it is the Jacobian ideal of the normal ring $k[\mathbf{t}]/I_2(B')$. Therefore, it remains to prove that the ideal $(I_2(B), I_2(B'), I) = (I, I_2(B))$ has codimension at least $n + 1$ or, equivalently, that $I_2(B)$ has codimension at least $n - 1$. But this is the statement of Proposition 9.38. $\qquad\square$

Remark 9.40

(i) The shape of the generators in item (b) of the above theorem looks sort of vague. However, any possibility with $1 \le r \le n - 1$ actually happens and in fact even when I is a monomial ideal; for a typical example, one can take

$$I = (xy^{n-2}, xy^{n-3}z, \ldots, xy^{n-(r+1)}z^{r-1}, y^{n-(r+1)}z^r, y^{n-(r+2)}z^{r+1} \ldots, yz^{n-2}, z^{n-1}).$$

(See Section 9.3.5.3 for a more systematic way of dealing with the monomial case.)

(ii) As to the two alternatives in item (c), both can take place. For an emblematic example of the case where the special fiber is a cone over the Veronese embedding of $\mathbb{P}^1$, consider the ideal $I = (x(y, z)^{n-2}, zf)$, where $f \in k[y, z]_{n-2}$, $f \ne 0$. An easy calculation shows that the special fiber is the ring of a cone over the ring of the rational normal curve in $\mathbb{P}^{n-2}$.

9.3.5 Paradigm Classes

In this part we introduce three classes of linearly presented codimension 2 ideals along with their chaos invariants.

9.3.5.1 Linearly Presented Ideals of Plane Fat Points

We assume that k is an algebraically closed field.

An ideal of plane fat points has the form

$$I = p_1^{m_1} \cap \cdots \cap p_r^{m_r},$$

where $p_i \subset R = k[x, y, z]$ is a codimension 2 prime ideal. In particular, it is Cohen–Macaulay, hence a perfect ideal of codimension 2.

Note that since k is algebraically closed, each p_i is generated by two independent linear forms; m_i is a positive integer, called the *multiplicity* of the point associated to the prime ideal p_i. We assume that $m_1 \geq \cdots \geq m_r$.

The following establishes the value of the chaos invariant for these ideals.

Proposition 9.41 *Let $I = p_1^{m_1} \cap \cdots \cap p_r^{m_r}$ denote an ideal of fat points as above. Suppose that I is linearly presented, with syzygy matrix φ of size $n \times (n-1)$. Then $u(\varphi) = n - m_1 - 1$, where m_1 is the highest multiplicity.*

Proof On the one hand, for every i, one clearly has $\mu(I_{p_i}) = \mu(p_i^{m_i}{}_{p_i}) = m_i + 1$.

On the other hand, we know from Lemma 9.27 (b) that $\mu(I_{p_i}) = n - u_{p_i}$, where u_{p_i} denotes the chaos invariant of φ at p_i. This gives $u_{p_i} = n - m_i - 1$, for every i. By (9.22), $u(\varphi) = \min\{n - m_i - 1 \mid 1 \leq i \leq r\} = n - m_1 - 1$, as contended. $\square$

Corollary 9.42 *Let $I = p_1^{m_1} \cap \cdots \cap p_r^{m_r}$ denote an ideal of fat points as above. Suppose that I is linearly presented, with syzygy matrix φ of size $n \times (n-1)$. Then:*

(a) *The minimal primes of $R/I_{u+1}(\varphi)$ are the primes of I having the highest multiplicity.*

(b) *The following conditions are equivalent:*

    (i) $u(\varphi) = n - 2$.
    (ii) *I has only simple points (i.e., I is a radical ideal).*
    (iii) *I is generically a complete intersection.*
    (iv) *I satisfies property G_3.*

Proof

(a) This follows from the proof of the previous proposition.

(b) By the previous proposition, (i) and (ii) are obviously equivalent. Since a radical ideal is generically a complete intersection, (ii) implies (iii). Conversely, if I has a prime p with multiplicity $m \geq 2$, then $\mu(I_p) = \mu(p_p^m) = m + 1 \geq 3$. Therefore, (iii) implies (ii).

    Finally, quite generally for any codimension 2 perfect ideal I in a Cohen–Macaulay ring of dimension 3, (iii) and (iv) are equivalent. $\square$

A major question in this universe is how broad is the class of ideals of fat points allowing for an application of Corollary 2.16. Here we deal with a few special results.

Definition 9.43 Let $\mu = \{\mu_1, \ldots, \mu_r\}$ be a set of nonnegative integers satisfying the following condition: there exists an integer $s \geq 2$ such that

$$\sum_{i=1}^{r} \mu_i = 3(s-1) \quad \text{and} \quad \sum_{i=1}^{r} \mu_i^2 = s(s-1). \tag{9.33}$$

Note that the first of the above relations implies that s is uniquely determined. We will say that μ is a *sub-homaloidal multiplicity set in degree s*. We note that such an integer s is necessarily odd.

Fixing a set of points of cardinality r, the fat ideal with these points and multiplicities $\mu_1, \ldots, \mu_r$ will be called an ideal of sub-homaloidal type in degree s.

Proposition 9.44 *Let $I \subset R = k[x, y, z]$ denote an ideal of fat points which is of sub-homaloidal in degree $s \geq 3$ and generated in degree s. Then:*

(a) *The image of the rational map defined by the linear system I_s is a variety of degree $s/\eth$, where $\eth := (k(I_s) : k(R_s))$ is the field extension degree.*

Suppose in addition that the linear system I_s has the expected dimension $(s + 5)/2$ and that the special fiber $\mathcal{F}(I)$ is Cohen–Macaulay of dimension 3. Then:

(b) *$\eth = 1$, i.e., the rational map defined by I_s is birational onto the image.*
(c) *$r(I) = 2$, and the coefficient of the highest term of the numerator of the Hilbert series of $\mathcal{F}(I)$ is $(s-1)/2$.*

Proof

(a) Let $(s; \mu)$ denote the given sub-homaloidal type, and let p_i $(1 \leq i \leq r)$ denote the respective defining prime ideals of the given points.

   We apply the formula of [22, Theorem 6.6 (a)].

   For this, we first claim that $e_{J_{p_i}}(R_{p_i}) = \mu_i^2$, for every i – here, for an $\mathfrak{n}$-primary ideal $\mathfrak{a}$ in a local ring $(A, \mathfrak{n})$, the symbol $e_{\mathfrak{a}}(A)$ denotes the Samuel multiplicity of A with respect to $\mathfrak{a}$. This follows by recalling that $I_{p_i} = p_i p_i^{\mu_i}$ and applying Samuel's formula (see, e.g., [2, Proposition 4.5.2 (b)]), with $\lambda(\_)$ denoting length:

$$e_{I_{p_i}}(R_{p_i}) = \lim_{n \to \infty} \frac{2}{n^2} \lambda \left(\frac{R_{p_i}}{p_i^{(n+1)\mu_i}} \right) = \lim_{n \to \infty} \frac{2}{n^2} \binom{(n+1)\mu_i + 1}{2}$$

$$= \lim_{n \to \infty} \frac{(n+1)^2 \mu_i^2}{n^2} = \mu_i^2. \tag{9.34}$$

   Now in the notation of [22, Theorem 6.6 (a)], we take $R = k[x, y, z]$ to be our ground ring and I an ideal of height $g = 2$. Then p runs through the primes p_i, i.e., the minimal primes of R/I. According to that result, there is actually an equality

$$e(k[I_s]) = \frac{1}{r}\left(e(R)s^2 - \sum_i e_{I_{p_i}}(R_{p_i})\, e(R/p_i)\right) = \frac{1}{r}(s^2 - \sum_i \mu_i^2),$$

by (9.34) above and because p_i is generated by linear forms.

By the second equation of condition for I, one has $\sum_i \mu_i^2 = s^2 - s$. It follows that $e(k[I_s]) = s/r$.

(b) Let $h := 1 + h_1 t + h_2 t^2 + \cdots$ stand for the numerator of the Hilbert series of $\mathcal{F}(I)$. From (a) and the assumption on the expected dimension of I_s, one has

$$s/\partial = 1 + ((s+5)/2 - 3) + \sum_{t \geq 2} h_t = (s+1)/2 + \sum_{t \geq 2} h_t.$$

Since $\mathcal{F}(I)$ is Cohen–Macaulay, $\sum_{t \geq 2} h_t \geq 0$. This is impossible unless $\partial = 1$.

(c) Since the rational map defined by I_s is birational onto the image, the latter is a rational variety. By hypothesis, it is arithmetically Cohen–Macaulay; hence condition (iii) of Corollary 2.16 is satisfied (see, e.g., [9, The proof of Corollary 2.6]), thus implying that $r(I) \leq 2$.

In the notation of the proof of (b), we then obtain $h_t = 0$ for $t \geq 3$. This gives $h_2 = s - (s+1)/2 = (s-1)/2 \geq 1$ and, in particular, $r(I) = 2$. $\square$

Remark 9.45 Regarding the proof of item (c) in the above proposition, the reason to draw on Corollary 2.16, instead the more direct Theorem 9.33, is that I is not linearly presented. In [18] an appropriate quadratic transform of I is introduced that keeps the essential properties of the latter and, moreover, is a linearly presented ideal – for this Theorem 9.33 could safely apply.

9.3.5.2 Reciprocal Ideals of Hyperplane Arrangements

For the basic material of this part, we refer to [8] and the references thereof.

Let $\mathcal{A} = \{H_1, \ldots, H_n\} \subset \mathbb{P}^{d-1}$ be a central hyperplane arrangement of size n and rank r. Here $H_i = \ker(\ell_i)$, $i = 1, \ldots, n$, where each ℓ_i is a linear form in $R := k[x_1, \ldots, x_r]$ and $k[\ell_1, \ldots, \ell_n] = R$. From the algebraic viewpoint, there is a natural emphasis on the linear forms ℓ_i and the associated ideal theoretic notions.

We consider the ideal of $(n-1)$-fold products of the linear forms $\ell_1, \ldots, \ell_n$:

$$I := (\ell_1 \cdots \hat{\ell}_i \cdots \ell_n \mid 1 \leq i \leq n), \tag{9.35}$$

where the hat means omission.

As it turns out, I is a linearly presented codimension 2 ideal. Taking the generators in the above order, the corresponding syzygy matrix has the shape

$$\varphi = \begin{bmatrix} \ell_1 & & & & \\ -\ell_2 & \ell_2 & & & \\ & -\ell_3 & \ell_3 & & \\ & & \ddots & \ddots & \\ & & & -\ell_{n-1} & \ell_{n-1} \\ & & & & -\ell_n \end{bmatrix} = \begin{bmatrix} \ell_1 & & & & \\ & \ell_2 & & & \\ & & \ell_3 & & \\ & & & \ddots & \\ & & & & \ell_{n-1} \\ -\ell_n & -\ell_n & -\ell_n & \cdots & -\ell_n \end{bmatrix},$$

where the blank slots have null entries.

It is easy to see that, for any $1 \leq t \leq n - 1$, the ideal of t-minors $I_t(\varphi)$ is generated by the t-fold products of $\ell_1, \ldots, \ell_n$. Note that since the arrangement is of maximal rank, the height of the ideal $I_1(\varphi)$ is 3. This entails the question as to what values can the chaos invariant of φ take. The basic parts of code theory over an arbitrary field give the following proposition:

Proposition 9.46 *Let $r = 3$, and let $A(\boldsymbol{\ell})$ stand for the $r \times n$ scalar matrix over k whose columns are the respective coefficients of $\ell_1, \ldots, \ell_n$. Then the minimal distance of the linear code defined by $A(\boldsymbol{\ell})$ coincides with the chaos invariant of φ.*

Proof Quite generally, by [4] (see also Subsection 1.1), it turns out that the minimal distance is the largest t such that $\dim R/I_t(\varphi) = 0$. For $r = 3$, this clearly coincides with the chaos invariant of φ. $\qquad\square$

The special fiber $\mathcal{F}(I)$ of this ideal is the so-called *Orlik–Terao algebra*. It is known to have dimension $r = \dim R$ (see [24]). It has been proved in [16] that $\mathcal{F}(I)$ is Cohen–Macaulay. Assuming this result, we prove that if $\dim R = 3$, then the Rees algebra $\mathcal{R}(I)$ of I is Cohen–Macaulay. This is a particular case of the general fact that $\mathcal{R}(I)$ is Cohen–Macaulay in any dimension, as proved in [8] by different methods.

Theorem 9.47 *Let $\dim R = 3$, and let I be as in (9.35). Then the Rees algebra $\mathcal{R}(I)$ is Cohen–Macaulay.*

Proof As remarked above, $\mathcal{F}(I)$ is Cohen–Macaulay. By Theorem 9.33, it suffices to show that $r(I_p) \leq 1$ for every associated prime p of R/I – this fact alone will hold in an arbitrary dimension of R.

Letting $\{\ell_1, \ldots, \ell_n\}$ be as above, where no two are proportional, we easily see that any associated prime p of R/I is generated by two such linear forms. Say, $p = (\ell_1, \ell_2)$ without loss of generality. Now consider all circuits of order 3 involving ℓ_1 and ℓ_2. Let c_p denote the number of these circuits, and consider the c_p additional linear forms intervening in these circuits, one for each such circuit. Without loss of generality, assume that the additional forms are $\ell_3, \ldots, \ell_t$, where we have set $t := c_p + 2$. Focusing on the sub-arrangement $\{\ell_1, \ell_2, \ldots, \ell_t\}$, we consider the ideal J of R generated by its $(t - 1)$-fold products. Clearly, $J \subset p^{t-1}$. But since these ideals are generated in the same degree and the products generating J are k-linearly independent, we conclude that $J = p^{t-1}$.

Now, as one easily sees, $I_p = J_p$. Therefore, $I_p = p_p^{t-1}$.

Since R_p is a regular local ring of dimension 2, the subideal $(\ell_1^{t-1}, \ell_2^{t-1})_p \subset p_p^{t-1}$ is a reduction with reduction number at most 1 (exactly 1 if $c_p \geq 1$).   □

The case of chaos invariant one is fairly simple:

Proposition 9.48 *Let $r = 3$ and write $R = k[x, y, z]$. Let I be as in (9.35). The following are equivalent:*

(i) *Up to a change of variables in $R = k[x, y, z]$, one has*

$$\ell_1 = x, \ell_2 = y + c_2 z, \ell_3 = y + c_3 z, \ldots, \ell_n = y + c_n z,$$

  where the coefficients $c_j \in k$ are mutually distinct.

(ii) $u = 1$.

(iii) *The reduction number of I is at most 1.*

Proof (i) $\Rightarrow$ (ii). By the shape of the syzygy matrix φ as depicted earlier, under the present values, one sees that $I_1(\varphi) = 3$ and $I_2(\varphi) \subset (y, z)$. Therefore, $u = u(\varphi) = 1$.

(ii) $\Rightarrow$ (iii). This reads off Corollary 9.39 (c).

(iii) $\Rightarrow$ (i). By [8, Proposition 27], quite generally for hyperplane arrangements of maximal rank in $k[x_1, \ldots, x_r]$, the reduction number of I is $r - q$, where q is the number of partitions of $\mathcal{A}$ into sets of linear forms in non-overlapping variables. In the present case, with $r = 3$, one gets $q \geq 2$. Therefore, up to a change of variables, it is clear that $\mathcal{A}$ has the required shape as in (i).   □

Quite generally, by the same argument as in (iii), the reduction number of I is at most 2. Thus, its exact value has no direct impact on u (except in the above case). At the other end, for arbitrary $u \geq 2$, we have no fluid characterization similar to the one in (i), except in the generic case.

We recall that an arrangement of maximal rank is *generic* provided any three of the defining linear forms are k-linearly independent. We are indebted to Ş. Tohăneanu for the argument in the following result.

Proposition 9.49 *Let $r = 3$, and let the arrangement be generic. Then the chaos invariant of the syzygy matrix φ is maximal, i.e., $u(\varphi) = n - 2$.*

Proof By Proposition 9.46, it suffices to prove that the minimal distance d of the linear code defined by the associated matrix $A(\ell)$ is $n - 2$. Since the arrangement is generic, the code is MDS (maximum distance separable); hence d achieves the Singleton bound $d \leq n - r + 1 = n - 2$.   □

9.3.5.3   Linearly Presented Ternary Monomial Ideals

Assume throughout that k is an algebraically closed field.

A linearly presented codimension 2 primary ideal $I \subset R = k[x, y, z]$ which is minimally generated by n monomials, all of degree $n - 1$, is up to the change of variables the power $(y, z)^{n-1}$. With the natural order of the monomials in the latter, the syzygy matrix of I has the shape

$$
\begin{pmatrix}
z & 0 & 0 & \ldots & 0 & 0 & 0 \\
-y & z & 0 & \ldots & 0 & 0 & 0 \\
0 & -y & z & \ldots & 0 & 0 & 0 \\
\vdots & \vdots & \vdots & \ddots & \vdots & \vdots & \vdots \\
0 & 0 & 0 & \ldots & z & 0 & 0 \\
0 & 0 & 0 & \ldots & -y & z & 0 \\
0 & 0 & 0 & \ldots & 0 & -y & z \\
0 & 0 & 0 & \ldots & 0 & 0 & -y
\end{pmatrix}.
\tag{9.36}
$$

Dealing with non-primary such ideals is somewhat more delicate. We next deal with the case of two minimal primes, while the case of three minimal primes requires a mixed approach elsewhere.

Proposition 9.50 *Let $I \subset R = k[x, y, z]$ denote a linearly presented codimension 2 perfect ideal minimally generated by n monomials of degree $n - 1$. If R/I has exactly two minimal primes, then, up to conjugation, the syzygy matrix of I has the form*

$$
\varphi =
\begin{pmatrix}
z & 0 & 0 & \ldots & 0 & 0 & 0 \\
-c_1 & z & 0 & \ldots & 0 & 0 & 0 \\
0 & -c_2 & z & \ldots & 0 & 0 & 0 \\
\vdots & \vdots & \vdots & \ddots & \vdots & \vdots & \vdots \\
0 & 0 & 0 & \ldots & z & 0 & 0 \\
0 & 0 & 0 & \ldots & -c_{n-3} & z & 0 \\
0 & 0 & 0 & \ldots & 0 & -c_{n-2} & z \\
0 & 0 & 0 & \ldots & 0 & 0 & -c_{n-1}
\end{pmatrix},
\tag{9.37}
$$

where $c_i \in \{x, y\}$ for each $1 \leq i \leq n - 1$.

Proof Up to change of variables in R, we may assume that the minimal primes of R/I are (x, z) and (y, z); hence $\sqrt{I} = (xy, z)$. Therefore, a power of z lies in I, and since I is generated by monomials of degree $n - 1$, it follows that $z^{n-1} \in I$. By the same token, I has a monomial generator of the form $x^a y^b$, with $a + b = n - 1$. Fixing such a generator, we claim that a remaining set of $n - 2$ monomial generators of I is

$$
\{x^{a_i} y^{b_i} z^i,\ 1 \leq i \leq n - 2,\ a_i + b_i = n - 1 - i\}.
\tag{9.38}
$$

That is, no two minimal monomial generators have in their support the same power of z.

Indeed, assuming otherwise, there would be a syzygy $\mathfrak{s}$ of I with only two nonzero coordinates which are monomials in x and y only. Since by assumption the syzygy matrix φ of I is linear, $\mathfrak{s}$ would have to be a combination of linear syzygies. Now, because I is minimally generated by monomials, a generating linear syzygy can be assumed to have only two nonzero coordinates, and these are then necessarily different (signed) variables. On the other hand, the generator z^{n-1} being a maximal minor of φ implies that every column of φ has an entry $\pm z$ in exactly one row except one. It follows that $\mathfrak{s}$ cannot be a combination of syzygies represented by the columns of φ.

Having granted (9.38), let us write the complete set of monomial generators thus obtained in the following ordering:

$$\{x^{a_0}y^{n-1-a_0},\ x^{a_1}y^{n-2-a_1}z,\ \ldots,\ x^{a_{n-2}}y^{1-a_{n-2}}z^{n-2},\ z^{n-1}\},\ a_i \geq 0. \tag{9.39}$$

One immediately sees that a generating linear syzygy between these generators can only involve contiguous ones in the above ordering because of the sequential powers of z. Moreover, its nonzero coordinates are z and necessarily a signed x or y. Therefore, I admits at least those many syzygies that appear in the matrix (9.37). But this partial matrix has rank $n-1$ and hence must be a full syzygy matrix of I since the latter is perfect of codimension two. $\qquad\square$

In the assumption of the above proposition, the sequence of entries $c_1\, c_2\, \ldots\, c_{n-1}$ will be referred to as the *basic entry sequence* of the ideal I.

We next give the structural properties of two remarkable cases of the basic entry sequence.

Theorem 9.51 *Let the assumptions and notation be those of* Proposition 9.50, *and let the basic entry sequence be an alternating sequence of x and y, i.e., of the shape $\ldots x\, y\, x\, y\, x\, y \ldots$ Then:*

(a) *The special fiber of the ideal I is Cohen–Macaulay and normal.*
(b) *The Rees algebra of I is Cohen–Macaulay and normal.*

Proof We assume that the sequence has the shape $x\, y\, x\, y\ \ldots\ x\, y\, x\, y$, i.e., that $n-1$ is even - the case where $n-1$ is odd is handled similarly with minor changes. Denoting $\{f_0, f_1, \ldots, f_{n-1}\}$ the generators in (9.39), one has

$$f_i = \begin{cases} x^{(n-1)/2}y^{(n-1)/2}, & \text{if } i = 0 \\ (f_{i-1}/x)z, & \text{if } i \text{ is odd} \\ (f_{i-1}/y)z, & \text{if } i \text{ is even.} \end{cases} \tag{9.40}$$

(a) With the generators in the form (9.40), it is straightforward to verify that the homogeneous defining ideal of the special fiber of I, on variables $\mathbf{t} = \{t_0, \ldots, t_{n-1}\}$, contains the 2-minors of the 2-leap catalecticant

$$C = \begin{pmatrix} t_0 & t_1 & t_2 & \ldots & t_{n-3} \\ t_2 & t_3 & t_4 & \ldots & t_{n-1} \end{pmatrix}.$$

By Corollary 7.4, $k[t_0, t_1, \ldots, t_{n-1}]/I_2(C)$ is a Cohen–Macaulay domain of codimension $n - 3$. Therefore, this algebra is the defining ideal of $\mathcal{F}(I)$.

Since an r-leap catalecticant of order $2 \times s$ is up to columns permutation the concatenation of independent Hankel matrices – e.g., in the case of C it suffices to permute the second and third columns – then it is the homogeneous coordinate ring of a rational normal scroll, hence normal.

(b) By (a), the special fiber of I is Cohen–Macaulay and $r(I) \leq 1$ (see Theorem 9.39 (c)).

We first show that the Rees algebra $\mathcal{R}_R(I)$ is Cohen–Macaulay. Since the fiber $\mathcal{F}(I)$ is Cohen–Macaulay, it suffices by Theorem 9.33 to prove that $r(I_p) \leq 1$ and $r(I_q) \leq 1$, where $p = (x, z)$ and $q = (y, z)$. As in (9.40), note that

$$I = (x^m y^m, x^{m-1} y^m z, x^{m-1} y^{m-1} z^2 \ldots, x^i y^i z^{n-1-2i}, x^{i-1} y^i z^{n-2i}, \ldots, xyz^{n-3}, yz^{n-2}, z^{n-1}),$$

with $m := (n - 1)/2$. Then, localizing and cancelling superfluous generators yields

$$I_p = (x^m, x^{m-1}z, \ldots, xz^{n-3}, z^{n-2}) \quad \text{and} \quad I_q = (y^m, y^{m-1}z^2, \ldots, yz^{n-3}, z^{n-1}).$$

By [10, Proposition 2.15], I_p and I_q are normal ideals of R_p and R_q, respectively. Thus, the Rees algebras $\mathcal{R}_{R_p}(I_p)$ and $\mathcal{R}_{R_q}(I_q)$ are Cohen–Macaulay. Therefore, $r(I_p) \leq 1$ and $r(I_q) \leq 1$.

We next prove that $\mathcal{R}_R(I)$ is normal, by showing that it satisfies Serre's condition (R_1). Let $\mathcal{J}$ stand for its presentation ideal on $R[\mathbf{t}]$. We note that, quite generally, the codimension of $\mathcal{J}$ is 1 less than the minimal number of generators of I. Now, we need to bound below the codimension of the ideal $(I_{n-1}(\Theta(\mathcal{J}), \mathcal{J})$. Thus, we are to show that the codimension of $(I_{n-1}(\Theta(\mathcal{J}), \mathcal{J})$ is at least $n - 1 + 2 = n + 1$.

On the other hand, $\mathcal{J}$ contains the ideal $\mathcal{I} := (t\varphi, I_2(C))$, and the generators of the latter can be taken to be a subset of a full set of generators of $\mathcal{J}$. Accordingly, the Jacobian matrix $\Theta(\mathcal{I})$ of the generators of $\mathcal{I}$ is a submatrix of the Jacobian matrix $\Theta(\mathcal{J})$ of the generators of $\mathcal{J}$, with the same number of columns. This gives $I_{n-1}(\Theta(\mathcal{I})) \subset I_{n-1}(\Theta(\mathcal{J}))$, and hence

$$(I_{n-1}(\Theta(\mathcal{I}), \mathcal{I}) \subset (I_{n-1}(\Theta(\mathcal{J}), \mathcal{J}),$$

so it suffices to bound below the codimension of the leftmost ideal.

Note that

$$\Theta(\mathcal{I}) = \begin{pmatrix} B & 0 \\ \varphi & \Theta' \end{pmatrix},$$

where Θ' denotes the Jacobian matrix of the 2-minors of C and B denotes the Jacobian dual matrix of φ. Therefore, the ideal $(I_2(B) I_{n-3}(\Theta'), I_2(C), I)$ is contained in $(I_{n-1}(\Theta(\mathcal{I}), \mathcal{I})$.

So far, the discussion has been kept on a fairly arbitrary level. Let us bring in the particulars of the alternating case via the special shape of B. Namely, the transpose of B has the shape

$$\begin{pmatrix} -t_1 & 0 & t_0 \\ 0 & -t_2 & t_1 \\ -t_3 & 0 & t_2 \\ 0 & -t_4 & t_3 \\ \vdots & \vdots & \vdots \\ 0 & -t_{n-1} & t_{n-2} \end{pmatrix}.$$

Therefore, $(t_0 t_{n-1}, t_1^2, t_2^2, \ldots, t_{n-2}^2) \subset I_2(B)$, implying that the codimension of $(I_2(B), I)$ is at least $(n-1) + 2 = n + 1$. As to the "component" $(I_{n-3}(\Theta'), I_2(C), I)$, one is through as well since $(I_{n-3}(\Theta'), I_2(C))$ is the Jacobian ideal of the normal domain $k[\mathbf{t}]/I_2(C)$ of codimension $n - 1$.   □

Conjecture 9.52 In the above case of an alternating basic entry sequence, the Rees algebra of the ideal I is of fiber type.

The second remarkable case is as follows.

Theorem 9.53 *Let the assumptions and notation be those of* Proposition 9.50, *and let the basic entry sequence be a separating sequence of x and y, i.e., of the shape* $c_1 = x, c_2 = x, \ldots, c_r = x, c_{r+1} = y, c_{r+2} = y, \ldots, c_{n-1} = y$. *Then:*

(a) *The special fiber of the ideal I is Cohen–Macaulay.*
(b) *The Rees algebra of I is Cohen–Macaulay and of fiber type.*

Proof

(a) The argument in this case is more involved, and it will appeal to a "deformation" argument.

   Set $s := n - 1 - (r + 1) + 1 = n - 1 - r$. The ideal I is generated by the monomials

$$\begin{cases} x^r y^s, & \text{if } i = 0 \\ x^{r-i} y^s z^i, & \text{if } 1 \le i \le r \\ y^{n-1-i} z^i, & \text{if } r + 1 \le i \le n - 1. \end{cases} \tag{9.41}$$

Consider the following Hankel matrices:

$$H_1 = \begin{pmatrix} t_0 \ t_1 \ \cdots \ t_{r-1} \\ t_1 \ t_2 \ \cdots \ t_r \end{pmatrix} \quad \text{and} \quad H_2 = \begin{pmatrix} t_r \ \ t_{r+1} \ \cdots \ t_{n-2} \\ t_{r+1} \ t_{r+2} \ \cdots \ t_{n-1} \end{pmatrix},$$

and set $P := (I_2(H_1), I_2(H_2))k[t_0, t_1, \ldots, t_{n-1}]$.

The content of the item is a consequence of the following claims:

CLAIM 1. Let $\mathcal{L}$ denote the homogeneous defining ideal of the special fiber of I on variables $t_0, \ldots, t_{n-1}$ mapping to generators as in (9.41). Then $P \subset \mathcal{L}$.

CLAIM 2. $k[t_0, t_1, \ldots, t_{n-1}]/P$ is a Cohen–Macaulay domain of dimension 3.

The first claim is a straightforward verification, similar to the one in the alternating case, so is left to the reader.

For the second claim, we first show that the ring is Cohen–Macaulay. Introduce a new variable T, and consider the following Hankel deformation of H_2

$$H_2' := \begin{pmatrix} T \ \ t_{r+1} \ \cdots \ t_{n-2} \\ t_{r+1} \ t_{r+2} \ \cdots \ t_{n-1} \end{pmatrix}.$$

Set $P' := (I_2(H_1), I_2(H_2')) \subset k[t_0, \ldots, t_{n-1}, T]$ and $S := k[t_0, \ldots, t_{n-1}, T]/P'$. Then

$$S \simeq k[t_0, \ldots, t_r]/I_2(H_1) \otimes_k k[t_{r+1}, t_{r+2}, \ldots t_{n-1}, T]/I_2(H_2').$$

Since k is algebraically closed, the k-algebra S is a Cohen–Macaulay domain of dimension $2 + 2 = 4$. In particular, $T - t_r$ is regular on S; hence

$$k[t_0, t_1, \ldots, t_{n-1}]/P \simeq S/(T - t_r)$$

is a Cohen–Macaulay ring of dimension 3.

It remains to show that P is a prime ideal.

We first show that t_r is regular on $k[t_0, t_1, \ldots, t_{n-1}]/P$. If not, let $Q \subset k[t_0, t_1, \ldots, t_{n-1}]$ denote an associated prime of $k[t_0, t_1, \ldots, t_{n-1}]/P$ containing t_r. We assert that necessarily $\{t_1, \ldots, t_{n-2}\} \subset Q$, and hence Q has codimension at least $n - 2$, which gives a contradiction since $k[t_0, t_1, \ldots, t_{n-1}]/P$ is Cohen–Macaulay of codimension $n - 3$.

To prove the assertion, if $t_r \in Q$, since both minors $\det \begin{pmatrix} t_{r-2} \ t_{r-1} \\ t_{r-1} \ \ t_r \end{pmatrix}$ and $\det \begin{pmatrix} t_r \ \ t_{r+1} \\ t_{r+1} \ t_{r+2} \end{pmatrix}$ belong to Q, it follows that $t_{r-1}, t_{r+2} \in Q$. But again, since $\det \begin{pmatrix} t_{r-3} & t_{r-2} \\ t_{r-2} & t_{r-1} \end{pmatrix}$ and $\det \begin{pmatrix} t_{r+1} & t_{r+2} \\ t_{r+2} & t_{r+3} \end{pmatrix}$ both belong to Q it follows that $t_{r-2}, t_{r+2} \in Q$. Continuing, eventually $\{t_1, \ldots, t_{n-2}\} \subset Q$, as claimed.

Finally, we argue that the localization of $k[t_0, t_1, \ldots, t_{n-1}]/P$ at t_r is a domain. Indeed, inverting t_r in each of the matrices H_1 and H_2 by the well-known procedure, the resulting 2-minors become dependent only on the three variables t_r, t_{r-1}, and t_{r+1}. More exactly,

$$\frac{k[t_0, \ldots, t_{n-1}][t_r^{-1}]}{Pk[t_0, \ldots, t_{n-1}][t_r^{-1}]} \simeq k[t_r, t_{r-1}, t_{r+1}][t_r^{-1}].$$

This wraps up the item.

(b) Consider the following scroll matrices:

$$S_1 = \begin{pmatrix} x & t_0 & t_1 & \ldots & t_{r-1} \\ -z & t_1 & t_2 & \ldots & t_r \end{pmatrix} \quad \text{and} \quad S_2 = \begin{pmatrix} y & t_r & t_{r+1} & \ldots & t_{n-2} \\ -z & t_{r+1} & t_{r+2} & \ldots & t_{n-1} \end{pmatrix},$$

and set $\mathcal{P} := (I_2(S_1), I_2(S_2)) \subset k[x, y, z, t_0, \ldots, t_{n-1}]$.

We will show that $\mathcal{P}$ coincides with a presentation ideal $\mathcal{J}$ of $\mathcal{R}(I)$. For this it suffices to show that $\mathcal{P} \subset \mathcal{J}$ and that $\mathcal{P}$ is a prime ideal of dimension 4. The inclusion is clear since the 2-minors on each side are either defining relations of the symmetric algebra of I or polynomial relations of I (the latter by part (a)). Moreover, once verified that $\mathcal{P} = \mathcal{J}$, then it follows immediately that $\mathcal{R}(I)$ is of fiber type.

We proceed to show that $\mathcal{P}$ is a Cohen–Macaulay prime ideal.

Let T and U be new variables, and consider the following deformation of S_2 to "separate" variables:

$$S_2' := \begin{pmatrix} y & U & t_{r+1} & \ldots & t_{n-2} \\ T & t_{r+1} & t_{r+2} & \ldots & t_{n-1} \end{pmatrix}.$$

Note that all three S_1, S_2, and S_2' define rational normal scroll surfaces in their respective projective ambients.

Set $\mathcal{P}' := (I_2(S_1), I_2(S_2')) \subset k[x, y, z, t_0, \ldots, t_{n-1}, T, U]$.

CLAIM 1: The ring $S := k[x, y, z, t_0, \ldots, t_{n-1}, T, U]/\mathcal{P}'$ is a Cohen–Macaulay domain of dimension 6.

Note that

$$k[t_0, \ldots, t_{n-1}, x, y, z, T, U]/\mathcal{P}' \simeq \frac{k[t_0, \ldots, t_r, x, z]}{I_2(S_1)} \otimes_k \frac{k[t_{r+1}, t_{r+2}, \ldots t_{n-1}, y, T, U]}{I_2(S_2')}$$

is Cohen–Macaulay of dimension $3 + 3 = 6$. Since k is assumed to be algebraically closed, it is a domain.

Specializing back only one of the two deformation variables

$$S_2'' := \begin{pmatrix} y & t_r & t_{r+1} & \cdots & t_{n-2} \\ T & t_{r+1} & t_{r+2} & \cdots & t_{n-1} \end{pmatrix}$$

still gives a matrix defining a rational normal scroll in the respective projective ambient.

Set $\mathcal{P}'' := (I_2(S_1), I_2(S_2''))$.

CLAIM 2: The ring $S' := k[x, y, z, t_0, t_1, \ldots, t_{n-1}, T]/\mathcal{P}''$ is a Cohen–Macaulay domain of dimension 5.

Since this is an obvious specialization with $U - t_r$ regular on S, the assertion on the Cohen–Macaulayness and the dimension are immediate.

To show that $\mathcal{P}''$ is a prime ideal, we will argue that t_r is regular on S' and that the localization of the latter at t_r is a domain.

Let $Q \subset k[x, y, z, t_0, t_1, \ldots, t_{n-1}, T]$ denote an associated prime of S' containing t_r. We claim that necessarily $\{t_1, \ldots, t_{n-2}, zt_0, yt_{n-1}\} \subset Q$, and hence Q has codimension at least n; but this is impossible since S' is Cohen–Macaulay of codimension $n - 1$.

To see the claim, if $t_r \in Q$, since both minors $\det\begin{pmatrix} t_{r-2} & t_{r-1} \\ t_{r-1} & t_r \end{pmatrix}$ and

$\det\begin{pmatrix} t_r & t_{r+1} \\ t_{r+1} & t_{r+2} \end{pmatrix}$ belong to Q, it follows that $t_{r-1}, t_{r+2} \in Q$. But again,

since $\det\begin{pmatrix} t_{r-3} & t_{r-2} \\ t_{r-2} & t_{r-1} \end{pmatrix}$ and $\det\begin{pmatrix} t_{r+1} & t_{r+2} \\ t_{r+2} & t_{r+3} \end{pmatrix}$ both belong to Q, it follows that $t_{r-2}, t_{r+2} \in Q$. Continuing, eventually $\{t_1, \ldots, t_{n-2}\} \subset Q$. Moreover, both

minors $\det\begin{pmatrix} x & t_0 \\ -z & t_1 \end{pmatrix}$ and $\det\begin{pmatrix} y & t_{n-2} \\ T & t_{n-1} \end{pmatrix}$ belong to Q, and it follows that

$zt_0, yt_{n-1} \in Q$. Therefore, $\{t_1, \ldots, t_{n-2}, zt_0, yt_{n-1}\} \subset Q$, as claimed.

Finally, inverting t_r yields an isomorphism

$$\frac{k[x, y, z, t_0, t_1, \ldots, t_{n-1}, T][t_r^{-1}]}{\mathcal{P}'' k[x, y, z, t_0, t_1, \ldots, t_{n-1}, T][t_r^{-1}]} \simeq k[y, z, t_r, t_{r-1}, t_{r+1}][t_r^{-1}].$$

CLAIM 3: The ring $k[x, y, z, t_0, t_1, \ldots, t_{n-1}]/\mathcal{P}$ is a Cohen–Macaulay domain of dimension 4.

Again, this is a specialization with $T + z$ regular on S'; hence the assertions about the Cohen–Macaulayness and the dimension are immediate.

To see that $\mathcal{P}$ is a prime ideal, we follow the same path as above, namely, we will show that t_r is regular on S'' and that the localization of the latter at t_r is a domain.

Let $Q \subset k[t_0, t_1, \ldots, t_{n-1}, x, y, z]$ denote an associated prime of S'' containing t_r. In the same way as above, we have that $\{t_1, \ldots, t_{n-2}, zt_0, yt_{n-1}\} \subset$

Q, and hence Q has codimension at least n; but this is impossible since S'' is Cohen–Macaulay of codimension $n - 1$.

Finally, inverting t_r yields an isomorphism

$$\frac{k[x, y, z, t_0, \ldots, t_{n-1}][t_r^{-1}]}{\mathcal{P}'' k[x, y, z, t_0, \ldots, t_{n-1}][t_r^{-1}]} \simeq k[y, t_r, t_{r-1}, t_{r+1}][t_r^{-1}],$$

thus proving the statement. $\square$

Remark 9.54 The special fiber of the separating case is not normal even when it is balanced (i.e., when the number of x entries and of y entries is equal). The basic entry sequences picked in random order of x and y have even less chance of having a normal special fiber since the latter may not even be Cohen–Macaulay. The question remains as to whether there are other events in the present context where normality takes place besides the alternating (scroll) case.

Exercises

9.55 Consider the following 5×4 matrix in five variables:

$$\varphi = \begin{bmatrix} x_1 & 0 & 0 & 0 \\ x_3 & x_1 & 0 & 0 \\ 0 & x_4 & x_2 & 0 \\ 0 & 0 & x_5 & x_2 \\ 0 & 0 & 0 & x_3 \end{bmatrix},$$

and set $I = I_4(\varphi)$.

1. Prove that I is not of linear type.
   (HINT: I does not satisfy condition G_5.)
2. Show that the rational map defined by the maximal minors of φ is a Cremona map of $\mathbb{P}^4$ (of degree 4).

9.56 Consider the matrix of (9.4)

$$\varphi = \begin{pmatrix} x_1 & x_2 & x_3 \\ x_2 & x_3 & 0 \\ x_3 & 0 & x_1 \\ 0 & x_1 & x_2 \end{pmatrix},$$

and let $N = \text{Im}(\eta)$ stand for the matrix of (9.2) associated to φ. This exercise asks for the details of this example:

1. Prove that the k-vector space dimension of the linear forms in N is 8.
2. Show that, by changing six out of the nine nonzero entries of φ into general 1-forms, the resulting matrix has $\dim_k I_1(N)_1 = 9$, the maximal possible value.
   (HINT: Computer assistance may be needed.)

9.57 This problem is intended to work the details of Example 9.16.
   Let $I := I_3(\varphi) \subset R = k[x_1, x_2, x_3]$ denote the ideal of 3-minors of the matrix φ in the previous problem.

1. Show that $I^{(2)}/I^2$ is a cyclic R-module generated by the residue class of a form $f \in I^{(2)}$ of degree $4 < d(d-1) - 1 = 5$, thereby showing that Proposition 9.15 fails.
2. Prove that the map defined by the 3-minors is still birational onto the image, with inversion factors $x_1 f$, $x_2 f$, and $x_3 f$; in particular, the latter are not minimal generators of $I^{(2)}$.
3. Argue that a slight "perturbation" of the entries of φ may not be enough to make Proposition 9.15 hold.
   (HINT: Consider the following perturbation:

$$\begin{pmatrix} x_1 & x_2 & x_3 \\ x_2 & x_3 & 0 \\ x_3 & 0 & x_1 - x_2 \\ 0 & x_1 - x_3 & x_2 - x_3 \end{pmatrix} .)$$

4. Show that the richer perturbation of φ

$$\begin{pmatrix} x_1 & x_2 & x_3 \\ x_2 & x_3 & x_1 - x_2 \\ x_3 & x_1 - x_3 & x_2 - x_3 \\ x_1 + x_2 & x_2 + x_3 & x_1 + x_3 \end{pmatrix}$$

makes Proposition 9.15 hold and so Theorem 9.13 as well.
5. Show however that, under the perturbation in the previous item, now the statement in Theorem 9.17 (i) below fails as those generators have a nontrivial common divisor.
6. Show that changing the lower right corner entry of the first of the above matrices into $x_1 - x_2 + x_3$ causes the ideal of maximal minors to be no longer radical.
   (HINT: Computer assistance may be needed throughout.)

9.58 Consider the matrix

$$\varphi = \varphi \begin{pmatrix} 0 & x & 0 \\ 0 & -z & -y^2 \\ x & 0 & 0 \\ -z & 0 & x^2 \end{pmatrix}$$

with entries in $k[x, y, z]$, and let $I := I_3(\varphi)$.

1. Verify that I has height 1, and retrieve its factor $\widetilde{I}$ of degree 3.
2. Prove that the ideal $\widetilde{I}$ of the previous item has height 2 but is not the ideal of 3-minors of a 4×3 matrix of linear entries.
3. Produce the free linear resolution of $\widetilde{I}$ over $k[x, y, z]$.

9.59 In the terminology and notation of Section 9.3.5.3, let the basic entry sequence be $x\,y\,x\,x\,y\,y$, obtained by a single transposition of the alternating case. Let $I \subset k[x, y, z]$ denote the ideal of the maximal minors of the associated matrix φ.

1. Set $n = 7$, the embedding dimension of the special fiber $\mathcal{F}(I)$. Show that the chaos invariant is $u = 3$, hence $n < 2(u + 1)$, and that $I_{u+1}(\varphi) = I_4(\varphi)$ has two minimal primes. This should be confronted with Proposition 9.30 (b).
2. Let $J \subset k[t_1, \ldots, t_7]$ denote the ideal generated by the sum of the ideals of maximal minors of the following matrices:

$$\begin{pmatrix} t_1 & t_2 & t_5 \\ t_2 & t_3 & t_6 \end{pmatrix}, \quad \begin{pmatrix} t_2 & t_4 & t_5 \\ t_4 & t_6 & t_7 \end{pmatrix} \quad \text{and} \quad \begin{pmatrix} t_3 & t_4 \\ t_6 & t_7 \end{pmatrix}.$$

Verify that $J \subset P$, where $P \subset k[t_1, \ldots, t_7]$ denotes the defining ideal of the special fiber $\mathcal{F}(I)$.
3. Prove that J is Cohen–Macaulay of height 4.
4. From the previous item, the ideal $Q := (t_2, t_3, t_4, t_6)$ is a minimal prime of J. Give enough substance to the assertion that $J : Q = (J, f)$, where f is a form of degree 3 totally contained in the subring $k[t_1, t_5, t_7]$.
5. Prove that $\mathcal{F}(I) = (J, f)$ and is not Cohen–Macaulay.
     (HINT: Moderate intervention of computer assistance may help.)

References

1. L. Avramov, J. Herzog, The Koszul algebra of a codimension 2 embedding. Math. Z. **175**, 249–260 (1980) 239, 241, 261
2. W.Bruns, J.Herzog, *Cohen–Macaulay Rings*. Cambridge Studies in Advanced Mathematics, vol 39 (Cambridge University Press, Cambridge, 1993) 42, 248, 279
3. T. Cortadellas, S. Zarzuela, Burch's inequality and the depth of the blow up rings of an ideal. J. Pure Appl. Algebra **157**, 183–204 (2001) 269

4. M. De Boer, R. Pellikaan, Gröbner bases for codes, in *Some Tapas of Computer Algebra* (Springer, Berlin, 1999), pp. 237–259 281

5. D. Eisenbud, On the resiliency of determinantal ideals, Proceedings of the U.S.-Japan Seminar, Kyoto 1985, in *Advanced Studies in Pure Math. II, Commutative Algebra and Combinatorics*, ed. by M. Nagata, H. Matsumura (North-Holland, Amsterdam, 1987), pp. 29–38 263

6. D. Eisenbud, Linear sections of determinantal varieties. Amer. J. Math **110**, 541–575 (1988) 261, 263

7. D. Eisenbud, B. Mazur, Evolutions, symbolic squares, and fitting ideals. J. Reine Angew. Math. **488**, 1810–201 (1997) 242

8. M. Garrousian, A. Simis, Ş. Tohăneanu, A blowup algebra for hyperplane arrangements. Algebra Number Theory **12**, 1401–14210 (2018) 280, 281, 282

9. A.V. Geramita, A. Gimigliano, Y. Pitteloud, Graded Betti numbers of some embedded rational n-folds. Math. Ann. **301**, 363–380 (1995) 280

10. P. Gimenez, A. Simis, W. Vasconcelos, R. Villarreal, On complete monomial ideals. J. Commun. Algebra **8**, 207–226 (2016) 285

11. J. Harris, *Algebraic Geometry—A First Course*. Graduate Texts in Mathematics, vol. 133 (Springer, Berlin, 1992) 272, 276

12. C. Huneke, Hilbert functions and symbolic powers. Michigan Math. J. **34**, 293–318 (1987) 269, 276

13. N.P.H. Lan, On Rees algebras of linearly presented ideals in three variables. J. Pure Appl. Algebra **221**, 2180–2191 (2017) 268, 272

14. P. Mantero, C. Miranda-Neto, U. Nagel, A formula for symbolic powers (2022). arXiv:2112.12588v2 [math.AC] 243

15. S. Morey, B. Ulrich, Rees algebras of ideals with low codimension. Proc. Amer. Math. Soc. **124**, 3653–3661 (1996) 266

16. N. Proudfoot, D. Speyer, A broken circuit ring. Beiträge Algebra Geom. **47**, 161–166 (2006) 281

17. Z. Ramos, A. Simis, Symbolic powers of perfect ideals of codimension 2 and birational maps. J. Algebra **413**, 153–197 (2014) 243, 258, 259, 261

18. Z. Ramos, A. Simis, Homaloidal nets and ideals of fat points II. Intern. J. Algebra Comput. **27**, 677–715 (2017) 280

19. Z. Ramos, A. Simis, *Graded Algebras in Algebraic Geometry*. Expositions in Mathematics, vol. 70 (De Gruyter, Berlin, 2022) 240, 250, 252, 253, 266, 267, 269, 270

20. A. Simis, N.V. Trung, Divisor class group of ordinary and symbolic blow-ups. Math. Z. **198**, 479–491 (1988) 242, 260

21. A. Simis, B. Ulrich, W. Vasconcelos, Jacobian dual fibrations. Amer. J. Math. **115**, 47–75 (1993) 242

22. A. Simis, B. Ulrich, W. Vasconcelos, Codimension, multiplicities and integral extensions. Math. Proc. Camb. Philos. Soc. **130**, 237–257 (2001) 279

23. A. Simis, W.V. Vasconcelos, The Syzygies of the Conormal Module **103**, 203–224 (1981) 239, 241, 261

24. H. Terao, Algebras generated by reciprocals of linear forms. J. Algebra **250**, 549–558 (2002) 281

25. B. Ulrich, W. Vasconcelos, The equations of Rees algebras of ideals with linear presentation. Math. Z. **214**, 79–92 (1993) 266

26. W. Vasconcelos, *Arithmetic of Blowup Algebras*. London Mathematical Society, Lecture Notes Series, vol. 195 (Cambridge University Press, Cambridge, 1994) 242, 247

Chapter 10
Apocryphal Classes

Abstract In this chapter one surveys a few exceptional situations not fitting the known classes. There is an effort to collect some of these examples into plausible classes along with basic properties they enjoy as such. Some of these will involve a mixing of types of linear sections making the contents quite anomalous as compared to previous sufficiently well-behaved classes. An additional feature is the idea of looking at the properties of certain intermediate stages as when a linear section gradually degenerates into another – the case of the generic symmetric matrix and its Hankel like degeneration treated with some detail. By and large, the material is conjectural, while quite a bit of the examples resort to computer assistance. In spite of this, it is the hope that this survey like chapter will stimulate the interested reader that has thus far copped with the book contents.

10.1 Mixed Catalecticants

What we have in mind in this section is an $m \times n$ matrix that has the structure of a coherent stack of catalecticant matrices of various leaps. The one thing common to them is that their generic versions are 1-generic (same proof as in Proposition 7.3). Thus, even in the non-square case, the ideal of its maximal minors is a Cohen–Macaulay prime ideal with the expected codimension $n - m + 1$. As before, also here there may be interest in looking at some of their linear sections.

Let $R = k[x_1, \ldots, x_n]$ be a polynomial ring over a field k. Consider an $m \times m$ matrix such that $1 \leq m \leq n - 1 \leq 2m - 1$, having the following semi-hollow like shape:

$$\mathbf{C}(m, n-m) := \begin{pmatrix} x_1 & x_2 & x_3 & \cdots & x_{m-1} & x_m \\ x_{n-m+1} & x_{n-m+2} & x_{n-m+3} & \cdots & x_{n-1} & x_n \\ x_{n-m+2} & x_{n-m+3} & x_{n-m+4} & \cdots & x_n & 0 \\ \vdots & \vdots & \vdots & \vdots & \vdots & \vdots \\ x_{n-2} & x_{n-1} & x_n & \cdots & 0 & 0 \\ x_{n-1} & x_n & 0 & \cdots & 0 & 0 \end{pmatrix}.$$

Note the boldface $\mathbf{C}$ to distinguish from the previous notation for the $(n-m)$-leap catalecticant of Chapter 7. The matrix has a catalecticant $(n-m)$-leap at the first step and then ensues as a Hankel matrix for the following rows. When $m = n-1$, we retrieve the ordinary sub-Hankel matrix, but otherwise the behavior may often be different.

One advantage of considering this format is that, like the sub-Hankel matrix, the number of ground variables is not so much larger than the degree of the determinant. In this regard, we recall one of the main results of [?] as a search for homaloidal polynomials whose degree is arbitrarily large as compared to the dimension of the ground projective space, while no examples are known of such a homaloidal polynomial that is moreover the determinants of a square matrix – the sub-Hankel case providing the nearest known example.

Throughout the section, we let $\mathfrak{f} := \det \mathbf{C}(m, n-m)$ and let $J = J_{\mathfrak{f}} \subset R = k[x_1, \ldots, x_n]$ denote the gradient ideal of $\mathfrak{f}$. We assume throughout that $\mathrm{char}(k) = 0$.

10.1.1   Basic Results

Lemma 10.1 *With the above notation, assume that $m \geq 3$. Then:*

(i) *$\mathfrak{f}$ is irreducible.*
(ii) *The Hessian determinant of $\mathfrak{f}$ has the form $cx_n^{(m-2)n}$, for a certain nonzero $c \in k$. In particular, the partial derivatives of $\mathfrak{f}$ are algebraically independent over k.*
(iii)

$$\mathrm{codim}(J) = \begin{cases} 3 & \text{if } m = 3 \text{ and } n = 5, 6 \\ 2 & \text{otherwise.} \end{cases}$$

Proof

(i) One can write $\mathfrak{f} = x_1 x_n^{m-1} + \mathfrak{g}$, where $\mathfrak{g}$ includes a nonzero term in $x_m x_{n-1}^{m-1}$. Therefore, $\mathfrak{f}$ is a degree one primitive polynomial in x_1 since x_n cannot divide $\mathfrak{g}$.

(ii) This follows from the following:

CLAIM. Letting $\mathfrak{f}_i := \partial\mathfrak{f}/\partial x_i$, one has

$$\partial\mathfrak{f}_i/\partial x_j = \begin{cases} 0 & \text{if } j \leq n - i \\ c_{i,n} x_n^{m-2} & \text{if } j = n \end{cases}$$

for a suitable nonzero $c_{i,n} \in k$.

An argument in the sub-Hankel case is provided in [? , Theorem 4.4 (iii)].

To see it in the general case, use Proposition 2.10 to deduce that, for $1 \leq i \leq n$, up to a nonzero scalar multiplier, $\mathfrak{f}_i = x_{n-i+1} x_n^{m-2} +$ terms involving only lower powers of x_n and terms involving x_t, with $t > n - i + 1$. Clearly, then the second derivative with respect to $j \leq n - i$ has the claimed values.

(iii) First, consider the case where $m = 3$. If $n = 4$, the matrix is the ordinary 3×3 sub-Hankel matrix

$$\begin{pmatrix} x_1 & x_2 & x_3 \\ x_2 & x_3 & x_4 \\ x_3 & x_4 & 0 \end{pmatrix}.$$

By inspection of the partial derivatives, one easily sees that $J \subset (x_3, x_4)$.

If $n = 5$, the matrix is

$$\begin{pmatrix} x_1 & x_2 & x_3 \\ x_3 & x_4 & x_5 \\ x_4 & x_5 & 0 \end{pmatrix}. \tag{10.1}$$

Then

$$J = (x_5^2,\ x_4 x_5,\ x_4^2 - 2x_3 x_5,\ x_3 x_4 - 1/2 x_2 x_5,\ x_3^2 + x_2 x_4 - 2x_1 x_5).$$

An immediate verification gives that a prime containing J has to contain (x_3, x_4, x_5).

A similar verification holds for $n = 6$.

Now, assume that $m \geq 4$, and hence $n \geq 5$.

CLAIM. $J \subset (x_{n-1}, x_n)$.

Clearly, it suffices to argue that in the Leibniz expansion of $\mathfrak{f}$ every nonzero monomial term is divisible by a monomial $x_{n-1}^i x_n^j$, with $i + j \geq 2$. In fact, it suffices to check this behavior for the single case where $n = 2m$ (i.e., when the first two rows form a generic $2 \times m$ submatrix) because any other case can be viewed as a specialization of the former along the first row. The argument is left to the reader.   □

Proposition 10.2 *With the current notation, let $m = 3$. Then $\mathfrak{f}$ is homaloidal, and $J_{\mathfrak{f}}$ has maximal linear rank $(= n - 1)$.*

Proof Since $m = 3$ is assumed, we have $4 \leq n \leq 6$:

- $n = 4$. This is the ordinary sub-Hankel case (Section 6.4).
- $n = 5$. Then we are in the case of (**??**). This case is very nearly like the sub-Hankel case in that one has the following linear syzygies of $J_{\mathfrak{f}}$:

$$
\begin{matrix}
8x_1 & x_2 & 2x_3 & x_4 \\
5x_2 & 2x_3 & x_4 & x_5 \\
2x_3 & 0 & x_5 & 0 \\
-x_4 & x_5 & 0 & 0 \\
-4x_5 & 0 & 0 & 0
\end{matrix}
$$

  as one readily checks. Clearly, the matrix of these linear syzygies has rank 4. By Lemma **??** (ii), the partial derivatives of $\mathfrak{f}$ are algebraically independent over k; hence, the polar map of $\mathfrak{f}$ is dominant. Therefore, $\mathfrak{f}$ is homaloidal by Theorem 2.27.
- $n = 6$. The corresponding matrix is

$$
\begin{pmatrix}
x_1 & x_2 & x_3 \\
x_4 & x_5 & x_6 \\
x_5 & x_6 & 0
\end{pmatrix}.
\tag{10.2}
$$

The following linear syzygies were obtained via computer assistance, but they are easily verified:

$$
\begin{matrix}
2x_1 & 0 & 0 & x_4 & x_5 \\
x_2 & x_2 & x_3 & x_5 & x_6 \\
0 & 2x_3 & 0 & x_6 & 0 \\
x_4 & -2x_4 & -x_5 & 0 & 0 \\
0 & -x_5 & 0 & 0 & 0 \\
-x_6 & 0 & 0 & 0 & 0
\end{matrix}.
$$

Clearly, their matrix has rank 5, maximal possible. Thus, the assertion follows as in the case $n = 5$. $\qquad\square$

Remark 10.3 Further computation yields that, for $n = 4, 5$, the linear syzygies span a free submodule of maximal rank, while this fails for $n = 6$, where there are 7 minimal linear syzygies. In all cases, J is of linear type.

10.1.2   Basic Conjectures

Keeping the notation of the previous subsection, the main facet of the theory is the following:

Conjecture 10.4 $\mathfrak{f}$ is homaloidal.

Less ambitiously, we pose:

Conjecture 10.5 With the current notation, let $m \geq 3$. Then:

(i) *The following conditions are equivalent:*

   (1) $J_{\mathfrak{f}}$ *is of linear type (that is, since $\mathfrak{f} \in J_{\mathfrak{f}}$, $\mathfrak{f}$ is a regular element on the symmetric algebra $S_R(J_{\mathfrak{f}})$).*
   (2) $J_{\mathfrak{f}}$ *has maximal linear rank.*

   In particular, under any one among the above conditions, $\mathfrak{f}$ is homaloidal.

(ii) *The following conditions are equivalent:*

   (a) *The linear syzygies of $J_{\mathfrak{f}}$ span a free submodule of rank $n - 1$.*
   (b) $\mathfrak{f}$ *is homaloidal, and the base ideal of the inverse map to its polar map is Cohen–Macaulay.*
   (c) *Either $n = m + 1$ (sub-Hankel) or $n = 2m - 1$.*

Elements of evidence: (i) If Conjecture **??** holds, then (1) and (2) are equivalent.

   (ii) In the case where $n = m + 1$, the implications (c) $\Rightarrow$ (a) $\Rightarrow$ (b) come out of Theorem 6.44 and its proof.

Example 10.6 Let the ground ring be $k[x_1, x_2, x_3, x_4, x_5, x_6]$ and consider the matrix

$$
\mathbf{C}(4, 2) = \begin{pmatrix}
x_1 & x_2 & x_3 & x_4 \\
x_3 & x_4 & x_5 & x_6 \\
x_4 & x_5 & x_6 & 0 \\
x_5 & x_6 & 0 & 0
\end{pmatrix}.
$$

Computer-assisted calculation gives that the following syzygies generate the submodule of minimal linear syzygies of $J_{\mathfrak{f}}$:

$$
\begin{matrix}
15x_1 & 2x_3 & 2x_4 & x_5 \\
-11x_2 & -2x_4 & -x_5 & -x_6 \\
7x_3 & x_5 & x_6 & 0 \\
-3x_4 & -x_6 & 0 & 0 \\
3x_5 & -x_5 & 0 & 0 \\
-10x_6 & 0 & 0 & 0
\end{matrix}.
$$

Thus, the linear rank of $J_{\mathfrak{f}}$ is 4, one below the maximum possible. A consequence of item (i) of Conjecture **??** would say that $J_{\mathfrak{f}}$ is not of linear type. And, in fact, another computer-assisted calculation shows the existence of three minimal relations of $J_{\mathfrak{f}}$ of bidegree $(1, 2)$ not belonging to the subideal generated by the linear syzygy relations. Anyhow, the Jacobian matrix rank is maximum $(= 5)$, establishing that $\mathfrak{f}$ is homaloidal.

Remark 10.7 The computation also yields that the linear rank of the base ideal of the inverse map is not maximal, and, quite generally, since J is not linearly presented, this base ideal is not of linear type either, regardless whether it is the gradient ideal of a form.

10.1.3   The Dual Variety

Keeping the standing notation of the section, we consider the dual variety $V(\mathfrak{f})^*$ of the hypersurface $V(\mathfrak{f})$.

Lemma 10.8 $V(\mathfrak{f})^*$ *is non-deficient. In particular, the multiplicity of $\mathfrak{f}$ as a factor of the Hessian determinant $h(\mathfrak{f})$ is the trivial expected one $(= 0)$.*

Proof By Lemma **??** (ii), the Hessian matrix has maximal rank $(= n)$ modulo $(\mathfrak{f})$ as well. By Lemma **??** (i), the dimension value follows from (2.5).   □

Remark 10.9 We suspect that $\mathfrak{f}$ is self-dual (i.e., $V(\mathfrak{f})^* \simeq V(\mathfrak{f})$ up to a change of variables) if (and only if) $\deg V(\mathfrak{f})^* = \deg \mathfrak{f}$. If $n = m + 1$, the self-duality is proved in Proposition 8.11. If $n = 2m$, i.e., the first two rows form a generic $2 \times m$ matrix, then both the degree equality and the self-duality look quite plausible. Yet, e.g., for $m = 4$, the degrees are different if and only if $n = 2m - 1 = 7$. It looks within reach to guess $\deg V(\mathfrak{f})^*$ in terms of n, m.

10.2   3 × 3 Matrices: Unmapped Cases

There are further apocryphal classes of homaloidal determinants that have not been included in the previous accounts at large. We first wrap up a fairly complete list of 3×3 matrices whose determinants have been shown in this book to be irreducible homaloidal polynomials:

1. The generic matrix

$$\begin{pmatrix} x_1 & x_2 & x_3 \\ x_4 & x_5 & x_6 \\ x_7 & x_8 & x_9 \end{pmatrix} \qquad \text{(Section 3.3.1)}$$

2. The one entry hollow-filled generic matrix

$$\begin{pmatrix} x_1 & x_2 & x_3 \\ x_4 & x_5 & x_6 \\ x_7 & x_8 & x_5 \end{pmatrix} \qquad \text{(Section 4.21)}$$

3. The generic symmetric matrix

$$\begin{pmatrix} x_1 & x_2 & x_3 \\ x_2 & x_4 & x_5 \\ x_3 & x_5 & x_6 \end{pmatrix} \qquad \text{(Section 3.3.1)}$$

4. The one entry hollow-filled generic symmetric matrix

$$\begin{pmatrix} x_1 & x_2 & x_3 \\ x_2 & x_4 & x_5 \\ x_3 & x_5 & x_4 \end{pmatrix} \qquad \text{(Section 5.6)}$$

5. The generic 2-leap catalecticant

$$\begin{pmatrix} x_1 & x_2 & x_3 \\ x_3 & x_4 & x_5 \\ x_5 & x_6 & x_7 \end{pmatrix} \qquad \text{(Section 7)} \tag{10.3}$$

6. A semi-hollow like linear section of the generic 2-leap catalecticant

$$\begin{pmatrix} x_1 & x_2 & x_3 \\ x_3 & x_4 & x_5 \\ x_5 & x_6 & 0 \end{pmatrix} \qquad \text{(Section ??)}$$

7. A semi-hollow like linear section of the Hankel matrix (sub-Hankel)

$$\begin{pmatrix} x_1 & x_2 & x_3 \\ x_2 & x_3 & x_4 \\ x_3 & x_4 & 0 \end{pmatrix} \qquad \text{(Section 6.4)}$$

8. A semi-hollow like linear section of the mixed catalecticant

$$\begin{pmatrix} x_1 & x_2 & x_3 \\ x_3 & x_4 & x_5 \\ x_4 & x_5 & 0 \end{pmatrix} \qquad \text{(\textbf{C}(5, 2) of Section ??)}$$

Example 10.10 The matrix

$$\begin{pmatrix} x_1 & x_2 & x_3 \\ x_2 & 0 & x_4 \\ x_3 & x_4 & x_5 \end{pmatrix}$$

lies on the other extreme as regards homaloidness. Here, though its determinant is even irreducible, its partial derivatives are algebraically dependent. In fact, this example is essentially the Gordan–Noether counter-example to Hesse's well-known flawed statement, expressed in a determinantal fashion. This example, up to permutation of rows and columns, can also be viewed as the semi-hollow like linear section $\mathbf{L}(\mathcal{S})[1]$ of the 3×3 generic symmetric matrix dealt with in Section 5.3.2, which has a vanishing Hessian.

There are also examples of 3×3 homaloidal determinants obtained by relaxing the irreducibility requirement. Some of them are linear sections of the generic symmetric matrix. We can disregard trivial examples where the determinant is the product of the variables up to a scalar coefficient or products of linear forms, such as those of the matrices

$$\begin{pmatrix} x_1 & 0 & 0 \\ 0 & x_2 & 0 \\ 0 & 0 & x_3 \end{pmatrix}, \quad \begin{pmatrix} 0 & x_1 & x_2 \\ x_1 & 0 & x_3 \\ x_2 & x_3 & 0 \end{pmatrix}, \quad \begin{pmatrix} x_1 & x_2 & x_3 \\ x_2 & x_3 & x_1 \\ x_3 & x_1 & x_2 \end{pmatrix},$$

where the last one is the circulant discussed in Section 3.3.3.

The following example is a deformation of the semi-hollow like catalecticant sections $\mathbf{C}(m, n - m)$ considered in previous subsections:

Proposition 10.11 (Mixed Generic Hankel) *The determinant of the matrix*

$$\begin{pmatrix} x_1 & x_2 & x_3 \\ x_4 & x_5 & x_6 \\ x_5 & x_6 & x_7 \end{pmatrix} \tag{10.4}$$

is an irreducible homaloidal polynomial.

Proof Let f denote the determinant. Up to a sign it can be written as $f = \Delta_{1,1} x_1 + D$. In the reverse lex order, the initial term of $\Delta_{1,1}$ is x_6^2, while that of D is $x_3 x_5^2$. Therefore, $\Delta_{1,1}$ and D are relatively prime, and since f is a primitive polynomial in x_1, it must be irreducible.

Let $f_1, \ldots, f_7$ be the partial derivatives of f. By Proposition 2.10, f_1, f_2, f_3, f_4, f_7 are the respective cofactors, while f_5, f_6 are sums of two cofactors each. The syzygies of f_1, f_2, f_3 are thus the Hilbert–Burch syzygies of the last two rows of the matrix, and these give two independent linear syzygies of J_f. No other Hilbert–Burch syzygies are at sight, but one can use certain linear syzygies of the ideal I of 2-minors, by slightly changing the latter to have the partial derivatives generate a subset of a minimal set of generators of I. By this method, one eventually obtains 4 new independent linear syzygies, and the total of 6 linear syzygies is linearly independent. Of course, this method depends on having a good grip on the syzygies of I, itself a major burden.

Resorting to a computer-assisted calculation, as a bonus one obtains that these six linear syzygies already generate the entire submodule of linear syzygies of J_f, which consequently is free.

Next one checks that the derivatives are algebraically independent. For this one applies to the entries of the Hessian matrix of f the endomorphism of R that fixes each one of the variables x_2, x_4, x_7 and maps the remaining ones to zero. The resulting Hessian determinant will be a nonzero monomial with support $x_2 x_4 x_7$.

Therefore, f is homaloidal by [? , Proposition 3.2.26]. $\qquad\qquad\qquad\square$

Remark 10.12 It is curious that if one changes the order of row mixing in (??), then the determinant is no longer homaloidal – this follows from [? , Corollary 3.2.27], by showing that the gradient ideal has linear rank one and is of linear type (the calculation was done with computer assistance).

Examples of 3×3 matrices with weirder behavior are available.
As an invitation, consider the following example:

$$\mathcal{M} := \begin{pmatrix} x_1 & x_2 & x_3 \\ x_2 & 0 & x_4 \\ x_3 & x_4 & 0 \end{pmatrix}. \tag{10.5}$$

One can see that the gradient ideal J_f of $f = \det \mathcal{M}$ is linearly presented and that the polar map is an involution (up to a trivial change of coordinates). However, f is not irreducible. Geometrically, the determinantal variety is the union of the smooth quadric $x_1 x_4 - 2 x_2 x_3 = 0$ in $\mathbb{P}^3$ and the coordinate hyperplane $x_4 = 0$ cutting the quadric in two intersecting lines – the union of these lines is the set-theoretic singular locus of f, i.e., its defining ideal is the radical of J_f.

Remark 10.13 (4 × 4 and Beyond)
It would seem that after a reasonably understood account in the case of 3×3 matrices, one has walked some way to look at homaloidal determinants of degree four and higher. Alas, in higher size, the models may face a variety of directions. Thus, for example, 4×4 analogues of matrix (??) have been inspected, most of which turned up to be of linear type with insufficient linear syzygy rank. The nearest homaloidal candidate is given by the following mixing:

$$\begin{pmatrix} x_1 & x_2 & x_3 & x_4 \\ x_5 & x_6 & x_7 & x_8 \\ x_9 & x_{10} & x_{11} & x_{12} \\ x_{10} & x_{11} & x_{12} & x_{13} \end{pmatrix}.$$

The linear syzygy rank here is 11 (would need 12).

By and large, it will be the case that the examples come up naturally within a more structured theory, as in the next section.

10.3 Stepwise Hankelization

Fix integers m, r with $0 \leq r \leq m - 2$ and consider the linear section $\mathbf{L}(\mathcal{S})[r]$ of the $m \times m$ generic symmetric matrix as established in Section 5.3.2.

The total number of null entries in $\mathbf{L}(\mathcal{S})[r]$ is obviously $\binom{r+1}{2}$, whereas its codimension in the generic symmetric matrix is given by Lemma 5.14:

$$\operatorname{codim}_{\mathcal{S}} \mathbf{L}(\mathcal{S})[r] = \mathfrak{o}(r) := \begin{cases} \frac{(r+1)^2}{4} & \text{if } r \text{ is odd} \\ \frac{r(r+2)}{4} & \text{if } r \text{ is even} \end{cases} . \tag{10.6}$$

We wish to focus on linear sections of $\mathbf{L}(\mathcal{S})[r]$ of a special kind, some of whose anti-diagonals become of the Hankel type, i.e., equal entries all along. However, this process should be so that one can speak about each such section as defined by a series of *Hankelization steps*, culminating with the Hankel matrix $\mathcal{H}_m[r]$, where the number of steps is exhausted. For this, we decree as rules for the steps:

   (i) Orderly, along the anti-diagonals, starting from the leftmost anti-diagonal.

  (ii) Moving over to the next anti-diagonal only after the present anti-diagonal has been fully symmetrized (i.e., all of its entries coincide).

 (iii) Inwardly along a given anti-diagonal – e.g., $x_{1,j}$ moves into $x_{2,j-1}$ and replaces it (repeating automatically for $x_{j,1}$ into $x_{j-1,2}$). Next step, $x_{1,j}$ moves into $x_{3,j-2}$, and so on.

The third step above is vacuous if one only deals with full steps, i.e., a single step for every anti-diagonal.

Example 10.14 Consider the 4×4 linear section $\mathbf{L}(\mathcal{S})[2]$

$$\begin{pmatrix} x_1 & x_2 & x_3 & x_4 \\ x_2 & x_5 & x_6 & x_7 \\ x_3 & x_6 & x_9 & 0 \\ x_4 & x_7 & 0 & 0 \end{pmatrix} .$$

It can be verified that, like $\mathbf{L}(\mathcal{S})[2]$, each of the following linear sections

$$\begin{pmatrix} x_1 & x_2 & x_3 & x_4 \\ x_2 & x_3 & x_6 & x_7 \\ x_3 & x_6 & x_9 & 0 \\ x_4 & x_7 & 0 & 0 \end{pmatrix}, \begin{pmatrix} x_1 & x_2 & x_3 & x_4 \\ x_2 & x_5 & x_4 & x_7 \\ x_3 & x_4 & x_9 & 0 \\ x_4 & x_7 & 0 & 0 \end{pmatrix}, \begin{pmatrix} x_1 & x_2 & x_3 & x_4 \\ x_2 & x_5 & x_6 & x_7 \\ x_3 & x_6 & x_7 & 0 \\ x_4 & x_7 & 0 & 0 \end{pmatrix}, \begin{pmatrix} x_1 & x_2 & x_3 & x_4 \\ x_2 & x_3 & x_6 & x_7 \\ x_3 & x_6 & x_7 & 0 \\ x_4 & x_7 & 0 & 0 \end{pmatrix}, \begin{pmatrix} x_1 & x_2 & x_3 & x_4 \\ x_2 & x_5 & x_4 & x_6 \\ x_3 & x_4 & x_6 & 0 \\ x_4 & x_6 & 0 & 0 \end{pmatrix}$$

has non-dominant polar map. Here, only the first of them has a first step obeying the above rules of process. Thus, from this example, not abiding by (i) or (ii) may lead to "unwelcome" behavior, in the sense, e.g., that the corresponding polar map is not even dominant.

In the opposite, the following two steps Hankelizations obey the rules and have homaloidal determinants.

$$
\begin{pmatrix}
x_1 & x_2 & x_3 & x_4 \\
x_2 & x_3 & x_4 & x_7 \\
x_3 & x_4 & x_8 & 0 \\
x_4 & x_7 & 0 & 0
\end{pmatrix},
\quad
\begin{pmatrix}
x_1 & x_2 & x_3 & x_4 \\
x_2 & x_3 & x_4 & x_7 \\
x_3 & x_4 & x_7 & 0 \\
x_4 & x_7 & 0 & 0
\end{pmatrix}.
\tag{10.7}
$$

The second of these is the sub-Hankel case of Section 6.4. The first is more involved and may require computer assistance (Exercise **??**).

Let $\mathbf{L}(S)[r](t)$ denote the matrix obtained at the full Hankelization step of order t. Thus, for example, one has in the notation of Section 6.3:

Lemma 10.15 $\mathcal{H}_m[r] = \mathbf{L}(S)[r]\left(\binom{m-1}{2} - \lceil (\mathfrak{o}(r) - 2)/2 \rceil \right).$

Proof Suppose first that $r = 0$. One has $\binom{m+1}{2}$ distinct entries in S. By rule (iii), along each anti-diagonal with three or more entries, we count the number of entries minus one. Counting the number of these diagonals adds up to $2(m-3)+1$. We still have to subtract the four distinct entries of the upper left and lower right corners. Therefore, we arrive at

$$
\binom{m+1}{2} - 4 - 2(m-3) - 1 = (m^2 - 3m + 2)/2 = (m-1)(m-2)/2 = \binom{m-1}{2}.
$$

Thus, $\mathcal{H}_m = \mathbf{L}(S)\left(\binom{m-1}{2}\right)$. To get the step value of $\mathcal{H}_m[r]$ for arbitrary r, we subtract the contribution of the sparse region, which follows the same principle. $\square$

Set $h(r) := \binom{m-1}{2} - \lceil (\mathfrak{o}(r) - 2)/2 \rceil$. We enact the following additional notation: For any $0 \leq t \leq h(r)$, let $f(t)$ denote the determinant of $\mathbf{L}(S)[r](t)$ and let $J(t)$ denote the gradient ideal of $f(t)$.

10.3.1   The Especial Case $r = m - 2$

In this part we assume that $r = m - 2$, for which we state a few conjectured statements concerning invariants along the Hankelization process.

Lemma **??** gives

$$
h(m-2) = \begin{cases}
\frac{(m-1)^2}{8} + 1 & \text{if } m \text{ is odd} \\[2mm]
\frac{(3m-4)((m-2)}{8} + 1 & \text{if } m \text{ is even}
\end{cases}
\tag{10.8}
$$

Theorem 10.16 (Conjectured) *Let t_0 denote the least index in the interval $0 \leq t \leq h(m-2)$ such that the Hessian determinant of $\det \mathbf{L}(S)[r](t)$ does not vanish. Then the analytic spread of $J(t)$ is stable in the interval $0 \leq t \leq t_0$, while in the interval $t_0 + 1 \leq t \leq h(m-2)$ the analytic spread is maximal and drops by 1 at each step.*

In other words, from $h(m-2)$ back all the way to $h(m-2) - t_0$, the analytic spread has values $m + 1, m + 2, \ldots, m + h(m-2) - t_0$ correspondingly; from $h(m-2) - t_0$ back to 0 the value of the corresponding analytic spread is stably $m + h(m-2) - t_0$.

Theorem 10.17 (Conjectured) *If $m \geq 4$, then $f[t]$ is homaloidal if and only if $t = h(m-2) - 1$ or $t = h(m-2)$.*

In other words, homaloidness only occurs for the Hankel step and one before it. For the "if" implication, the Hankel case has been dealt with in Section 6.4. As for step $t = h(m-2) - 1$, one has:

Proposition 10.18 (Conjectured) *Let $t = h(m-2) - 1$. Then the Jacobian dual matrix is of size $(m+1) \times (m+2)$, necessarily of rank $m+1$. In particular, $f(t)$ is homaloidal, and the base ideal of the inverse map to the corresponding polar map is a perfect codimension 2 ideal generated by $m+2$ linear forms.*

Proof (Modicum) For $t = h(m-2) - 1$, $J(t)$ has only 2 linear syzygies, but the square $J(t)^2$ has $3(m+1) = 3m + 3$ linear syzygies of which $m - 1$ are "fresh," i.e., independent of the $2(m+2) = 2m + 4$ "old" syzygies coming from the two linear ones of $J[t]$ (proof?). In addition, the powers of order 3 and higher of $J(t)$ introduce no fresh linear syzygies (proof?). It follows that the Jacobian dual matrix is of size $(m+1) \times (m+2)$, necessarily of rank $m+1$.

The supplementary statement follows by [? , Theorem 3.2.22]. $\square$

10.4   Hankel Matrices Bordered by the Circulant

Let C_n denote the generic circulant matrix based on variables $x_1, \ldots, x_n$ over a field of characteristic zero (see Section 3.3.3). Let $H = H_{\lceil n/2 \rceil}$ denote the submatrix of C_n with the first $\lceil n/2 \rceil$ rows and $\lceil n/2 \rceil$ columns from upper left. This is the largest generic Hankel upper left submatrix of C_n.

Let BH_i denote the submatrix of C_n with the first i rows and i columns from upper left, for $\lceil n/2 \rceil \leq i \leq n$. Note that $BH_{\lceil n/2 \rceil} = H$.

Proposition 10.19 (Conjectured) *With the above notation, let $\mathfrak{f}_i := \det BH_i$ and let $\mathfrak{J}_i$ denote its gradient ideal. Then:*

(i) *$\mathfrak{J}_i$ is of linear type.*
(ii) *$\mathfrak{J}_i$ is Cohen–Macaulay if and only if $i = n$.*
(iii) *$\mathfrak{f}_i$ is homaloidal if and only if $i = n$.*

Proof (Modicum)

(i) For $i = n$ this is established in Section 3.3.3. For $i \neq n$ this is as yet an untold story even when $i = \lceil n/2 \rceil$ (Conjecture 6.30).

(ii) The "if" implication is provided in Section 3.3.3. The "only if" assertion starts on with the fact that the gradient ideal of an $m \times m$ ($m \geq 3$) generic Hankel matrix is not Cohen–Macaulay (Theorem 6.8 (d)). For $\lceil n/2 \rceil < i < n$, we conjecture similarly that the minimal primary part of $\mathfrak{J}_i$ is the ideal of submaximal minors $I_{i-1}(BH_i)$, in which case again $\mathfrak{J}_i$ is not Cohen–Macaulay.

(iii) By [? , Corollary 3.2.27], granted (i), it suffices to show that the linear rank of $\mathfrak{J}_i$ is maximal if and only if $i = n$. We conjecture, more strongly, that the initial standard degree of the syzygy matrix of $\mathfrak{J}_i$ is ≥ 2 for $\lceil n/2 \rceil < i < n$. $\square$

Exercises

By and large, the problems to follow will require computer assistance.

10.20 Let $f \in k[x_1, \ldots, x_7]$ denote the determinant of the matrix (**??**), where $\mathrm{char}(k) \neq 2, 3$, and let $J_f \in k[x_1, \ldots, x_7]$ denote its gradient ideal:

1. Show that J_f has maximal analytic spread.
   (HINT: Prove that the Hessian of f does not vanish by applying to its entries a convenient k-linear transformation of $k[x_1, \ldots, x_7]$.)
2. Justify that J_f has maximal linear rank.
   (HINT: You may need computer assistance.)
3. Deduce that f is homaloidal and prove that J_f is an ideal of linear type.

10.21 Perform the details of Remark **??**.

10.22 Compute and discuss the details of Example **??**.

10.23 Consider the leftmost matrix in Example **??** and its determinant f:

1. Show that the corresponding gradient ideal J_f has two independent linear syzygies and three independent relations of bidegree $(1, 2)$ not coming from the linear syzygies.
2. Deduce that the Jacobian dual matrix of J_f has rank 5 (maximal); hence, f is homaloidal.
3. Argue that the inverse map to the polar map of f is defined by the maximal minors of a linear Hilbert–Burch type of matrix. (This follows from conjectured Proposition **??**.)

10.24 Discuss the details of the entire Section **??** for $m = 3$.

References

1. C. Ciliberto, F. Russo, A. Simis, Homaloidal hypersurfaces and hypersurfaces with vanishing Hessian. Adv. Math. **218**, 1759–1805 (2008) 296, 297
2. Z. Ramos, A. Simis, *Graded Algebras in Algebraic Geometry*. Expositions in Mathematics, vol. 70 (De Gruyter, Berlin, 2022) 303, 306, 307

Appendix A
Complement

A.1 Non-vanishing Hessian

We give a proof of Theorem 6.31.

The method consists in sufficiently "degenerating" the Hessian matrix to allow direct calculation with some of the submatrices to eventually get a non-vanishing expression. Namely, consider the ring endomorphism φ of R mapping any variable in $\mathbf{v} := \{x_1, x_{m-r-1}, x_{2m-r-1}\}$ to itself and mapping any variable of $\mathbf{v}$ to zero. We will show that by applying φ to the entries of the Hessian matrix $H(f)$ the resulting matrix $H(f)(\mathbf{v})$ has non-vanishing determinant. This is an easy calculation for low values of m with computer assistance, and yet the exact resulting sparse picture of $H(f)(\mathbf{v})$ gets blurred for arbitrary values of m and $1 \leq r \leq m - 2$.

For visualization we depict the matrix $\mathcal{H}_m[r]$ for arbitrary $r \leq m - 3$:

$$
\left(
\begin{array}{ccccc|ccccc}
x_1 & x_2 & \cdots & x_{m-r-2} & x_{m-r-1} & x_{m-r} & x_{m-r+1} & \cdots & x_{m-1} & x_m \\
x_2 & x_3 & \cdots & x_{m-r-1} & x_{m-r} & x_{m-r+1} & x_{m-r+2} & \cdots & x_m & x_{m+1} \\
\vdots & \vdots & \cdots & \vdots & \vdots & \vdots & \vdots & \ddots & \vdots & \vdots \\
x_{m-r-1} & x_{m-r} & \cdots & x_{2m-2r} & x_{2m-2r-3} & x_{2m-2r-2} & x_{2m-2r-1} & \cdots & x_{2m-r-3} & x_{2m-r-2} \\ \hline
x_{m-r} & x_{m-r+1} & \cdots & x_{2m-2r-3} & x_{2m-2r-2} & x_{2m-2r-1} & x_{2m-2r} & \cdots & x_{2m-r-2} & x_{2m-r-1} \\
x_{m-r+1} & x_{m-r+2} & \cdots & x_{2m-2r-2} & x_{2m-2r-1} & x_{2m-2r} & x_{2m-2r+1} & \cdots & x_{2m-r-1} & 0 \\
\vdots & \vdots & \cdots & \vdots & \vdots & \vdots & \vdots & \ddots & \vdots & \vdots \\
x_{m-1} & x_m & \cdots & x_{2m-r-4} & x_{2m-r-3} & x_{2m-r-2} & x_{2m-r-1} & \cdots & 0 & 0 \\
x_m & x_{m+1} & \cdots & x_{2m-r-3} & x_{2m-r-2} & x_{2m-r-1} & 0 & \cdots & 0 & 0
\end{array}
\right).
$$

The goal is to isolate terms of the partial derivatives of f that have in their support a product of at least *two* variables of $\mathbf{v}$, since such terms will produce at least one variable of $\mathbf{v}$ in the entries of $H(f)$ and hence will vanish upon applying φ.

© The Author(s), under exclusive license to Springer Nature Switzerland AG 2024

Z. Ramos, A. Simis, *Determinantal Ideals of Square Linear Matrices*,

https://doi.org/10.1007/978-3-031-55284-7

The expression "terms of degree at least 2 off **v**" will next appear recurrently in the sense just explained. In order to avoid tedious repetition, we replace the expression by the letter T.

Now, recall that the partial derivatives of f are sums of signed $(m-1)$-minors (see Proposition 2.6). More precisely, for $k = 1, \ldots, 2m - r - 1$, we have

$$f_k = \sum_{i+j=k+1} M_{i,j}, \tag{A.1}$$

where $M_{i,j}$ is the cofactor of the (i, j)th entry.

Let us pick up the shape of such a partial derivative of f as we go through the various relevant intervals for the sum $i + j$:

(a) $i + j \le m - r$

Expanding the minor $M_{i,j}$ by the Laplace rule along its first $m - r - 1$ rows yields $M_{i,j} = D_{i,j} x_{2m-r-1}^{r+1} + T$, where $D_{i,j}$ is the cofactor of the (i, j)th entry of the submatrix

$$D = \begin{pmatrix} x_1 & x_2 & \cdots & x_{m-r-1} \\ x_2 & x_3 & \cdots & x_{m-r} \\ \vdots & \vdots & \cdots & \vdots \\ x_{m-r-1} & x_{m-r} & \cdots & x_{2m-2r-3} \end{pmatrix}. \tag{A.2}$$

Expanding $D_{i,j}$ for $i + j < m - r$ yields

$$D_{i,j} = \pm z_{m-r-i,m-r-j} x_{m-r-1}^{m-r-3} + T = \pm x_{2m-2r-(i+j)-1} x_{m-r-1}^{m-r-3} + T.$$

Then (A.1) becomes

$$f_k = \sum_{i+j=k+1} D_{i,j} x_{2m-r-1}^{r+1} + T, \tag{A.3}$$

which then gives, for $k + 1 = i + j < m - r$,

$$f_k = \pm \sum_{i+j=k+1} x_{2m-2r-(i+j)-1}\, x_{m-r-1}^{m-r-3}\, x_{2m-r-1}^{r+1} + T$$

$$= \pm k \cdot x_{2m-2r-(k+1)-1}\, x_{m-r-1}^{m-r-3}\, x_{2m-r-1}^{r+1} + T.$$

Therefore

$$\varphi\left(\frac{\partial^2 f}{\partial x_l \partial x_k}\right) = \begin{cases} \pm k \cdot x_{m-r-1}^{m-r-3} x_{2m-r-1}^{r+1}, & \text{if } l = 2m - 2r - (k+1) - 1 \\ 0, & \text{if } l \ne 2m - 2r - (k+1) - 1 \end{cases}.$$

As for $i + j = m - r$ one has

$$D_{i,j} = \begin{cases} \pm x_{m-r-1}^{m-r-2} + T, & \text{if } i = 1 \text{ or } j = 1 \\ \pm x_{m-r-1}^{m-r-2} \pm x_1 x_{2m-2r-3} \, x_{m-r-1}^{m-r-4} + T, & \text{if } i \neq 1 \text{ and } j \neq 1 \end{cases}.$$

Observe that, when $i \neq 1$ and $j \neq 1$, then we have necessarily $m - r \geq 4$.

By (A.3), it obtains

$$f_{m-r-1} = \pm \sum_{i+j=m-r} D_{i,j} x_{2m-r-1}^{r+1} + T = \pm (m - r - 1) x_{m-r-1}^{m-r-2} x_{2m-r-1}^{r+1}$$

$$\pm (m - r - 3) x_1 x_{2m-2r-3} \, x_{m-r-1}^{m-r-4} x_{2m-r-1}^{r+1} + T.$$

It follows that

$$\varphi\left(\frac{\partial f}{\partial x_l \partial x_{m-r-1}}\right) = \begin{cases} \pm (m - r - 1)(m - r - 2) x_{m-r-1}^{m-r-3} x_{2m-r-1}^{r+1}, & \text{if } l = m - r - 1 \\ \pm (m - r - 1)(r + 1) x_{m-r-1}^{m-r-2} x_{2m-r-1}^{r}, & \text{if } l = 2m - r - 1 \\ \pm (m - r - 3) x_1 x_{m-r-1}^{m-r-4} x_{2m-r-1}^{r+1}, & \text{if } l = 2m - 2r - 3 \\ 0, & \text{if } l \neq m - r - 1, 2m - r - 1, 2m - 2r - 3 \end{cases}.$$

So far we have discussed the first $(m - r - 1)$ columns of $H(f)(\mathbf{v})$ and hence, by symmetry, its first $(m - r - 1)$ rows as well. In particular, the columns $m - r, \ldots, 2m - 2r - 3$ of $H(f)(\mathbf{v})$ are partially obtained. For the concluding argument on the entire shape of $H(f)(\mathbf{v})$, it will suffice to move all the way to the interval $2m - 2r - 1 \leq i + j \leq 2m - r - 1$, which will give the shape of columns $2m - 2r - 2, \ldots, 2m - r - 1$ of $H(f)(\mathbf{v})$.

(b) $2m - 2r - 1 \leq i + j < 2m - r - 1$

Write $M_{i,j} = (-1)^{i+j} \det C_{i,j}$, where $C_{i,j}$ is the submatrix of $\mathcal{H}_m(r)$ obtained by omitting its ith row and its jth column. If $i > m - r - 1$ and $j > m - r - 1$, then $C_{i,j}$ misses the entry x_{2m-r-1} sitting on the $(i, 2m - r - i)$th and $(2m - r - j, j)$th slots $\mathcal{H}_m(r)$. Therefore, in this case $C_{i,j}$ has only $(m - 2)$ columns with some entry in $\mathbf{v}$, and hence, $M_{i,j}$ cannot have any term supported on $\mathbf{v}$; in addition, by the same token, any of its terms having degree 1 in variables of $\mathbf{v}$ involves necessarily the $m - r - 1$ entries equal to x_{m-r-1}, the $r - 1$ entries equal to x_{2m-r-1} on the matrix $C_{i,j}$ and the entry of $\mathcal{H}_m(r)$ in the $(2m-r-j, 2m-r-i)$th slot. But the latter is zero since $(2m-r-j)+(2m-r-i) > 2m-r$ when $2m - 2r - 1 \leq i+j < 2m-r-1$.

Summing up, we have shown that, for $i > m - r - 1$ and $j > m - r - 1$, the cofactor $M_{i,j}$ has neither terms supported on $\mathbf{v}$ nor terms having degree 1 in variables of $\mathbf{v}$.

Thus, we are left with the next two possibilities:
(i) $i < m - r - 1$ or $j < m - r - 1$

By symmetry, it suffices to consider the case where $i < m - r - 1$ and we do so. Then $C_{i,j}$ misses the entries x_{m-r-1} and x_{2m-r-1} in the $(i, m - r - i)$th and $(2m - r - j, j)$th slots of $\mathcal{H}_m(r)$, respectively. Clearly, $M_{i,j}$ does not have terms supported in $\mathbf{v}$. Moreover, any term of $M_{i,j}$ involving x_1 cannot simultaneously involve variables of $\mathbf{v}$ in slots $(m - r - 1, 1)$ and $(1, m - r - 1)$ on $C_{i,j}$, and hence ought to have degree at least 2 in the variables of $\mathbf{v}$. Then one has

$$M_{i,j} = \pm x_{m-r-1}^{m-r-2} x_{2m-r-1}^{r} x_{3m-2r-(i+j)-1} + T.$$

(ii) $i = m - r - 1$ or $j = m - r - 1$

Again, by symmetry, it suffices to argue for the case where $i = m - r - 1$. A similar argument as above concerning slots $(m - r - 1, 1)$ and $(2m - r - j, j)$ of $\mathcal{H}_m(r)$ will do, and one gets

$$M_{i,j} = \pm x_{m-r-1}^{m-r-2} x_{2m-r-1}^{r} x_{2m-r-j} \pm x_1 x_{m-r-1}^{m-r-3} x_{2m-r-1}^{r} x_{3m-2r-j-2} + T.$$

A count of these cofactors gives for each $l = 2m - 2r - 2, \ldots, 2m - r - 2$:

$$f_l = \pm 2 x_1 x_{m-r-1}^{m-r-3} x_{2m-r-1}^{r} x_{4m-3r-l-4} \pm c x_{m-r-1}^{m-r-2} x_{2m-r-1}^{r} x_{3m-2r-l-2} + T,$$

where $c \in k$. Therefore, for $l = 2m - 2r - 2, \ldots, 2m - r - 2$, one gets

$$\varphi\left(\frac{\partial^2 f}{\partial x_k \partial x_l}\right) = \begin{cases} \pm 2\mathbf{q} = \pm 2 x_1 x_{m-r-1}^{m-r-3} x_{2m-r-1}^{r}, & \text{if } k = 4m - 3r - l - 4, \\ 0, & \text{if } k > 4m - 3r - l - 4. \end{cases}$$

(c) $i + j = 2m - r - 1$

Here expanding along the first $m - r - 1$ rows, one has $M_{i,j} = \det D \cdot x_{2m-r-1}^{r} + T$. Expanding $\det D$, one gets

$$M_{i,j} = \pm x_1 x_{m-r-1}^{m-r-3} x_{2m-r-1}^{r} x_{2m-2r-3} \pm x_{m-r-1}^{m-r-1} x_{2m-r-1}^{r} + T,$$

and therefore,

$$f_{2m-r-1} = \pm (r + 1) x_1 x_{m-r-1}^{m-r-3} x_{2m-r-1}^{r} x_{2m-2r-3} \pm (r + 1) x_{m-r-1}^{m-r-1} x_{2m-r-1}^{r} + T.$$

Thus,

$$\varphi\left(\frac{\partial^2 f}{\partial x_{2m-r-1} \partial x_{2m-r-1}}\right) = \pm (r + 1) r x_{m-r-1}^{m-r-1} x_{2m-r-1}^{r-1}.$$

Collecting the information gathered so far, we see that by applying φ to the entries of the Hessian matrix $H(f)$, one obtains a matrix in the form:

$$H(f)(\mathbf{v}) = \left(\begin{array}{c|c} A & B^t \\ \hline B & A' \end{array}\right).$$
(A.4)

Here A and B are matrices of sizes $(2m - 2r - 3) \times (2m - 2r - 3)$ and $(r + 2) \times (2m - 2r - 3)$, respectively, and the stack $\frac{A}{B}$ has the shape

$$\left(\begin{array}{ccccccccc}
0 & 0 & \cdots & 0 & & 0 & & 0 & \cdots 0 \;\; \pm\mathbf{p} \\
0 & 0 & \cdots & 0 & & 0 & & 0 & \cdots \pm2\mathbf{p} \;\; 0 \\
\vdots & \vdots & \cdots & \vdots & & \vdots & & \vdots & \ddots \;\; \vdots \;\; \vdots \\
0 & 0 & \cdots & 0 & & 0 & \pm(m-r-2)\mathbf{p} \cdots & 0 & 0 \\
0 & 0 & \cdots & 0 & \pm(m-r-1)(m-r-2)\mathbf{p} & & 0 & \cdots & 0 \;\; \mathbf{d} \\
0 & 0 & \cdots \pm(m-r-2)\mathbf{p} & & 0 & & * & \cdots & * \;\; * \\
\vdots & \vdots & \ddots & \vdots & & \vdots & & \vdots & \cdots \;\; \vdots \;\; \vdots \\
0 & \pm2\mathbf{p} & \cdots & 0 & & 0 & & * & \cdots \;\; * \;\; * \\
\pm\mathbf{p} & 0 & \cdots & 0 & & \mathbf{d} & & * & \cdots \;\; * \;\; * \\
\hline
0 & 0 & \cdots & 0 & & 0 & & * & \cdots \;\; * \;\; * \\
\vdots & \vdots & \ddots & \vdots & & \vdots & & \vdots & \cdots \;\; \vdots \;\; \vdots \\
0 & 0 & \cdots & 0 & & 0 & & * & \cdots \;\; * \;\; * \\
0 & 0 & \cdots & 0 & \pm(m-r-1)(r+1)x_{m-r-1}^{m-r-2}x_{2m-r-1}^{r} & & * & \cdots \;\; * \;\; *
\end{array}\right),$$

where $\mathbf{p} = x_{m-r-1}^{m-r-3}x_{2m-r-1}^{r+1}$ and $\mathbf{d} = \pm(m - r - 3)\, x_1\, x_{m-r-1}^{m-r-4}\, x_{2m-r-1}^{r+1}$. This part follows from the result in item (a).

As for the matrix A', its shape follows from items (b) and (c):

$$A' = \left(\begin{array}{ccccc}
* & * & \cdots \pm2\mathbf{q} & & 0 \\
\vdots & \vdots & \ddots \;\; \vdots & & \vdots \\
* & \pm2\mathbf{q} & \cdots \;\; 0 & & 0 \\
\pm2\mathbf{q} & 0 & \cdots \;\; 0 & & 0 \\
0 & 0 & \cdots \;\; 0 & \pm(r+1)r x_{m-r-1}^{m-r-1} x_{2m-r-1}^{r-1} &
\end{array}\right)$$

with $\mathbf{q} = x_1 x_{m-r-1}^{m-r-3} x_{2m-2r-1}^{r}$.

Now expand the determinant of (A.4) along its first $2m - 2r - 3$ rows. Note that the complementary minor to a $(2m - 2r - 3)$-minor of the first $2m - 2r - 3$ rows, avoiding the first $m - r - 2$ columns, vanishes as any of its columns is null. At the other end, the submatrices consisting of non-vanishing minors of the first $(2m - 2r - 3)$ rows that involve the first $m - r - 2$ columns are submatrices of A itself and the following matrix X:

$$
\begin{pmatrix}
0 & 0 & \cdots & 0 & 0 & 0 & 0 & \pm\mathbf{p} & 0 \\
0 & 0 & \cdots & 0 & 0 & & 0 & \pm2\mathbf{p} \; 0 & 0 \\
\vdots & \vdots & \cdots & \vdots & \vdots & \cdot{}^{\cdot} & \vdots & \vdots & \vdots \\
0 & 0 & \cdots & 0 & \pm(m-r-2)\mathbf{p} & \cdots & 0 & 0 & 0 \\
0 & 0 & \cdots & 0 & 0 & & \cdots \; 0 & & \pm(m-r-1)(r+1)x_{m-r-1}^{m-r-2}x_{2m-r-1}^{r} \\
0 & 0 & \cdots & \pm(m-r-2)\mathbf{p} & * & & \cdots \; * & * & * \\
\vdots & \vdots & \cdot{}^{\cdot} & \vdots & \vdots & & \cdots \; \vdots & \vdots & \vdots \\
0 & \pm2\mathbf{p} & \cdots & 0 & * & & \cdots \; * & * & * \\
\pm\mathbf{p} & 0 & \cdots & 0 & * & & \cdots \; * & * & *
\end{pmatrix},
$$

obtained upon replacing the $(m - r - 1)$th column of A with the last column of B^t (i.e., the transpose of the last row of B). Their complementary matrices are, respectively, A' and

$$
X' =
\begin{pmatrix}
 & * & & & & * & * & \cdots & \pm2\mathbf{q} \\
 & \vdots & & & & \vdots & \vdots & \cdot{}^{\cdot} & \vdots \\
 & * & & & & * & \pm2\mathbf{q} & \cdots & 0 \\
 & * & & & & \pm2\mathbf{q} & 0 & \cdots & 0 \\
(m-r-1)(r+1)x_{m-r-1}^{m-r-2}x_{2m-r-1}^{r} & & & & & 0 & 0 & \cdots & 0
\end{pmatrix}.
$$

Thereof, we obtain $\det H(f)(\mathbf{v}) = \pm \det A \det A' \pm \det X \det X'$. Expanding the various determinants in this expression gives

$$
\det H(f)(\mathbf{v}) = 2^{r+2}(r + 1)(m - r - 1)!\,\mathbf{p}^{2m-2r-4}\mathbf{q}^{r+1}
$$

$$
\left(\pm r(m - r - 2)\mathbf{p}x_{m-r-1}^{m-r-1}x_{2m-r-1}^{r-1} \pm (m - r - 1)(r + 1)x_{m-r-1}^{2m-2r-4}x_{2m-r-1}^{2r} \right).
$$

The first factor above is a term in $\mathbf{p}$ and $\mathbf{q}$ and hence does not vanish. The second factor is a sum of distinct terms, so does not vanish either. Therefore, the expression is nonzero. $\square$

Remark A.1 It may seem that there might exist an easy argument for the non-vanishing of the Hessian determinant of an intermediate $\det \mathcal{H}_m[r]$ since it is "squeezed" between the extreme situations where $r = 0$ and $r = m - 2$, where we know the Hessian does not vanish. Unfortunately, we may need the specifics of the present setup as for arbitrary threads of coordinate sections some intermediate Hessian determinants may vanish or not. On the other hand, a more fundamental argument would be very welcome.

Index

A

Affine space, 5
Algebra
  dimension, 59
  fiber
    submaximal minors (Hankel), 174
  fiber algebra
    Hankel, 167
  Grassmann, 58
    dimension, 41
    regularity, 168
  k-algebra, 32
  Orlik–Terao, 281
  Rees, 41, 46
    normal, 10
    submaximal minors (Hankel), 174
  special fiber, 10, 41
    submaximal minors (Hankel), 174
  symbolic Rees, 48, 50
    Noetherian test, 48
  symmetric, 44
    Cohen–Macaulay, 46
Analytically independent, 44
Analytic spread, 41
  maximal, 41, 96, 113, 171

B

Birkhoff–Grothendieck theorem, 11
Buchsbaum, D.
  Buchsbaum–Eisenbud criterion, 80

C

Cauchy, A.-L., 29
Cofactor, 4, 37
Condition
  F_0, 44
  F_1, 44
  G_s, 44
  G_∞, 44, 45, 67, 82
  sliding-depth, 82
Conjecture
  analytic spread
    stepwise Hankelization, 306
  catalecticant determinant
    parabolic, 209
  dual variety
    Hankel hollow sparse, 224
  gradient
    hollow determinant, 93, 148
  Hankel bordered by circulant, 306
  homaloidal
    apocriphal, 299
    Hollow sparse Hankel, 183
    stepwise Hankelization, 306
  linear rank
    gradient ideal, 101
    Hankel hollow sparse, 181
  linear type
    Hankel hollow sparse, 182
  Rees algebra
    fiber type, 286
    Hankel hollow sparse, 174
    vs. special fiber, 42

Conjecture (*cont.*)
  special fiber
    Hankel hollow sparse, 174

D
Determinant, 4, 29
  Hessian, 35, 54, 68
  inverse problem, 69
    Dickson theorem, 70
    Dixon theorem, 70
    Edge method, 70
Dual
  Jacobian matrix, 90, 144
  variety, 55, 63, 64

E
Eagon, J., 15

F
Field, 32, 45
  algebraic extension, 33, 57
  algebraic independence over, 45
  extension, 33
  perfect, 16
  residue, 41
  separable extension, 56, 57
  separably generated, 56
Fitting, H., 4

H
Hessian
  determinant, 296
Homaloidal, 54, 298
  apocriphal, 300
  mixed Hankel (irreducible), 302

I
Ideal
  associated primes
    hollow-like (Hankel), 177
  codimension of, 40
  Cohen–Macaulay, 23, 149
    strongly, 82
  determinantal, 5
  Fitting, 4
  free presentation of, 43, 44
  free resolution
    graded, 249
    mapping cone, 192
  Gorenstein, 83, 96, 113

  normal, 102
  tight, 83
grade of, 40
gradient, 45, 54, 93, 97, 115, 135, 148
  almost Cohen–Macaulay, 192
  associated primes, 120
  associated primes (sub-Hankel), 194
  Cohen–Macaulay, 123
  embedded prime, 135
  hollow-like (Hankel), 175, 179, 182
  linear type, 182
  minimal primes, 123, 179
  primary component (sub-Hankel), 193
  is reduction, 167
  reduction number, 168
  sub-Hankel, 195
  unmixed part, 135, 179
height of, 6, 40
homogeneous, 41
  graded resolution, 42
initial, 33
Jacobian, 45, 54
ladder, 77
  Cohen–Macaulay, 78
  Gorenstein, 78
  one-sided, 77, 78, 113
linearly presented, 43, 134
linear presentation, 188
linear rank of, 43
of linear type, 44, 46, 49, 82, 93
minimal prime of, 6
of minors, 4
  Plücker relations of, 164, 165
  strategic, 121
  submaximal, 131, 135
monomial, 33
normally torsionfree, 47
perfect, 45
  codimension two, 188
Pfaffians, 79
  is prime, 79
presentation
  graded free, 98
prime, 5, 23, 25
reduction, 41, 102, 135, 179
  minimal, 41
  number, 41
resolution
  graded free, 96, 112, 189
  mapping cone, 84
symbolic power, 47
syzygies, 43, 80
transform, 48
unmixed part, 98, 103

L

Linear
  determinantal hypersurface, 63
  general forms, 19
  presentation, 43
  rank, 43, 49, 106
    maximal, 115, 138, 298
    maximal (sub-Hankel), 186
  section, 18, 24, 77
    codimension one, 93
    general, 19
    of Hankel matrix, 161
    hollow, 86, 87, 141
    hollowfilled, 87, 92, 141
    hollow like (Hankel), 168
    one-term, 101
    semi-hollow, 87, 112, 142
    symmetry preserving, 129
  system, 8
  syzygies, 43, 108

M

Map
  birational, 49, 96, 102, 113, 123, 133, 153
    inversion factor, 49
    source inversion factor, 49, 63
  Cremona, 49, 54
    plane, 49
  linear projection, 8
  polar, 54, 105, 167
    Hankel, generic, 166
    image, 54, 123, 153
    involution, 63
  rational, 8, 49, 96, 102, 133
  Segre, 14
  Veronese, 8, 10
Matrix, 3
  adjugate, 30, 59, 80, 89, 134, 139, 143, 177
  alternating, 79, 81
  catalecticant, 25
  circulant, 66
  conjugation of, 22
  generic, 32, 77
  1-generic, 21, 94
  t-generic, 21
    symmetric, 64, 129
  Hankel, 10, 23, 161
    generic, 24, 25, 162
  Hankel bordered by circulant, 306
  Hankelization steps, 304
  Hessian, 35, 54, 65
    rank, 57, 92, 147
  of indeterminates, 4

indeterminate trick, 6
Jacobian, 35, 49, 56, 59, 88, 133, 142, 147
  dual, 90, 144
  rank, 56, 59
linear, 18
linear part of, 43
linear section, 18, 86
minor, 4
  maximal, poset of, 163
  submaximal, 4, 25, 33, 95
(trace) orhogonal, 22
persymmetric, 10
rational normal scroll, 288, 289
r-leap catalecticant, 201
  mixed, 295
scroll, 24
section
  hollow, 87
  hollowfilled, 87, 92
  semi-hollow, 87
sparse, 87, 113
sub-Hankel, 171
  homology, 189
  partial derivatives (structure), 185
  theory, 185
symmetric, 9
syzygies of ideal, 98
syzygies of partial derivatives, 89
trace-orhogonal, 26, 95
Module
  Cohen–Macaulay, 45
  conormal of higher order, 47
  cyclic
    regularity of, 42
  finitely generated, 45
  grade of, 45
  homological dimension of, 45
  perfect, 45
  of pure grade, 45

N

Northcott, D., 15

P

Polynomial
  Euler formula of, 58
  gradient ideal, 93, 148
  higher polar of, 54
  homaloidal, 63, 64, 68, 88, 106, 142
  homogeneous, 35, 45
  hypersurface defined by, 53
  parabolic, 55

Polynomial (*cont.*)
    partial derivatives of, 35, 53, 54, 57, 65
        algebraically independent, 167
    polar of, 54
    totally Hessian, 55, 63

R
Regular sequence, 40
Ring
    associated graded
        torsion, 47
    Cohen–Macaulay, 40, 41, 45, 133
    determinantal, 5
        Hankel, Cohen–Macaulay, 172
        Hankel, parameters, 172
        Hankel, prime, 172
    extension
        integral, 167
        Noether normalization, 167
    ladder
        symmetric, 153
    Noetherian, 41
    normal, 133
        failure of (Hankel), 174
    polynomial, 4, 18, 19, 21–24, 32, 49, 55
    regular local, 16
    standard graded, 41

S
Segre, B., 35

embedding, 14
Serre, J.-P., 10
Special fiber, 41
Symbolic power, 49

T
t-minor, 4

V
Variety
    arithmetically Cohen–Macaulay, 10
    curve
        rational normal curve, 10
    determinantal, 5, 6
    double structure, 135
    dual, 35, 55, 63, 64
    Gauss map of, 35
    Grassmann, 6, 165
    incidence, 6
    non-deficient, 300
    Plücker embedding of, 35
    rational normal scroll, 11, 12
    Segre, 14
    singular locus of, 16, 54
    unirational, 5
    vector bundle
        Birkhoff–Grothendieck theorem, 11
    Veronese, 8, 12
Vector bundle, 11

If you have any concerns about our products,
you can contact us on
ProductSafety@springernature.com

In case Publisher is established outside the EU,
the EU authorized representative is:
Springer Nature Customer Service Center GmbH
Europaplatz 3, 69115 Heidelberg, Germany

Printed by Libri Plureos GmbH
in Hamburg, Germany